STATISTICS

STATISTICS

Thirteenth Edition

by

A. R. ILERSIC
M.Sc.(Econ)., B.Com., F.I.S.

Reader in Economic and Social Statistics at
Bedford College (University of London)

HFL (PUBLISHERS) LTD
9 BOW STREET, COVENT GARDEN
LONDON WC2

The first eleven editions of this book were published under the title of
Statistics and Their Application to Commerce.

Twelfth edition completely revised under the title of *Statistics* 1959.

Twelfth edition	*1959*
(Second impression)	*1962*
(Third impression)	*1963*
(Fourth impression)	*1963*
Thirteenth edition	*1964*
(Second impression)	*1968*
(Third impression)	*1969*
(Fourth impression)	*1970*

This edition © HFL (Publishers) Ltd 1964

I S B N 0 372 01632 4

Printed in Great Britain at
THE STELLAR PRESS, HATFIELD, HERTS

PREFACE TO THE THIRTEENTH EDITION

For those teachers and students who have used earlier editions of this book, a brief note of the changes in this edition may be helpful. Considerable sections of most chapters have been re-written, more especially the earlier chapters on elementary statistical methods. There has also been some re-arrangement of the 'method' chapters, so as to keep the main topics of the average examination paper together. More illustrative examples of calculations, which form the backbone of most examination papers, have also been introduced. The chapters on Economic and Social Statistics have been brought up to date and extended with the addition of a few new sections, as has been the chapter on Vital Statistics. The chapter on Business Statistics has been re-written, and the gist of the chapter on Quality Control has been incorporated therein. Finally, questions from past examination papers have been added to the 'method' chapters. These should enable the student to try his hand and also to see what sort of questions appeal to the examiners.

To the new reader some explanation of the scope and form of this book is due. The text is designed primarily for students taking external University or professional examinations in which statistics is a subsidiary subject. Many of these part-time students have limited opportunities of attending lecture courses in this subject, a large proportion relying entirely on a correspondence course. The absence of a teacher to clear up any difficulties places an additional responsibility on a writer to ensure that his ideas are clearly expressed and, more important, that the overall picture of the scope of the subject is not concealed behind a mass of apparently unconnected detail. This consideration also explains why points which may appear obvious to some readers are repeated, while some explanations tend to be longer and fuller than would be the case if the text were to be used solely in class work.

It is especially important that the reader should not gain the impression that statistical method comprises a series of incomprehensible formulae to be learnt by heart and applied automatically to any data available. There is an understandable tendency on the part of many students taking statistics as a subsidiary subject to 'cram' formulae and hope that the latter will turn up in the examination paper. If their luck holds, they may for ever after believe themselves to possess some knowledge of statistics. Perhaps it is as well that most of them promptly forget what they learn as being useless, as all too often they possess little appreciation of the ideas and principles underlying the methods learnt.

I have tried to explain as simply as possible the purpose of the various methods, largely in the hope that interest and curiosity will be aroused. Every effort has been made to consider those whose studies of arithmetic and algebra stopped, possibly more years ago than they care to remember and probably very much to their relief, at the General Certificate of Education Ordinary Level stage. This consideration has also determined the rather limited outline of sampling theory. This is, after all, a branch of mathematics and for the adequately equipped student there are numerous first-class texts covering this subject. I have tried merely to indicate the basic notions underlying significance tests which recur regularly in any statistical or survey report. The object is not that the student shall be able to carry out such tests outside the examination hall, but rather that he shall appreciate the limitations of sample data. The need for such understanding is amply evidenced by the discussion surrounding the results of recent pre-election opinion polls.

Consideration of the type of reader for whom the book is intended is closely linked with the question of what an introductory text such as this should cover. Its primary task, I believe, should be to arouse interest in the subject, rather than form a medium for passing examinations. Unfortunately examinations exist and even the seekers after knowledge for its own sake among the students are bound to be influenced to some extent by the prospect of the tests their hardly acquired knowledge must undergo. Accordingly, every effort to meet the requirements of the various syllabuses has been made. Considerable space has been devoted to method, although at this level it is the view of the writer that definitions and sources, together with an ability to comprehend and present factual information, are of infinitely greater importance than a mechanical application of memorised formulae to hypothetical and improbable data which the statistics papers in so many examinations demand. Commonsense is far more important in statistical work at all levels than a knowledge of formulae. I can only hope that as the student works his way through the succeeding pages he will find a greater appeal to his intelligence and imagination than to his memory. If I have succeeded in that respect I shall feel reasonably content.

It is with pleasure that I acknowledge the assistance given me by various people and organisations. My thanks are due to Mr. L. Moss, Director of the Social Survey, The Times Publishing Company, Dr Henry Durrant of Social Surveys (Gallup Poll) Ltd, the Institute of Practitioners in Advertising, and in particular the Controller of Publications at H.M. Stationery Office, as well as the Editor of

Barclays Bank Review for allowing me to reproduce copyright material. I am indebted to Professor Sir Ronald A. Fisher, Cambridge, and to Messrs Oliver and Boyd Limited, Edinburgh, for permission to reproduce an extract from Table No.3 in their book *Statistical Methods for Research Workers*. I acknowledge with grateful thanks the kind permission granted by the following bodies to reproduce questions taken from their past examination papers: The University of London Examinations Department; the Local Government Examinations Board; the Institute of Statisticians; the Chartered Institute of Municipal Treasurers and Accountants; the Institute of Cost and Works Accountants; the Rating and Valuation Association and the Institute of Hospital Administrators.

It is with great pleasure that I acknowledge the assistance given me by Mr. R. A. Pluck, B.Sc. (Econ.) in revising this edition, in particular the chapter on Economic Statistics, and the preparation of material for the questions and answers. I owe a particular debt to Mr. J. R. Hews, F.C.A. and Mr. W. Hummerstone, not least for their patience, and I have to thank Mrs Anna Hackel, B.Sc.(Econ.) for the preparation of the index and for her secretarial assistance.

Lastly, I take this opportunity of thanking all the teachers and students who have written to me drawing attention to both the virtues and limitations of previous editions. I shall be grateful if they will continue to do so.

A. R. ILERSIC

TABLE OF CONTENTS

PART II

ELEMENTARY SAMPLING THEORY AND PRACTICE

PART III

SOCIAL STATISTICS

PART IV

ECONOMIC & BUSINESS STATISTICS

PART I

ELEMENTARY STATISTICAL METHOD

CHAPTER I

THE NATURE AND PURPOSE OF STATISTICS

According to a memorandum prepared by a committee of the Royal Statistical Society,[1] 'the science and methods of *statistics* in the modern sense range from the mere recording and tabulation of numerical data to subtle processes of inductive reasoning based on the mathematical theory of probability'. To allay any fears that may be aroused in the reader's mind, it can be stated that this study stops well before the latter stage. Statistics as defined in the latter part of the above quotation is a relatively recent development. Most of the significant advances in statistical techniques are little more than 50 years old, and recent years have witnessed a substantial expansion in the employment of statistical techniques in industry. The term statistician has until recently been very widely defined, the more so since there existed no recognised professional statistical qualification. In lieu thereof, graduates with degrees in mathematics or economics with some training in statistics formed the main source of statistical workers. Since the war, however, a new professional body, the Institute of Statisticians, has established such a qualification while the universities have been extending the scope and scale of their teaching and granting degrees in which statistics is the dominant, *i.e.* honours, subject. The undoubted importance of statistics is reflected in the large number of professional bodies which set a paper in statistics at some stage of their qualifying examinations. For this purpose, the student is required to learn the basic ideas and principles of the subject and it is, in the main, with such fundamentals that this book is concerned.

Definitions of 'Statistics'

The term 'statistics' is somewhat loosely employed to cover two separate concepts: (1) descriptive statistics; (2) statistical methods. 'Descriptive statistics' covers the collection and summaries of numerical data, popularly known as the 'facts'. Examples of this type of statistic are encountered every day in the national Press; one reads, for example, that the 'total working population of Great Britain in June 1963 was some 24·8 million of whom 16·4 million were male and 8·4 million female'. Figures relating to United Kingdom overseas trade and balance of payments appear regularly in

[1] Memorandum on Official Statistics. By a committee of the Council of the Royal Statistical Society, 1943.

the press. For example: 'total imports in the third quarter of 1963 amounted to £1,109 million compared with exports of £1,083 million, so that for that quarter there was an excess of total imports over total exports of £26 million.

In the illustrations given above, the number of workers in Great Britain was given to the nearest hundred thousand, whereas the trade figures were stated to the nearest £ million. Actually, it would have been possible to have given the exact number of male and female workers in Great Britain as counted by the Ministry of Labour, just as it would have been possible to give the trade figures to the nearest £. Two considerations determine the form in which such data are presented. First, and less important, is the purpose to which the figures are to be put; for simple comparisons it is often sufficient to state, *e.g.* 'in 1958-59 the United Kingdom government collected £5,480 million of revenue as against £6,794 million in 1962-63. The significant fact here is that tax revenues have risen substantially, not their precise yields to the nearest pound.

The second consideration is more important. How reliable are the figures themselves? Descriptive statistics are very often not exact or accurate in the arithmetic or accounting sense. Some data may be; for instance in Great Britain the number of Civil Servants and their salaries can be given precisely, but on the other hand no one can state *exactly* how many people there are in Great Britain at any date. This is true even for the census day itself, because errors due to omissions and miscountings are inevitable when enumerating some 50 million people living in over 14 million households. An exact figure is certainly not possible several years after that date, although the registration system, which should record every birth and death between the censuses, is very efficient. On the other hand the figures for emigrants from and immigrants into the United Kingdom, are only estimates.

Since statistics is concerned mainly with large aggregates, exactness to the nearest ton, person, penny or any other unit of measurement is, in most instances, quite unnecessary and, it must frankly be admitted, usually quite unattainable. The unemployment figures, for example, are given to the last unit, *i.e.* 474,373 persons were unemployed on 11th November, 1963, of whom 463,126 were wholly unemployed and 11,247 were temporarily stopped. This is unquestionably an underestimate since a number of individuals temporarily unemployed would not yet have registered with the local labour exchange and would be excluded from this total. Consequently, the above statement normally reads '474,373 persons were *registered* as unemployed . . .', in other words, the figure is dependent on the fact of registration. Thus, while it is perfectly all right to give an exact figure in such cases, it should not be forgotten that the appearance of accuracy can be misleading, especially if for some reason the registration system is at all unreliable or temporarily affected by extraneous factors. For example, in some of the under-developed countries registration of births is incomplete partly because not all births may be registered, but the result may also be misleading since female children are sometimes ignored and not registered. Persistent under-notification can be detected since the sex ratio of registered births should not differ significantly from the expected ratio of about 51 male to 49 female births. It is far more important, when handling and studying statistical data, to ensure that the source of the original data is reliable and that the data collected are relevant to the question to which an answer is sought. Whenever any statistics are consulted, the first question is not, 'To what unit or place of decimals are these figures reliable?' but 'Where were the figures obtained, by whom, and why?' If satisfactory answers to these questions can be given, few qualms as to the reliability of that data need be entertained.

Interpreting the data

Statistical data in their raw state are by themselves of little value. It may be interesting to the managing director of a company to learn that the sales of a particular product in one country total £500,000 last calendar month, but the information only becomes statistically significant when related, for instance, to the fact that a turnover of only £300,000 was achieved in the corresponding month of the preceding year. The essence of statistics is not mere counting, but *comparison* – and provided this fact is always borne in mind, careless and consequently useless compilation of data may be avoided. Comparisons are valid only when they are between quantities expressed in identical terms, *e.g.* a direct comparison of weekly earnings as an indication of living standards between two periods is invalidated if in the first the working week was 45 hours and in the second only 40 hours.

A good example of careless and consequently irrelevant comparison was provided by a Board of Trade Press notice concerning the tobacco shortage in 1949. After pointing out that more tobacco was reaching the shops, it continued: 'Against this is the fact that more persons are smoking than before the war, the United Kingdom population having risen from 47,700,000 in 1939 to 50,000,000 at the end of 1949'. Since the population can only expand by immigration and by new births, and the former was relatively insignificant in this period, it might be assumed that the infants were tobacco addicts at an early age! The relevant figures were, of course, those of the *adult* population which, due to the changing age structure of the whole population, had increased.[1]

---

[1] Even this needs qualification, since it assumes that the smoking habits of the average adults are unchanged, or at least that the *per capita* consumption has not decreased.

On the other hand, too much should not be read into a comparison of such data. The results should be scrutinised because there is a difference. The question to be asked is then not so much 'how big is the difference', but 'why is there a difference'. The following illustrations may bring home the point. An examination of the marriage rates in the United Kingdom show occasional but sharp fluctuations between the relative numbers of marriages registered in the first and second quarters of the calendar year. Such fluctuations are largely attributable to the fact that the date of Easter varies and occasionally falls within the first quarter, although a new influence in recent years has been the attraction of an income tax refund if the marriage takes place before 6th April. Comparisons of quarterly births in the United Kingdom have not always been straightforward. After 1941 the number of births registered in the final and first quarters of successive years had to be adjusted to provide a true comparison with previous years. Previously, births were often registered up to six weeks after the event, as allowed by law. In consequence December births were often registered in January, thereby inflating the following quarter's total. But, with the introduction of ration books, the arrival of a new baby – subject to registration – meant a new and extra ration book. The elimination of the delay between the two events markedly affected the official statistics, yet at the time there was a very simple explanation of the sudden change in the birth pattern! Similarly, the sharp increase in the number of divorce petitions in the immediate post-war years gave rise to much public comment and suggestions that family life was disintegrating. Further reflections indicated that although the number of petitions had greatly increased, the increase could be explained largely by the following facts

1. the war had led to an accumulation of petitions which would normally have been spread over the war-years thus avoiding the marked post-war accumulation;
2. that the enforced separation of husbands and wives during the war inevitably led to what now seems to be a once-and-for-all increase and the cumulative effect of wartime influences was only apparent with the return of peace.

Thus the collection of statistical data and their verification regarding source and definitions used, forms only the first part of the statistician's work. The statistics themselves prove nothing; nor are they at any time a substitute for logical thinking. There are, as pointed out above, many simple but not always obvious snags in the data to contend with. Variations in even the simplest of figures may conceal a compound of influences which have to be taken into account before any conclusions are drawn from the data. For example, it was contended by the civic authorities that the reduction in the number of street accidents in Birmingham during the first half of 1963 was due to the success of their experiment of persuading drivers to use dipped headlamps while driving in the city. No doubt the scheme was helpful, but how can one attribute the entire difference in the number of accidents to this single factor when other factors undoubtedly played a part? In particular, the exceptionally bad weather at the time would tend to keep people at home, hence there would be fewer pedestrians crossing the streets at night. Fewer cars would be out, particularly on pleasure, *e.g.* from cinemas, dance halls, and hotels, so that there would be less traffic, less dazzle from on-coming lights, etc. Also, the only drivers out in such weather would tend to be those whose profession necessitated driving a great deal and these would tend to be above-average drivers in skill whose accident rate would be lower than that for weekend drivers and others. Finally, the number of accidents fluctuates from time to time even if road conditions remain constant; to what extent was the difference accounted for by such fortuitous fluctuations? It is for such reasons that the Road Research Laboratory has been asked to analyse the results of the Birmingham experiment.

Another case of dubious use of statistics was provided by the controversy which followed the assertion that the high toll of road accidents in Christmas 1963 was attributable to drivers who had been drinking. It was argued by the motoring correspondent of one newspaper that it was false to assert that drink causes as many accidents as the public seemed to believe. Statistics of convictions for road accidents showed, he argued, only a small proportion of cases in which the driver was convicted for driving under the influence of drugs or alcohol. This is true, but the inference is not justified that drink is a minor factor in road accidents, any more than the number of convictions measures the extent to which drivers drink too much. It is common knowledge that most drivers charged with drunken driving elect to go before a jury, where their chance of getting off are very much better than with a magistrate, since too many jurors probably feel, 'there but for the grace of God, go I'. The police, aware of this problem, are thus very reluctant to charge drivers with this offence unless the evidence is incontrovertible and witnesses can be found to ensure a conviction. Thus, there are at least two important factors tending to reduce the recorded figures of accidents in which drink was a contributory factor.

Statistical Methods

As indicated in the quotation in the opening paragraph, statistical methods range from simple numerical processes to 'subtle processes of inductive reasoning based on the mathematical theory of probability'. These processes are also covered by the term *statistics*, but to distinguish this meaning of the term from our first definition – descriptive statistics – they are usually referred to as *statistical method*. The simpler methods which involve little more than elementary arithmetic are discussed in the chapters which follow. The really fascinating developments in technique, which have made statistical science one of the most powerful tools in the hands of the modern research worker, are of relatively recent origin, mainly in the last thirty years. In industry, particularly since the war, statistical methods have been extensively employed to control the quality of the product and to ensure that faulty goods are not sent out. A brief introduction to this technique, known as *quality control*, is given later.[1] Another important development has been that dealing with the design of experiments and their subsequent analysis, which is of course determined by the actual design. Modern designs are much more efficient in that they enable much more information to be extracted from a given amount of data. These techniques, however, require a mathematical ability of no mean order and are only mentioned in this text to emphasise the potentialities of statistical method. A brief outline of these techniques is given in Chapter XIX.

Perhaps the most useful and generally applied technique devised by the statistician is known as *sampling*. Most readers will be familiar with the election forecasts of the public opinion polls; many will have heard about market research whereby manufacturers learn what the consumer thinks of a product by posing specially designed questions to a selected few people. Fortunately, although this branch of statistical method is complex in application, the lay reader can acquire an understanding of the principles underlying sampling techniques which will enable him to appreciate what the statistician terms the 'significance' of his results from an analysis of sample data.[2] Information and conclusions based upon sample data, although they are reported with all their limitations are later often misquoted and bandied about with a complete disregard for their limitations which the statistician has so carefully emphasised. It is hardly surprising that the statistician so often hears the dictum, 'lies, damned lies and statistics'quoted at him. Yet it is not the statistician who is at fault; it is the individual who quotes – often selectively to support his own tenuous arguments – the statistical results without any of the qualifications with which the statistician has surrounded his conclusions. For this reason the quip has been made that some people use statistics in the same way as a drunk uses a lamp-post. more for support than for illumination.

[1] Chapter XIX.
[2] Explained in Chapter XIII.

Statistics in Business

Every business man today appreciates the value of accurate and regular financial statements in the conduct of his business. The widespread expansion in management accounting in recent years is attributable to the growing realisation on the part of the industrial community that without facts concerning output, costs, turnover, expenses, etc., it is impossible to conduct the affairs of a business so as to obtain maximum efficiency.

Whatever the legal form of a business, be it public or private, a public corporation or a unit of nationalised industry, management today has become highly complex. The administration must make its decisions in the light of facts prepared by the executives rather than – as was the case in the smaller businesses of the past – from intimate personal knowledge of the concern. Although no volume of statistics can replace the knowledge and experience of the executives, it supplements it with more precise facts than were hitherto available. The point has been well made that far too many managers and directors are inclined to base their judgements on a 'sample of one'. In other words, they generalise from their own very limited experience of the matter under discussion. The value of statistics in the large concern is indicated by the comments of Lord Heyworth, Chairman of Unilever, in his Presidential Address to the members of the Royal Statistical Society[1]: 'We also have an interest in statistics of a general nature. We operate in most of the countries of the world. We frequently have to decide whether it would be, say, a better proposition to erect a new factory in Malaya or extend our ice-cream business in England. To come to a satisfactory decision we have to know all about not only the specific conditions of the trade in which we are proposing to engage in Malaya and England, but also the whole general health of their economies . . . So for such purposes a whole range of outside statistics may be relevant, from the amount of debt in the countryside to the number of votes obtained by communists in municipal elections.'

It is noteworthy that in recent years the Federation of British Industries has been publishing the results of its Industrial and Export Trends Enquiry. A copy of the schedule used is reproduced on page 27. The purpose of the survey, which takes place every four months, is to ascertain variations in the level of industrial activity and manufacturers' exports, together with industrialists' expectations regarding the immediate prospects for their firms and industries. Increasing attention is being paid by both industry and the government to such 'forward-looking' statistics' as well as in reducing the time-lag between the event and publication of relevant statistics which, in the

---

[1] 'The Use of Statistics in Business' – Lord Heyworth: Presidential Address to Fellows of the Royal Statistical Society, read 31st October, 1949.

case of so many series, has been such as to reduce the usefulness of the data. The National Association of British Manufacturers carries out an enquiry similar to that of the F.B.I., while the Association of British Chambers of Commerce collects data relating to the level of orders, stocks and expectations of merchants and export agencies. In all these cases the data are supplied by the member firms of the organisations.

It is much easier and probably more important for the larger undertaking to assemble and prepare statistical data relevant to their problems. This does not mean, however, that the smaller firm can derive no benefit from the simpler statistical techniques outlined in later pages. The mere assembly of data relating to the financial and production activities of the firm in tabular form and simple charts can often bring to light facts and trends which were hitherto not fully appreciated or not apparent from a scrutiny of the bare figures themselves. Given some knowledge of even the simpler methods, errors in the interpretation of such statistical data can often be avoided.

Finally, in government publications there is a wealth of information relating to virtually every industry as well as to the economy at large, accessible to all. One government committee found that in contrast with American business, British industrialists and businessmen made little use of such data as are compiled from the periodic censuses of production and distribution. Yet it is from such information that the business man can make a better assessment of the economic situation when forecasting the prospects of his business and provide him with information relating to other problems directly concerning him; for example, prospective supplies of labour, raw materials, machinery, building, etc. To ignore such information is at best unwise and at worst may be likened to a ship's captain who ignores his barometer in assessing the weather prospects.

Government Statistics

Most of these data are the by-product of government activity in the social and economic spheres which has brought with it the need for positive and constructive legislation in economic affairs. The acceptance by all parties of responsibility for the maintenance of economic stability, which was acknowledged in the White Paper on Full Employment,[1] necessitates the collection and preparation by the responsible Government departments of statistics covering the whole vast complex of the nation's economy. Many of these data can only be obtained from the business and industrial community, and this involves the generally disliked task of form filling. There is scope for much education on both sides in this matter. The Government

[1] Employment Policy. Cmd. 6527.

statisticians responsible for the preparation of the forms and the presentation of the results will perform their tasks more efficiently given reasonable co-operation from their informants just as the simplification of the questionnaires will facilitate their completion.

Statistical techniques enable the Government Actuary to estimate the potential demand for pensions, sickness benefit and unemployment allowances. The B.B.C. has, since October 1936, maintained a statistical department for audience research in order that programmes should as nearly as possible meet the demands of the public.[1] The Budget, which influences every side of the economic life of the community, is itself based on estimates of the probable consumption of taxed commodities, prospective income levels and mortality rates (for estate duty receipts), all of which owe their accuracy to statistical techniques. There is little doubt as to the importance attached by the government to reliable and up-to-date statistical data on the economic scene. The Central Statistical Office is expanding continuously its output of economically significant series such as hire-purchase statistics and estimates of investment outlays by large firms in the United Kingdom. Much is being done to improve the economic industrial and commercial statistics needed for economic planning. In 1957 Mr Macmillan, then Chancellor, likened the then available data to 'looking up the trains in last year's Bradshaw'. Economic planning is impossible without accurate statistics published with the minimum of delay. The material condensed into Chapters XVIII and XIX will later give the reader some impression of the tremendous volume of statistical information produced in connection with the economic situation. Yet without it, a body such as the National Economic Development Council would be quite unable to function effectively.

The contents of the following chapters may at times appear to be far removed from these important and fascinating problems; but just as a prospective mathematician has first to learn the multiplication tables, so the potential statistician must first acquaint himself with the elementary principles of statistical analysis. If after finishing this book, the reader is still interested in the subject, then – his mathematics permitting – a brief bibliography offers him additional material for study.

---

[1] *Methods of Listener Research employed by the B.B.C.* – R. J. E. Silvey, J.R.S.S., Vol. CVII, pts. III/IV, 1944, p. 190. A brief description of this work is given in Chapter XV.

CHAPTER II

STATISTICAL DATA:
DEFINITIONS AND SOURCES

Statistical method consists of two main operations; counting and analysis. The analysis may entail no more than a simple comparison, e.g. at last Saturday's football matches there were 60,000 spectators at Chelsea and 40,000 at Manchester United. From these data at least one conclusion can be drawn, i.e. there were 50 per cent more spectators at one match than at the other. The ensuing question, 'why this was so?' requires more information than is given here. At more advanced levels statistical analysis is very much more complicated and is based on specially designed mathematical techniques. But to whatever type of analytical technique the data are to be subjected, the first stage of any statistical enquiry is the collection of the facts and this means counting. The statistician has no use for information that cannot be expressed numerically, nor generally speaking, is he interested in isolated events or examples. The term 'data' is itself plural and the statistician is concerned with the analysis of aggregates.

The data themselves may be of any kind as long as they can be counted. They must, however, be susceptible to classification as well as to counting. In the statement that there are 50 million people in Great Britain, the unit of counting is a person, just as in the figure of unemployed quoted earlier the unit is a person registered at the local Employment Exchange as unemployed. The actual process of counting, whether it be the census enumeration of heads of households and the persons in them, or the number of insurance cards lodged at the Employment Exchange on a given day each month, is easy enough, although with very large aggregates mistakes will occur. But the latter are, generally speaking, insignificant in relation to the absolute figures in the totals. A much more serious consideration for the statistician is to be certain that not merely have all the units been counted, but that only the units relevant to his enquiry are included. For example, the characteristic of the unit in a count of the unemployed is the fact of being both unemployed and registered as such. The Ministry of Labour publishes, among other analyses, totals sub-divided by sex and age (actually as between men, youths, women and girls) by the duration of unemployment as well as the location. All these analyses depend on classifications which in turn depend on the definitions employed. Clearly, some definitions are easier than others, e.g. male

10

and female, men and youths – the latter being defined as under 18 years of age.

It is a good rule to remember that the first step in analysing any statistical data, whether it be culled from an official publication or a report prepared by someone else, is to check the definitions used for classification. How often do we hear radio news items recounting the number of road accidents. For most listeners this conjures up a scene of two cars colliding head-on. Yet this collision is not classified as an accident unless personal injury has been sustained. When we learn a few days after the Christmas holiday that sixty people have been 'killed' on the road, we think of the people who died at the time of the accident. In the road accident statistics, however, a person is classified as 'killed' if he dies within thirty days of the accident; thus early announcements of the Bank Holiday road deaths are usually under-statements. To take another example, most of us have a clear idea of what 'murder' is. The official statistics classify murders according to the number of victims, which is not the same thing as the number of murderers. Several people, *e.g.* a gang, may commit one murder; one person may kill several people, *i.e.* multiple murderers. The distinction is very important if one is researching into the causes of 'murder'.

Some classifications, however, are more arbitrary since the definition of the unit is itself arbitrary. For example, suppose we sought to classify a group of women by the colour of their hair or eyes. Inevitably there will be some colours which will be difficult to determine although if the same women were classified by height, the classification would be simple. The basis of classification depends on the nature of the characteristic by which units are being identified and counted, *e.g.* women over 16 years of age in the United Kingdom are either married, single, widowed, separated or divorced. The characteristic in this case is described as an *attribute*. The same women can be classified by age; in this case the classification is based on a characteristic which is known as a *variable*. The distinction between 'attribute' and 'variable' is important but quite simple. The former describes a characteristic which is not capable of numerical definition, *e.g.* colour of eyes, houses condemned as 'unfit for habitation', recruits classified as 'grade one'. A variable is a characteristic which can be expressed in quantitative terms, *e.g.* height in inches, salary in pounds, or marks in an examination.

The point of the foregoing paragraph is to emphasise the fact that the statistician is usually concerned with groups and 'populations' consisting of units possessing a common characteristic, although he may compare such groups which are dissimilar in one particular respect. For example, in the pre-election polls it is usual to classify the respondents (*i.e.* those who have been interviewed) by their political party to ascertain how far allegiance to a party affects their attitudes to particular problems of the day. But, it should be noted, all the units in the various groups share a common characteristic, *i.e.* the right to vote, and it is for this reason that they have been interviewed.

The significance of precise definition becomes apparent when information is being collected. This is especially true of what are termed 'secondary' statistics, *i.e.* statistical data available in published form such as the Annual Abstract of Statistics.[1] Especial care must be taken in using published data to ensure that over the relevant period of time the definitions or coverage of a series of data have not been altered. For example, it is impossible to compare pre- and post-war living costs by simply comparing the pre-1947 Cost of Living Index with the current Index of Retail Prices. Similarly, the published official estimates of the 'working population' at the present time are compiled in a different way from the figures published at the end of the war so that comparisons of 'total working populations' in 1947 and 1963 can only be made after adjusting certain figures to a common basis.

Such considerations are especially relevant when it is learnt that by far the most important and prolific source of published statistical data is the government. Each department produces a great deal of statistical data, primarily as a by-product of its administrative functions. For example, the Home Office is responsible for the preparation of the Annual Report on 'Criminal Statistics' which gives information on the extent and type of crimes committed, the activities of the police in clearing up such crimes and the punishment meted out to the convicted offenders by the Courts. The published statistics on crime are especially interesting in view of their chequered history.[2] For example, the only crimes which officially exist are those known to the police. Thus, it is probable that the extent of blackmail and sexual assault is understated in the annual returns, since the victims are often unwilling to run the risk of publicity if they were to charge the offender. Fluctuations from year to year in the published figures may reflect not only changes in the incidence of a particular crime, but mainly the fact that the police have at intervals undertaken a major drive against it. This, for example, probably explains the apparent increase in homosexual offences of recent years. It is unlikely that the extent of such practices among the male population has changed as much as recent publicity given to court cases would suggest. The increased number of convictions merely indicates that the police have

[1] A compendium of official statistics compiled by the Central Statistical Office and published annually. The Abstract for 1963 is the 100th in the series.
[2] See Chapter XVI for references

become more active in trying to suppress these practices. Thus, a *Police Review* article commenting on the extent of homosexual practices notes that in Manchester in 1955 there was only 1 prosecution for importuning, 0 in 1956 and 1957, 2 in 1958. In 1959 the number rose to 30, with 105 in 1960, 135 and 216 in the next two years. A new Chief Constable had been appointed at the end of 1958 and the *Police Review* rightly comments that the local Watch Committee ought to have considered the figures for 1955-58 more closely.[1] Contrasting rates of crime, *i.e.* of certain types of offence in different areas of the country, may merely reflect local police directives or the prejudices of the local bench of magistrates. For example, where the bench is reputed to be lenient towards certain types of offence, it is probable that the local police will be more reluctant to charge the offender than in another area where the bench is not so inclined.

In brief, the greatest care must be exercised in using any statistical data, especially when it has been collected by another agency. At all times, the statistician who uses published data must ask himself, by whom were the data collected, how and for what purpose? Many official statistics, *i.e.* government produced, are reliable and complete. But this is not true of all published data, least of all that covering a long period of time. While there has been a great improvement in the quality of official statistics since the last war, it is not so many years since the 'working population' was differently classified by the Board of Trade, the Registrar-General and the Ministry of Labour, thereby making comparisons of similar data collected by these three government departments virtually impossible. Even with the substantial improvement that has undoubtedly taken place, published statistics can sometimes be quite misleading. For example, a few years ago a small town was largely dependent for employment on one major undertaking which, due to inadequate orders, was only working alternate weeks. Since the count of unemployed each month takes place on the second Monday, if that happened to be in a working week there were virtually no unemployed. If next month it coincided with the weekly lay-off, then unemployment would be substantial. Thus, anyone studying the monthly data without a knowledge of this background would be sorely puzzled by the periodic sharp jump in registered unemployed in that town.

With these provisos and warnings regarding the use of published data, some indication of the scope and extent of published statistics can be given. Published sources can be divided into two categories. The *first* and most important are the official statistics prepared by the government. These are of two kinds: those which are the administrative by-product of a department's daily work and those which

[1] Quoted in the *Observer*, Sept. 1, 1963.

are collected at intervals for specific purposes. An example of the first type are the figures prepared by the Customs and Excise department. The bulk of these appear in the annual report of the department. Some of these annual reports, *i.e.* Blue Books, contain a wide range of statistical information. For example, apart from such routine statistics as the amount of tax collected and number of assessments made, the annual report of H.M. Commissioners of Inland Revenue has been considerably expanded in recent years and is now a mine of information regarding the size and distribution of incomes and direct taxation. Examples of the second type of official statistics are the various censuses, *e.g.* the decennial population census and the census of production. Apart from information of a statistical character contained in Blue Books and special census reports, there are the periodic publications such as those on the Balance of Payments and the Economic Survey. These papers are usually prepared by the appropriate department, but since the war there has been created a Central Statistical Office which has two main functions. First, it is responsible for co-ordinating the statistical information produced by the various departments so that, for example, with a Standard Industrial Classification introduced in 1948 and revised in 1958, all departments now prepare data classified (where this is appropriate) under the same industrial headings and not as before the war, as was pointed out above, on their own classification. Second, the C.S.O. as it is usually termed, is responsible for bringing together the economic statistics needed for the formulation of policies designed to maintain economic stability. Many of them are published in the *Monthly Digest of Statistics* and *Economic Trends*. The more important of the United Kingdom economic statistics are discussed more fully in Chapter XVIII.

The *second* category of published statistics is not so large, nor is it so well known. Local authorities publish annually a wide range of statistical information relating to their financial and social activities, as do the various nationalised industries, *e.g.* the National Coal Board. Some of the City institutions produce statistical studies of matters which are of especial interest to them. For example, the Midland Bank has long published a series of statistics showing the amount of capital raised in the capital market by governments and companies, while Lloyds Bank publishes an index of the supply of money. In recent years the need for the independent collection of economic statistics has been greatly reduced by the government's recognition of its responsibilities in this field. There remains, however, one branch of statistical work of the utmost importance for the private investigator, either individual or, more usually, corporate. This is the carrying out of surveys among special groups of the population to obtain information which is not otherwise available. Thus the Department of Applied Economics in Cambridge has conducted local surveys into family budgets and the Institute of Economic and Social Statistics at Oxford was responsible among other surveys for a pioneer enquiry into personal savings in the United Kingdom. Another private organisation conducted a national enquiry into the conditions of life among the 'over-seventies', while a professional body annually carries out sample enquiries into the reading habits of the community. Although initiated and developed by non-government bodies, the technique of sample enquiries has been adopted by the government since the war. It has its own survey organisation known as the Social Survey and large scale sample surveys have formed the basis of the Retail Price Index which serves as a cost of living index in the United Kingdom. Using similar methods, there is a continuous survey into the food consumption habits of the community, the findings of which are published in the annual reports of the National Food Survey.

If it is proposed to use statistical data which have been obtained from a sample survey carried out by some other party, then it is imperative that the report of that survey be read, special attention being paid both to the sample design, i.e. the informants and whether they were representative of the larger population, and to the questionnaire. Any survey report which lays claim to serious attention will give the reader such information within its covers.

Similar considerations arise when the statistician is forced to collect his own data direct from a group of respondents. Questions must be carefully phrased so that they are unambiguous and mean the same to all respondents. It will often be necessary to define certain terms. For example, 'household income' covers all wages and salaries before tax, as well as pensions and tax-free spare-time earnings. Individuals in the Household Expenditure Survey of 1953-4 and its successors, known as the Family Expenditure Survey in 1959 and 1962, sometimes gave their net incomes after deduction of tax and national insurance contribution, as well as omitting cash receipts from any spare-time activities.

The importance of paying attention to the precise definition of statistical units and sources of data collected by others becomes evident when it is recalled that the conclusions derived from the enquiry can only be as good as the original data upon which they are based. You cannot make, runs the old adage, a silk purse out of a sow's ear. It is equally true that an ill-classified collection of inaccurately defined data from misguided informants will not provide reliable data for a useful statistical study. It is for such reasons as these that the Central Statistical Office has prepared a small booklet of terms and definitions of the units and data published in the

Monthly Digest of Statistics.[1] The student would be well advised to look through this, not for the purpose of learning any of the definitions, but first so that he becomes aware of the care which must be devoted to this apparently elementary branch of statistics, and second, to ensure that when he abstracts figures from the *Monthly Digest* he will do it properly. He will also understand why it has been so necessary to devote an entire chapter to what appear to be commonsense propositions!

Even when definitions and sources have been checked, the mere extraction of the data must be done with care and a clear appreciation of what the published figures mean, quite apart from the question of what can be inferred from them. Thus, a government publication entitled *The Influenza Epidemic in England & Wales 1957-58*[2] analysed the extent and incidence of that disease between June 1957 and April 1958. It concludes that 'almost certainly at least $7\frac{1}{2}$ million persons suffered some incapacity from influenza during the course of the epidemic'. Since influenza is not a notifiable disease such as typhus, smallpox, etc., how can such a statement be made? Note, however, the careful qualifications to the figure, 'almost certainly' and 'at least', suggesting that some estimation is involved. It was. The total of $7\frac{1}{2}$ million was derived partly from estimates of the number of new claims for sickness benefit during this period in excess of the corresponding period in each of the preceding five years. Since not every adult is insured under the National Insurance scheme further estimates had to be derived for this very large non-insured section, predominantly female, of the population. It was in fact assumed that the incidence of the disease in this group was the same as in the first. Then, for children of school age, estimates were derived from records of local medical officers of health which related to absences of children from school. Lastly, the under five's and those over 64 years also required a further rough estimate. The official report explains these facts very clearly, but such figures tend often to be quoted as if they were accurate to the last unit, particularly in abbreviated press reports. While there is no excuse for such mistakes or misinterpretations in real statistical work, it is idle to pretend they do not arise.

---

[1] This Supplement on Notes and definitions is published annually in January by H.M. Stationery Office.
[2] H.M.S.O. 1960.

CHAPTER III

COLLECTING THE DATA

So far the emphasis has been laid on statistical information which is available in government reports and from other published sources. Whatever the subject matter of the statistician's current enquiry he can usually be certain that someone has done some work on that problem before. Such work, if it can be traced, should always be consulted. It may produce the answer for which the statistician is searching. More likely, it will give him a few ideas and even draw his attention to points which so far had escaped his notice. It may even contain some mistakes in technique or interpretation from which valuable and important lessons may be drawn for the enquiry projected by the statistician. In other words, before embarking upon any project which necessitates the collection of information and statistics, start by consulting all the known statistical sources and examining earlier survey material. This done, the statistician will know what additional or new information he must obtain. An important benefit from such work is that it helps him to clarify the issue, as well as the precise nature of the problem, in his own mind.

The emphasis on the need to clarify the problem, quite apart from obtaining all the relevant information, is justified by the fact that unless this is done, the survey may well prove unsatisfactory. Once the precise nature of the enquiry or survey has been defined, it becomes possible to consider how it may best be carried out. This will enable the statistician to decide how much information he needs to collect, from how many and which people, what questions will produce the information he needs. If it is important to be able to generalise about the sample data, *i.e.* the information collected from a sample of respondents, then he may need a larger sample than would be the case if he is merely concerned with collecting information for local use. Alternatively, his financial resources may be limited and he must consider the advantages of asking a smaller sample more intensive questions compared with asking a larger number of people fewer and simpler questions. After all, he cannot hope to interview all the people selected in his sample by himself; this is the task of specially trained interviewers and such work costs money.

In brief, too much time cannot be spent on these initial stages of considering all aspects of the enquiry. It is a commonplace among market research organisations and business consultants that when their client firms first come to them, very few have a really clear notion of the nature of their difficulties. The first part of such work is for the specialist to uncover the facts and help clarify the problem. Once that is done, they can consider how their organisation can best help the client. No survey organisation worth its salt will undertake a sample survey for a client company just because some senior director of that company thinks it would be a good thing. The first question is always, how and in what way is a survey likely to contribute to a solution of the problem as defined. This said, let it now be assumed that the statistician needs further information beyond that available to him.

Collection of Data

There are three basic methods of collecting the information needed:

1. The investigator may interview personally everyone who is in a position to supply the information he requires. Such a procedure will be possible in very few cases indeed, since most statistical enquiries cover a wider field than any single investigator could possibly examine personally within any reasonable time. An interesting example of such an enquiry is that conducted by Professor Zweig who personally interviewed 400 people.[1]

2. The task of interviewing informants may be delegated to selected agents who are provided with a standardised questionnaire and explicit instructions as to the mode of its completion and, quite often, the names and addresses of the persons to be questioned. The main problems in this case are the selection of suitable agents and the cost involved. It is not merely sufficient that they should be given routine instructions; they should be fully conversant with the purpose of the enquiry, since inefficient interviewing will seriously affect the value of the results obtained. Against the disadvantage of high cost and the difficulties of obtaining suitable agents must be set the very considerable advantage that the information received will probably be highly reliable. Such interviewers are nowadays employed by all research organisations such as the Government Social Survey and the market research offices.

3. The last method is by questionnaire addressed to individual informants. This method, at one time extensively employed, possesses the apparent advantage that a very large field of enquiry may be covered at relatively low cost, and the larger the coverage the less significant will be occasional errors in the filling up of individual forms. This method of collecting information is not often satisfactory due to the low proportion of returns. The

¹ *Labour, Life and Poverty*, F. Zweig, Gollancz, 1948.

government uses it for various census enquiries, *e.g.*, population, election, production, but in.these cases the return of the form is compulsory. If it is voluntary, only those individuals, generally speaking, who are particularly interested in the subject matter of the enquiry will trouble to return the questionnaire. Then the investigator may merely have a collection of biased data. Nevertheless, some recent enquiries have shown that good results may be achieved with postal surveys.[1]

Drafting the Questionnaire

The old adage about asking silly questions and getting silly answers is probably engraved on most statistician's memories, if not their hearts. Most practitioners will be able to recall one or more such incidents in their professional careers. The simple fact is that a survey is as good as its questionnaire; it cannot be better although it may be worse because of other additional defects in the design or organisation of the enquiry. Although the subject of questionnaire and schedule design is discussed in Chapter XV, some of the more obvious points are worth making even at this early stage in the study of statistics. First, the distinction between questionnaire and schedule. Usually the former describes the form sent *by post* or delivered some other way to the respondent and which is then completed by him unaided. The term 'schedule' is usually kept for the form which is filled up by an interviewer, whose job it is to contact selected persons and ask them to answer the questions which the interviewer then puts as set out on the schedule.

A number of questionnaires and schedules have been reproduced at the end of this chapter and in Chapter XV. They show the layout and the type of question asked, as well as indicating how the interviewer or respondent should complete the form. The best way of learning how to design such forms is by studying the work of the professionals. In the meantime, here are a few simple and apparently obvious points to bear in mind. It is worth remembering that the more obvious something is, the easier it is to overlook!

1. Few people enjoy form-filling or answering questions. Keep it as short as possible.

2. Complicated and long-winded questions sometimes irritate and often confuse the respondent and result in careless replies. Make the individual questions short and simple.

3. Answers such as 'Probably', 'Fairly good', 'Average', mean nothing to a statistician, since they signify different degrees to different individuals. Ensure that all questions may be answered

[1] See Chapter XV for a discussion of postal enquiries.

as far as possible by either 'Yes' or 'No', or by a name or figure. Some readers may, however, have encountered the type of questionnaire circulated by the Audience Research section of the B.B.C., in which the listener is asked to indicate whether he or she considered a particular programme excellent, good, fair, or poor. This survey is concerned with estimating the number of listeners who 'thought' the programme was good, *i.e.*, the listener's *subjective* assessment of the programme. If 90 per cent. of the listening public answered 'excellent', then as far as the B.B.C. producer is concerned, the programme was outstandingly successful. There is no attempt here to arrive at an impartial, *objective* assessment of the quality of the programme, since it is clearly impossible to do so. In short, the questions and their answers depend on the purpose to be served by the enquiry.

4. The questions should follow a logical sequence, so that a natural and spontaneous reply to each is induced. Thus it is clearly politic to enquire whether a woman informant is married before asking her how many children she has!

5. Few people willingly provide intimate facts about themselves. Many resent such questions, which should be avoided as far as possible. In some cases where private information of this nature is needed, the method of personal enquiry may be most likely to yield results. Generally speaking, highly personal questions should be kept to the end of the interview, when the informant may feel more at ease with the interviewer.[1]

6. When public opinion on a particular issue is being assessed, it is important to ascertain from the respondent whether he has any knowledge of the subject, before asking his opinion on it! Opinion polls, as compared with factual enquiries, give rise to a host of very complicated problems in questionnaire design, some of which are discussed in Chapter XV.

7. Remember that even the best questions, however well phrased they may appear, do not always elicit the true answer. This may be due to honest misunderstanding of the question, or a desire on the part of the respondent not to disclose the truth. Thus, in the 1951 census of population, householders were asked if they had the exclusive or shared use of amenities such as piped water, cooking stoves, baths and lavatories. All the evidence goes to show that this question was misunderstood and the tabulated answers worthless.[2]

---

[1] The extent to which people will supply information even of the most intimate nature, is demonstrated by the enquiry into attitudes and practices of birth control among persons married since the First World War by the Population Investigation Committee. See 'Birth Control in Britain' by G. Rowntree and R. M. Pierce in *Population Studies*, July 1961.

[2] 1951 Census. General Report.

In the Family Expenditure Surveys householders are asked to state their outlay on tobacco and alcohol. Comparing the aggregate answers with the information available from the Customs and Excise relating to consumption of these commodities, it is obvious that there is considerable under-statement.

It need hardly be added that once the questionnaire has been drafted, it should first be tested on a small number of individuals, to assess the probable reaction of the wider public to be covered later. The advantages are considerable. This policy was pursued by the Board of Trade in the 1948 Pilot Census of Distribution. It was proposed to carry out the first-ever full enumeration of shops and service establishments in 1951, but the Department undertook initially a sample survey of a few selected areas and trades representing the whole country. The lessons drawn from the many criticisms and suggestions made enabled improvements in the questionnaire to be introduced, thereby facilitating the task of the informant and ensuring more accurate and prompt replies.[1] Note that the mere fact that the census is enforced legally does not help the statistician very much; he is much more anxious to obtain a fair sample of accurate replies than a mass of carelessly compiled and often inaccurate information from a non-representative group of respondents.

Technique of Sampling

As already explained, it is frequently physically impossible to obtain a really wide coverage of information from all those individuals who might come within the scope of the enquiry. This is especially true where, owing to the nature of the enquiry or survey, it has been decided to use agents interviewing their subjects personally. Examples of such surveys are the 1953/4 Ministry of Labour enquiry into some 20,000 'working-class' household budgets, the Ministry of Labour enquiry into the attitudes of retired workers to retirement, the many surveys carried out by the Social Survey on behalf of various Government Departments, and finally the well-publicised political polls which are discussed regularly in the daily press. In these cases only a small part of the field may be covered; estimates of national public opinion are often based on a sample of less than 2,000 informants. Yet it is true to say that in most cases, provided that the survey has been conducted with due regard for statistical technique and principles, the results should differ only to an insignificant extent from those which would be derived from a complete census.[2] The validity of this contention has been proved in the past by careful

---

[1] In view of the widespread protests and even hostility encountered with the Pilot survey, it is clear that as a result of the lessons learnt, a great deal of expense was avoided.

[2] A complete enumeration of the entire field is known as a 'census'. A limited enquiry based on part of the field is usually described as a 'sample survey'. These are discussed more fully in Chapter XV.

checking, repeated sample enquiries, and in some cases by a full-scale enquiry where the field to be covered was not impossibly large.

This system of selection of a part of the whole, known in statistical method as 'sampling', may be explained by a simple analogy. The practice of sampling is frequently encountered in the world of commerce. Thus a small sample of tea or grain, of cloth, or of many other commodities, is frequently the only means whereby a prospective buyer can assess the quality of the bulk. The principle underlying the process is the assumption, generally borne out in practice, that the part is genuinely representative of the whole; thus the Public Analyst bases his report on the quality of a product on a few tested samples, and the assayer assesses the mineral content of an area on samples of ore selected at random.

To ensure that a sample is representative, it must be selected by what are known as *random* methods of selection. In this particular context 'random' has a rather special meaning. The simplest definition is that a sampling procedure is random if every item in the group from which the statistician is making his choice has an equal chance of selection. The group or aggregate of the sampling units, *e.g.* all the households in Great Britain, all the farms in England and Wales, all boys in the Sixth form of grammar schools, etc.; all these aggregates are usually referred to as *populations*. In other words a statistical population is the total number or aggregate of all the units which, by virtue of a common characteristic, may be classified as belonging to the aggregate. Thus, we can refer to the 'population' of people who regularly attend symphony concerts; all the women who do not use cosmetics, or those who drink 'Lyons' tea. Obviously, some of these populations can hardly be counted or listed without the expenditure of a great deal of time and money in enumerating them. However, given that the population in which the statistician is interested can be clearly defined, then provided his sample is selected by a random process it is likely to be representative of that population.

Note that there can be no certainty that the sample is truly representative. Sometimes this can be checked against other information. For example, if the distribution of the sample by say, social class, reflects the known social-class composition of the population as shown in the figures prepared by the Registrar-General's Office, then this is evidence strengthening the belief that the sample is a replica of the population. Without such information, however, and often such information does not exist, then there can be no real certainty that the sample, although drawn by random methods, is representative. This is where what is known as sampling theory comes in and in

Chapters XII and XIII the reader may learn how the statistician can nevertheless generalise his sample results for the population with considerable confidence. Samples drawn by methods which are not random cannot yield reliable data. So well understood is this maxim, *i.e.* all samples must be random, that the statistician seldom prefaces the word 'sample' with the adjective 'random'. For him, all statistical samples are random. If they are not, then they have no interest for him. The methods used to select samples in various types of surveys are discussed in Chapter XIV. At this stage it is pertinent to point out that 'sample design' as it is known to statisticians is a highly complex technique. The purpose of good sample design is to ensure the maximum of information from the sample drawn. Contrary to a widely held belief, the size of the sample in a survey is quite independent of the size of the population from which it is drawn. For example, a random sample of some 2,000 voters would give a reliable estimate of the probable outcome of an election involving either three or thirty million voters. The only relevance that the size of the sample has is that the larger it is, the more confident the statistician may be about his conclusions from the data. But, as will be seen later, this does not alter the fact that quite small samples are big enough for most purposes.

Sample Schedules and Questionnaires

Although the subject of schedule design is discussed in more detail in Chapter XV, the points and observations made already may be better impressed on the reader's memory by the opportunity to study some illustrations of good questionnaires and schedules. The first questionnaire was used in the Family Census in 1946, which was undertaken on behalf of the Royal Commission on Population. The form was sent to some 1·7 million married women together with the accompanying letter. The form is exceptionally well-designed to facilitate its correct completion; note for example the three clear subdivisions of the questions relating to the woman, her children and her husband. Note too the footnote stressing the need for an accurate description of the husband's occupation. Questions relating to work or occupation are often inaccurately answered; either the description is too vague for classification or the respondent up-grades himself! In view of the subject matter of the enquiry, the stress laid on its confidential nature in the letter is noteworthy, while the heavy black type at the head of the questionnaire 'strictly confidential' is almost a stroke of genius on the part of the designer. The letter is well phrased and the reader will observe how the benefits of the enquiry are conveyed to the respondent in the opening paragraph with its references

ROYAL COMMISSION ON POPULATION

FAMILY CENSUS

Dear Madam,

You will probably have seen in the newspapers, or heard on the wireless, that the Royal Commission on Population are trying to find out how family size has been changing during the past generation. At present it is not known how much childlessness there is, or how many families there are with one, two, three or other numbers of children. These and similar facts are needed urgently, so that the Royal Commission can give to the people and the Government a sound basis for understanding the population problem and its bearing on housing programmes, family allowances, social insurance and other measures of social welfare.

To collect the facts a 'Sample' Family Census is being held. *A strictly confidential inquiry* is being addressed to one tenth of the women in this country who are or have been married. If your name has incorrectly been included in our lists and you are not or have not been married, please disregard the questions on the form. Just write 'SINGLE' in the space against Question I and give the form back to the Royal Commission enumerator who calls to collect it.

If you are or have been married, the Royal Commission are very anxious for you to answer the questions on the form. If you wish to fill up the form yourself, please do so and give the completed form to the Royal Commission enumerator who calls for it. If you have difficulty with any of the questions, please tell the enumerator, who may be able to help you.

When your form is filled up and collected, it will be sent to the Royal Commission, where it will be dealt with together with many thousands of other forms. All these forms will be treated with the utmost confidence. The information on occupation and family size will be put on special cards for study, *but your name and address will not be used in any way.*

This Family Census is of the greatest importance to the country. If the results are complete and trustworthy, they will throw much light on problems of family and population, problems in which, as the Royal Commission know, men and women throughout the country are taking a keen interest. The Commission need your co-operation in this inquiry and they ask you to do all you can to ensure that it is a success.

Yours truly,

SIMON,

Chairman,

Royal Commission on Population.

ROYAL COMMISSION ON POPULATION
FAMILY CENSUS — Strictly Confidential

If you care to fill up this form yourself, please do so and give it completed to the Royal Commission Enumerator who will call to see you. If you prefer it, the Enumerator will be glad to fill up the form for you or help you with any difficulties you may have.

YOURSELF

Please write clearly and in full

1 Are you now **Married or Widowed** – or was your last marriage ended by Divorce? } *Please state which*

	M 1
	M 2

2 When were you born?

Month	Year

	MM 3
	MM 4

3 For those who have been **Married Once Only**:

Month	Year

  (a) When were you married?

  (b) If your marriage has ended – when did it end?
  (*By death of your husband, or divorce – NOT separation*)

	W 5
	W 6

4 For those who have been **Married More Than Once**:

Month	Year

  (a) When were you **First Married**?

  (b) When did your **First Marriage End**?

1 P	WM
2 E	7 F
3 OA	8 AW
4 S	9 L
5 WE	X AF

YOUR CHILDREN

Month	Year

5 (a) Number of **Children Born Alive.**

Beginning with your FIRST BORN child – enter, in order of birth, the date of birth of EVERY LIVE BORN CHILD you have had – *whether or not the child is still living.* Do NOT include still-births or miscarriages.

In the case of twins or triplets, use a separate line for every child born alive. Step-children or adopted children should NOT be counted.

  1st child
  2nd ,,
  3rd ,,
  4th ,,
  5th ,,
  6th ,,

	L.C.

(b) **No Children**
If you have NOT borne a living child, write NIL in this box. [ ]

  7th ,,
  8th ,,
  9th ,,

Note: For those who have had more than 10 children there are more spaces on the back. 10th ,,

	T.C.

6 Of your children Alive today, how many have NOT yet reached their Sixteenth birthday?
(*Only children BORNE BY YOU and under 16 – even if they are living away from you*)

YOUR HUSBAND   *If possible, discuss this section with your Husband*

7 (a) What is your **Husband's Occupation?**

(*If he is retired, out of work, or dead* – state his former occupation)
(*If he is temporarily in the Services* – state former occupation. If no former occupation – put '*Armed Forces*')
(*If he is a regular Sailor, Soldier or Airman* – state which, and his rank)
(*If you have been married more than once* – the answer should refer to your FIRST husband)

Note: Please describe the KIND of work your husband does in as much detail as possible. For example: if your husband is an Engineer, it will help us if you can say EXACTLY which kind he is.

(b) Is Your Husband –

1 An employer of 10 or more people?
  or
2 Working for himself or employing LESS than 10 people?
  or
3 Employed and earning a monthly salary?
  or
4 Employed and earning a weekly or other wage?
PLEASE PUT A RING ROUND THE NUMBER WHICH APPLIES

(c) **Employer's Business** –

(*If your husband is NOT himself an employer or working for himself*)

2*

to housing, family allowances and social welfare, all topics in which the average respondent would have a real interest. The ultimate response rate was 87 per cent., a very good result when it is recalled that unlike most government enquiries, the return of the questionnaire was voluntary.

The second illustration shows the questionnaire distributed by the Federation of British Industries to a sample of its member firms. The object of this enquiry, which takes place at four monthly intervals, is to obtain an indication of industrial opinion on the current state of the economy and industrialists' views regarding the immediate prospects, *e.g.* during the next four months. The form is well designed and quite simple to complete. Note that the answers required are not in terms of figures, but merely a broad indication of the respondent's views. Obviously, it is much easier to complete such a questionnaire; in any case most firms would be unwilling to give actual figures, even if they were prepared to go to the trouble of extracting them from their books. The whole point of this particular enquiry is simply to evaluate business and industrial opinion, which is a highly important factor in the economic situation at any time. This type of enquiry is a relatively new development in the field of what is usually termed 'forward-looking' statistics. In other words, instead of the economic statistician concerning himself solely with past events, he is here endeavouring to make some estimate of the future trend of the economy. Inevitably, any deductions drawn from such data must be somewhat tentative, but a recent study of the results derived from this enquiry during the past six years since its introduction concludes that the information has been helpful to government economists in formulating policy.[1] A particular virtue of the F.B.I. enquiry is the speed with which the data are processed and the results published.

The final example is of a schedule used by interviewers in a survey undertaken by Research Services Ltd on behalf of the Independent Television Authority and reproduced in the report *Parents, Children and Television*.[2] In this survey only parents aged between 30 and 49, with children aged 5 to 13, were interviewed. The reader should note the classificatory data relating to family composition on the first page of the schedule and the additional information on the same subject placed at the end of the schedule. Such information, usually collected at the end of the interview, permits a more thorough analysis of the information collected, in this case parents' attitudes by reference to their social class, occupation, education, etc. The codes on the right hand side of the questions are used when the information is being transferred from the schedules to punched cards, the appropriate

[1] National Institute Economic Review. Nov. 1963.
[2] H.M.S.O., 1959.

Industrial Trends Enquiry Number 20 : June 1964

Please tick appropriate answers: If question not applicable, tick N/A

A.I.   What is the MAIN activity covered by this return   (For definitions see back page)

1 ☐ Mechanical engineering
2 ☐ Electrical engineering
3 ☐ Metals and metal manufacture
4 ☐ Vehicles
5 ☐ Shipbuilding and marine engineering
6 ☐ Textiles, clothing, footwear and leather

7 ☐ Food, drink and tobacco
8 ☐ Chemicals, paints, petroleum
9 ☐ Building materials and components
10 ☐ Paper and printing
11 ☐ Other

II. How many EMPLOYEES are covered by this return   (a) 0–499 ☐   (b) 500—4,999 ☐   (c) 5,000 and over ☐

B. Total trade

1 Are you more, or less, optimistic than you were four months ago about the general business situation in your industry

More	Same	Less

2 Do you expect to authorise more or less capital expenditure in the next twelve months than you authorised in the past twelve months on :   a. buildings   b. plant and machinery

More	Same	Less	N/A

3 Excluding seasonal factors is your present level of output below capacity (i.e., are you working below a satisfactorily full rate of operation)

Yes	No	N/A

Excluding seasonal variations, what has been the trend over the PAST FOUR MONTHS, and what are the expected trends for the NEXT FOUR MONTHS, with regard to :

4 Numbers employed
5 Rate of total new orders
6 Rate of new orders received for capital equipment
7 Level of output
8 Stocks of (a) raw materials and brought in supplies
        (b) finished goods
9 Average costs per unit of output
10 Average selling prices

Trend over PAST FOUR MONTHS				Expected trend over NEXT FOUR MONTHS			
Up	Same	Down	N/A	Up	Same	Down	N/A
1	2	3	4	5	6	7	8

11 What factors are likely to limit your output over the next four months. Please tick the *most* important factor or factors

Orders or Sales	Skilled Labour	Other Labour	Plant Capacity	Credit or Finance	Materials or Components	Other

C. Export trade

Questions 12—17 *should not be completed unless your direct exports exceed £10,000 per annum*

12 Are you more or less optimistic about your EXPORT PROSPECTS for the next twelve months than you were four months ago

More	Same	Less

Excluding seasonal variations, what has been the trend over the PAST FOUR MONTHS and what are the expected trends for the NEXT FOUR MONTHS, with regard to :

13 Rate of new orders received for exports

14 Level of export deliveries

15 Average prices at which export orders are booked

	Trend over PAST FOUR MONTHS				Expected trend over NEXT FOUR MONTHS			
	Up	Same	Down	N/A	Up	Same	Down	N/A
	1	2	3	4	5	6	7	8

16 Are your prospects likely to improve, remain steady or worsen in the following markets during the next four months (Please reply only for markets that are significant to you.)

	North America	Central & S. America	" The Six "	" The Seven "	Sov. Union & E. Europe	Austral asia	Africa	Middle East	Indian Sub-Cont.	Rest of Asia
Improving										
Steady										
Worsening										
N/A										

17 What factors are likely to limit your ability to obtain export orders over the next four months. Please tick the *most* important factor or factors

Prices (compared with overseas competitors)	Delivery Dates	Credit or Finance	Tariff Quota or Licence Restrictions	Other

Signature...

Company ....

Note : If you wish your reply to remain anonymous, please detach this slip and return it under separate cover

The Questionnaire

J.1016   *April, 1958*

Research Services Ltd., 91 Shaftesbury Avenue, W.1

Family composition

	Sex		Age	Office use		Codes
Relationship to informant	M	F	(*years*)	a	b	
Informant	X	Y		c	d	
	X	Y				
	X	Y		e	f	
	X	Y		g	h	
	X	Y				
	X	Y		i	j	
	X	Y				
	X	Y				
	X	Y				
	X	Y				

If more than one child aged 5—13 inclusive, questions are to be asked in respect of the child whose birthday comes first after date of interview. *Tick* to show which child.

All informants

1. What time did your child go to bed last night ?   Yesterday ....
   (On MONDAYS ask about *both* SATURDAY and SUNDAY)
   Saturday ....

2. How much time did he/she spend yesterday playing out of doors ?   Yesterday .... hours
   (On MONDAY ask about *both* SATURDAY and SUNDAY)
   Saturday .... hours

3. How much time did he/she spend yesterday doing homework ?   Yesterday .... hours
   (On MONDAY ask about *both* SATURDAY and SUNDAY)
   Saturday .... hours

4. How often does he/she go to the cinema ?

Twice a week or more often	1
Once a week	2
Once a fortnight	3
Once a month	4
Less often than once a month	5
Never	6

5. (*a*) In your opinion, is it easier or harder to be a parent than it was fifty years ago ?

Easier	7
Harder	8
Don't know	9

   (*b*) Why ? ....
   ....
   ....

J.1016

6. Here are some things which some children occasionally do. (Show list). Supposing your
own child were to behave this way, would you please tell me :
(a) How you would deal with it ?
(b) What would you think might be the cause ?
(c) How would your parents have dealt with you if you had behaved that way ?

Behaviour	How informant would deal with	Cause	How inft's parents would have dealt with
(i) Child constantly shouting, breaking things, fighting			
(ii) Child lazy, refusing to help around house, untidy			
(iii) Started stealing money			
(iv) Regularly would not eat meals prepared for him/her			

7. (a) Do you feel that films which children usually see at cinemas are good entertainment
for children or not ?

Good	1
Not good	2
Don't know	3

(b) Why do you feel that way ? ....
....
....

8. Do you have a television set in your home ?

Yes	1
No	2

If Yes, ask Questions 9-23, then 24-28
If No, go to Question 24
All Parents with TV in Home

9. Does it receive both BBC and ITA
or just the BBC ?

BBC and ITA	3
BBC only	4

10. Which of these programmes have you watched in the past week ? (SHOW LISTS—
ITA and BBC list for ITA and BBC set owners : BBC list for BBC only set owners)

11. (About each programme watched). Were you watching it along with your child ?

12. Are there any programmes on the list which you have encouraged your child to watch—e.g.
by suggesting to him/her that it's good ?

13. (a) Are there any on the list which you have discouraged him from watching ?
(b) If yes: in what way ?
(c) Why ?
(d) (If not mentioned at 11). Did your child, in fact, watch this programme ?

Record Answers to Questions 10-13 on next two pages.

Codes

J. 1016

	Codes

14. How much time did your child spend watching TV yesterday? (On MONDAY ask about *both* SATURDAY and SUNDAY)

Yesterday................hours
Saturday................hours

15. During the past week has your child stayed up beyond his usual bed time to see any particular programme? (Here is a list of last week's main evening programmes as a reminder). *If yes :* which?

Yes 1
No 2

Day	Programme

16. (a) What is your personal opinion of programmes on Children's Television—those in the afternoon and up till about 6 p.m., (i) on ITA. (ii) on BBC?

	ITA	BBC
Very good	1	7
Reasonably good	2	8
Not up to much	3	9
Bad	4	0
Don't know	5	X

(b) In what way?
  (i) ITA..
  (ii) BBC...

17. (a) And what about the programmes from 6–7.30 p.m. (i) on ITA. (ii) on BBC.

	ITA	BBC
Very good	1	7
Reasonably good	2	8
Not up to much	3	9
Bad	4	0
Don't know	5	X

(b) In what way?
  (i) ITA..
  (ii) BBC...

18. Do your children stay at home more or less because of television?

More 1
Less 2
Same 3
Don't know 4

19. Do you spend more time together or less as a family because of television?

More 5
Less 6
Same 7
Don't know 8

20. Would you say that television has made your home life more interesting and happier, or do you think your family life would be better without it, or does it make no difference?

More interesting 1
Better without 2
No difference 3
Don't know 4

21. Would you say that having television has made it easier for you to bring up your children as you would like them to be, or harder? Or does it make no difference?

Easier 5
Harder ,6
No difference ,7
Don't know 8

J. 1016

	Codes

22. What would you say your child did before you got your set with the time he now spends watching television?

....

23. How long have you had a television set? ....

All informants

24. Supposing your child had a couple of free hours with nothing to do, which would you prefer him/her to do :

Go to the pictures	1
Watch television	2
Read the newspaper	3
None of these	4
Don't know	5

25. Why do you think that? ....

26. Would you say television is :

Good for children	1
Bad for children	2
Makes no difference	3
Don't know	4

27. In what way? ....

28. With which of these do you agree or disagree :
Watching television makes children :

	Agree	Dis-agree	Don't know
More sociable and friendly	1	2	3
Better informed and more knowledgeable	4	5	6
Mentally lazy	7	8	9
Better at their school work	0	X	Y
Noisy and disobedient	1	2	3
Better behaved	4	5	6
Violent and cruel	7	8	9

29. Which daily newspaper(s) do you yourself read regularly? (i.e., 3 issues out of 4)

Daily Express	1	Daily Sketch	7
Daily Herald	2	Daily Telegraph	8
Daily Mail	3	Times	9
Manchester Guardian	4	Daily Worker	0
Daily Mirror	5	Other (state)	X
News Chronicle	6	None	Y

Classification :

Sex of informant	Man 1	Age 30–39—3	Social grade	AB	5
	Woman 2	40–49—4		C1	6
				C2	7
				DE	8

Occupation of informant ....

Occupation of H/H ....

Education of informant :　　　**Type of school (child) :**

Elementary	1	Primary	5	Day ....
Secondary	2	Secondary modern	6	
University or college	3	Secondary grammar	7	Date ....
Other (state)	4	Other (state)	8	

Date of child's return to school ....

Sample	BBC only	1
	BBC and ITA	2
	Neither	3

Name of informant ....

Home address ....

If interviewed at work, business address ....

Quota district ....　　Inv. No. 　.　.　.　.

code for the answer having been ringed by the interviewer. The lists referred to in Q. 10 to 13 are not reproduced here, but they consist of two lists of children's programmes throughout the week; one for B.B.C. and the other for I.T.V. The main value of such lists is to enable the respondent to recall just which programmes her child or children did view. Note also the use of the pre-coded question, *e.g.* Q. 4, 5, 20, etc., sometimes referred to as 'multi-choice' or 'cafeteria' questions. The advantages of this are obvious. The interviewer merely marks off the appropriate response, thus saving time and facilitating the classification of the respondent's answers. To the extent that the respondent's answer may not be expressed so tersely as these pre-coded answers suggest, it is for the interviewer to ring the answer which she feels most closely corresponds to the respondent's opinions. This can be a fruitful source of error, and the subject is discussed in Chapter XV.

The outstanding characteristic of the foregoing questionnaires and schedules is their relative simplicity. By this is meant not that the questions are simple, but the forms and their layout convey the impression that they will not pose difficulties for the respondent. While this first impression upon the respondent is clearly of vital importance with a postal questionnaire, it is no less important with a schedule which is to be completed by an interviewer. A great deal of care is needed to help the interviewer complete the form so as not to interrupt the smooth flow of questions. Even the size, much less the layout of the form, and the way it can be folded and turned over is important, as anyone who has tried reading a newspaper on a windy corner well knows.

A particular defect on many questionnaires are the footnotes and what are termed 'instructions for completion'. These reach their nadir with the leaflets provided by the Inland Revenue, but it is unfortunately true that many censuses and surveys relying on a questionnaire are bad in this respect. The reader should look up the agricultural return known as the 4th June return. It is reproduced in *Guides to Official Sources; No.4* and it is a prime example of a misguided effort to ensure statistical accuracy with its welter of notes and definitions to be completed by some half a million occupiers of agricultural holdings over one acre. Even the population census form is not free from this defect; the notes cover most of the back of the schedule. The sad truth is that many people do not bother to read such notes. Hence, if they must be inserted, the best place for them is by the question itself. There is then a fair chance of them being read. The Census of Production forms suffer from this handicap too, and no doubt it is only the fact that the respondent is under a statutory obligation to return the form that prevents it being consigned to the waste-paper basket immediately upon receipt. In the case of voluntary enquiries and surveys such shortcomings cannot be permitted. Simplicity of design must be the keynote of the questionnaire. If the letter accompanying the form can indicate any benefits which may accrue to the respondent from completing the form, so much the better. Above all, remember the respondent is doing you a favour by answering your questions.

CHAPTER IV

TABULATION

The Purpose of Tabulation

In no investigation of any size is the volume of collected data or material so small that it may be rapidly or easily assimilated by a perusal of the completed forms. At best only the haziest impressions may be gathered of the ultimate results, and those impressions may well be the reverse of the truth for it is usually the unusual or freak cases that stick in the memory to the exclusion of the many more 'ordinary' replies. The statistician's first task is to reduce and simplify the detail into such a form that the salient features may be brought out, while still facilitating the interpretation of the assembled data. This procedure is known as classifying and tabulating the data; *i.e.* extracting from the individual questionnaires or schedules the answers to each question and entering the replies on separate summary sheets. These totals are then transferred to the relevant columns of prepared tables.

Before the summarising is commenced, the questionnaires should have been checked on receipt to ensure that they have been completed reasonably correctly. The person editing the returned forms cannot know if they are *correctly* completed, otherwise there would be no need for the enquiry. 'Reasonably correct' in this context means that there are no obvious mistakes in, or contradictions between, the various answers. Inevitably there will be uncompleted forms, and these it may be possible to complete by further enquiry; in other cases the replies will be useless, as they are either irrelevant or patently false. It requires little imagination on the reader's part to visualise the task involved in sorting several thousands of forms and tabulating their contents without mechanical assistance. The results of many large-scale enquiries would be available only long after the field work had been completed. Fortunately the introduction of mechanical punching and sorting machines has facilitated the task of the statistician. All that is now necessary is for the information on the questionnaires to be transferred to specially designed punched cards, the machines then sort the cards, tabulate and compute the totals at great speeds. There is, of course, the risk that the cards may be incorrectly punched by the operator, but this problem can be overcome with adequate supervision. The cost of such methods is considerable and careful thought must be given to the information to be entered

35

on the cards to ensure that the maximum information will be given out by the machines in the minimum space of time.[1] According to an article in the *Journal of the Royal Statistical Society*,[2] the representative of a machine accounting company was able to devise a procedure whereby the first results of the census of the population of Cyprus were available within a few weeks, instead of several months, as is usually the case. More recently, however, punched card installations have been superseded by computers in cases where the volume of work is very large. Even so the data may still be recorded initially on punched cards and then transferred to tape which is then fed into the computer. The .computer, is as explained in Chapter XIX, serves as a store for such information and is also, upon demand, capable of processing the data contained in its storage units at very high speed. Thus large numbers of combinations of the basic data can be produced in a short space of time. The 1961 Census of Population data are being processed on a large computer and it was hoped that this would mean an end to the customary long delay between collection and publication. In the event, a shortage of programmers, *i.e.* persons who prepare the data for the computer, as well as the fact that the Census authorities did not have the sole use of the computer, have together produced serious delays.

The Basis of Classification

Before the actual tabulation can be undertaken there is an intermediate stage, generally described as classification. The point has already been made that statistics is concerned with aggregates, the individual members of which are homogeneous. That is, all the items comprising the aggregate or what the statistician calls the 'population' are of one type, *i.e.*, they possess a characteristic in common. For example, retail businesses may be classified according to turnover, schoolboys according to their heights, shares quoted on the Stock Exchange according to the dividend paid. The 'characteristics' in these cases are: turnover, height and dividends respectively, *i.e.* these constitute the link or basis of comparison between all the items within each group. These groups of individual items are usually termed statistical *series* or *distributions*.

A more precise way of defining a series or distribution is that it comprises a group of items which are related one to the other by the possession of some common characteristic. The term 'series' is usually restricted to data which have been collected over time. The figures of the annual turnover of a firm for the past decade would be

[1] *J.R.S.S.*, 1946, Part III, p. 284. An article by O. Kempthorne illustrates the use of mechanical methods in some detail, with special reference to the National Farm Survey data.

[2] *J.R.S.S.*, 1947, Part II, p. 138. An experiment in census tabulation. – D. A. Percival.

described as a *time* series. Data which relate to any characteristic other than time may be classified as spatial or attributive distributions. Whereas the time series indicated the turnover of a group of departmental stores over a period of several years, a *spatial* distribution may be one which classifies the turnover of any period according to the location of the sales. Thus, the turnover may be classified according to the departments of the stores, or on the basis of comparing the annual turnover of the various stores in the different towns. In brief, spatial distributions are concerned with location.

The term *attributive* covers all distributions other than time series and spatial distributions. An attribute is simply a characteristic; data are classified according to their attributes. As already explained (p.11) these fall into two main types, those which are capable of numerical expression and those which are not. Thus, in the example of schoolboys classified according to height, the characteristic can be expressed numerically, *i.e.* so many inches. But if the boys are being graded by the school doctor on the basis of their general health and physique, they could only be classified somewhat as follows: excellent, good, fair, and poor. The first example of an attribute, which can be expressed in quantitative terms, *i.e.* height in inches, is termed a *variable;* the second type of classification is based on the *attribute* itself, *i.e.* a quality incapable of being measured in numerical terms, such as 'good health', and is given that name. These points have been repeated because they are relevant not merely to classification; they also affect the type of tabulation used and, as will be shown later, the form of diagrammatic representation of the data.

The Construction of Tables

The purpose of tabulation is primarily to condense and thereby facilitate comparison of the data. The form of the tables employed will vary according to the nature of the data and the requirements of the survey. In consequence, it is not possible to lay down hard and fast rules which may be applied in all cases. It may come as a surprise to the reader to learn that the tables are usually drawn up before the enquiry is actually started. More precisely, the frame of the tables is drawn up and this has two advantages. First, it enables the survey team to visualise the sort of data they want and are going to get and second, it sometimes draws attention to other information which would be of interest and provision for such questions can then be made on the questionnaire. As with so many matters, common sense dictates certain guides which should be borne in mind in the construction of any statistical table if it is to serve the purpose of revealing the basic structure of the data.

In no case should the table be overloaded with detail. A closely-printed and concentrated mass of figures may appear most impressive to the casual observer, but merely compels the reader to do what ought to have been done by the compiler of the table at the outset: namely, to reduce the mass into several sub-tables, each bringing out a separate aspect of the data. The purpose of the table should be immediately apparent, *i.e.* it should have a clear and concise title, although clarity and precision should not be sacrificed for brevity. Occasionally tables are encountered where the main title is amplified by a series of footnotes; wherever possible, this practice should be avoided. Where the individual figures are large, the table gains in clarity far more than it loses by eliminating the '000's' or even '00,000's', *i.e.* the final digits. This is especially true of summary tables which are often inserted in the text in the body of the main report or its conclusions; individuals seeking detailed figures can be referred back to the full tables which are best put into an appendix separate from the main report.

It is highly desirable in any report presenting data collected for that enquiry to precede the information presented in tabular form by a short summary of the methods of collection employed, in order that the reader may obtain some idea as to the probable realiability of the results given in the tables. If secondary data from other published sources are given, say for purposes of comparison, then a footnote should be appended indicating the source of that particular section of the table. Especial care should then be taken to leave the reader in no doubt as to the unit of measurement: whether it be £ sterling or £ Australian, long or short tons, ton-miles of passenger trains, or goods trains, etc. If any heading is at all liable to misinterpretation, a clear definition should be provided as to what information is included under that head. Thus, the statistics published by the Home Office of 'persons proceeded against for drunkenness' do not include those persons charged with 'driving under the influence of drink'; these are incorporated with offences against the Highway Act.

Simple Tabulation

To illustrate the normal procedure in tabulation, the data given in Table 1 relating to the individual outputs of 180 workers producing a certain manufactured article are set out with the smallest output at the beginning of the group, and the largest at the end, *i.e.* in order of magnitude. Such an arrangement of the data is known as an *array*. Inspection of the table reveals that the minimum and maximum outputs are 501 and 579 respectively. The difference between these two quantities is described as the *range*. Apart from the range, it is impossible without further careful study to extract any exact information

TABLE 1

GREAT PRODUCERS LTD.

INDIVIDUAL OUTPUTS OF 180 FEMALE OPERATIVES IN PLANT 1, IN THE WEEK ENDING 8TH NOVEMBER, 1963

501	520	534	540	547	555
503	522	535	542	547	557
503	522	535	542	547	557
504	523	535	542	547	557
506	523	535	542	547	559
507	524	536	542	548	559
507	525	536	542	548	559
509	525	537	543	548	559
510	526	537	543	548	559
511	526	537	543	549	561
511	527	537	543	549	561
513	527	537	544	549	561
515	527	537	544	549	563
515	528	538	544	550	563
515	528	538	544	550	563
515	528	538	544	550	564
515	528	538	544	550	565
515	528	538	545	551	565
517	528	539	545	551	565
517	530	539	545	551	567
518	530	539	545	551	567
518	532	539	546	552	567
519	532	539	546	552	569
519	532	539	546	552	569
519	532	539	546	553	569
519	532	539	546	553	572
520	532	540	546	553	574
520	534	540	547	553	575
520	534	540	547	555	577
520	534	540	547	555	579

of any value from the table. By breaking down the data into the form of Table 2 below, however, certain features of the data become apparent. Thus, by setting the number of workers producing each individual output against that figure, a more intelligible picture is provided. Even a superficial scrutiny of Table 2 reveals that the outputs from 537 to 547 inclusive occur most frequently. Such a table is known as a *frequency distribution*. It is so described because it indicates the frequency or number of times each individual output figure occurs. More precisely, it tabulates the frequency of occurrence of the different values of any given variable. Nevertheless, even after this simplification, since there are still too many figures to assimilate, the conventional procedure is to construct a *frequency table* as in Table 3.

TABLE 2
FREQUENCY DISTRIBUTION OF OUTPUTS DETAILED IN TABLE 1

Output	Frequency	Output	Frequency	Output	Frequency
501	1	527	3	550	4
503	2	528	6	551	4
504	1	530	2	552	3
506	1	532	6	553	4
507	2	534	4	555	3
509	1	535	4	557	3
510	1	536	2	559	5
511	2	537	6	561	3
513	1	538	5	563	3
515	6	539	8	564	1
517	2	540	5	565	3
518	2	542	6	567	3
519	4	543	4	569	3
520	5	544	6	572	1
522	2	545	4	574	1
523	2	546	6	575	1
524	1	547	8	577	1
525	2	548	4	579	1
526	2	549	4		

TABLE 3
GROUPED FREQUENCY DISTRIBUTION. DATA FROM TABLE 2

Output	No. of Operatives
(Units per operative)	
500— 9	8
510—19	18
520—29	23
530—39	37
540—49	47
550—59	26
560—69	16
570—79 '	5
	180

The data in this form are sometimes described as a *grouped* frequency distribution.[1] Instead of the 'frequencies' of each single output being shown separately, the range (difference between maximum and minimum outputs) is sub-divided into smaller groups, usually termed 'classes'. In this example each class comprises ten units of output. Thus the first class, 500-509, covers all ten values inclusive. In this class eight operatives had outputs of 500 units or above, but

[1] The term generally used is 'frequency distribution.' *i.e.* the same term is usually used whether the data are grouped or not.

none more than 509. In the fifth class, 540-549, there were 47 operatives, none of whose individual outputs was below 540 or exceeded 549. The reader can and should verify the figures in Table 3 by reference to the previous table.

By grouping the data into the form of such a frequency table, the basic structure of the information is prominently revealed. The main body of operatives have outputs falling within the middle classes, and only a few operatives come within the classes at either extreme. In passing, it can be stated that the frequency table or 'grouped' frequency distribution is the most common form of presentation of numerical data, and, as will be seen later, is the basis of most statistical analysis. By convention, the figures of output are usually referred to as the *independent* variable. The corresponding frequencies are known as the *dependent* variable.

The same data can be presented in cumulative form, *i.e.* 86 operatives each produced less than 540 units per week, 133 operatives less than 550; and so on up to the last stage, when 180 operatives each produced less than 580 units. Note that the 'cumulation' may be upward or downward; the upper half of Table 4 is read as '133 operatives produced less than 550 units per week . . .' etc.; the lower half as '94 operatives produced 540 or more units per week'.

TABLE 4
DATA FROM TABLE 3 PRESENTED IN CUMULATIVE FORM

Output (Units per operative)	500–9	510–9	520–9	530–9	540–9	550–9	560–9	570–9
No. of operatives	8	26	49	86	133	159	175	180
	180	172	154	131	94	47	21	5

This table yields the data to answer such questions as 'What percentage or proportion of the workers produce 550 or more units per week?' In this case the answer would be 26 per cent. (*i.e.* $\frac{47}{180} \times 100$).

The Selection of 'Classes'

The preparation of grouped frequency distributions from the raw data may give rise to difficulties. The greatest of these is deciding the number of *classes* into which the series may be divided: *e.g.* in the above table there are eight classes, 500-9, 510-9, and so on up to 570-9. This problem is directly linked with the second: what is the size, or, more precisely, the range, of each class to be, *i.e.* what is the *class-interval*? Thus, in the above example, the class-interval is ten units. There are no hard and fast rules on these points, but generally it is desirable that the number of classes should not exceed 10 or 15, depending on the range of the independent variable and the total frequencies in the distribution, otherwise the purpose of the table, the reduction of the data to manageable size may be defeated. As to the size of the class-interval, this will depend primarily on the number of classes and the distribution of the frequencies. Sometimes the class-interval is easily determined, *e.g.* if families are being classified according to the number of children in each, then the class interval is clearly one child. The frequency table heading would read:

| Number of children per family | No. of Families |

There are no set rules which if followed will ensure good tabulation in all cases. The student should at all times bear in mind the two main needs of tabulation work. First the table must be comprehensible in that it reduces the data to manageable proportions, and second, the content of the table should be clearly yet simply defined in the title and column headings. In brief, Table 1 on p. 39 may have its uses for detailed analysis of the factory's production performance; but as a means of conveying the state of affairs in the plant Table 3 is far more comprehensible and its content easier to grasp. A frequency distribution should resemble Table 3 rather than Table 1. Note that it is of the utmost importance to define the successive classes so that there is no doubt as to which class any single value should be allotted. In Table 3 the classes are 500–509, 510–519, etc., which avoids any uncertainty as to the allocation of terminal values such as 510, or 520, which would arise if the classes had been written incorrectly as 500–510, 510–520, etc. Note that where the variable is discrete, then there will be a clear gap between the lower limit of one class and the upper limit of the preceding class, *e.g.* 520 and 519. This natural break does not arise with continuous variables. For example, if men are classified by height, then the simplest classification is as follows: under 5ft. 6ins.; 5ft. 6ins. and under 5ft. 7ins.; etc. This again ensures that no confusion arises as to the allocation of any individual height to its appropriate class.

Where the individual values to be classified tend to group themselves around particular values, care should be taken when deciding upon the class-interval that such values of the independent variable coincide as far as possible with the mid-points of the classes. For example, if in Table 1 a large majority of the outputs ended with the digit 5, *e.g.* 545, 555, etc., the classification used in Table 3 is excellent. If, however, there were many outputs ending in 9, 0 or 1, then it would be preferable to draw up classes as follows: 505–514, 515–524, 525–534 and so on. The reason for this is that it is customary to regard the middle values of classes as the 'average' value of the items in that class. If, as will be shown in Chapter VI, calculations are to be carried out on the data, any inaccuracy or bias in the grouping within the class intervals may distort the final results. To illustrate the effect of using the same-sized class interval (10 units) with different limits, Table 5 has been drawn up showing the data from Table 1 classified in two ways. The differences can be observed by comparing the resultant distributions.

TABLE 5
EFFECTS OF DIFFERENT CLASSIFICATION ON DISTRIBUTION OF DATA FROM TABLE 1

First Grouping	Frequencies	Frequencies	Alternate Grouping
—	—	4	Under 505
500— 9	8	8	505—514
510—19	18	24	515— 24
520—29	23	25	525— 34
530—39	37	46	535— 44
540—49	47	41	545— 54
550—59	26	18	555— 64
560—69	16	11	565— 74
570—79	5	3	575— 84
	180	180	

Further reference will be made to this matter when the various averages are discussed, but keeping the above principles in mind, the numerous types of tabulation and classification can be examined. Table 3 above, *i.e.* the frequency distribution, is the basic and most simple form of presenting data. Even the most complex tabulation comprises little more than a number of such tables brought together under one head. Nevertheless, the more information which it is sought to bring into one table, the more important it becomes to ensure that the table remains intelligible and easily read.

Further Examples of Tabulation

A number of tables taken from official publications are reproduced in the following sections, together with comments upon their salient features.

Table 6 is an illustration of double tabulation describing the distribution of personal incomes in Great Britain in the fiscal year ended 31st March, 1961. The first two columns form a frequency distribution of incomes; the final column gives additional information regarding the share of the aggregate between the classes. The class interval, it will be noted, is not constant, but is adequate to illustrate the main features of the distribution of incomes in this country. When the same data for the fiscal year 1952-3 were published in the 97th Report, the class-interval was as follows: £250–500, 500–750, 750–1,000, etc.

TABLE 6

DISTRIBUTION OF PERSONAL INCOMES AFTER TAX 1960-61

Range of Total Income	No. of Incomes	Income before Tax
£	000's	£ million
180 – 249	1,407	307
250 – 499	6,292	2,499
500 – 749	6,672	4,462
750 – 999	4,400	4,127
1,000 – 1,999	2,493	3,606
2,000 – 3,999	297	1,190
4,000 – 5,999	33·7	365
6,000 and over	4·1	134
	21,600	16,690

Source: Based on Table 70 in 105th Report of the Commissioners of H.M. Inland Revenue, Cmnd. 1906, H.M.S.O. 1963.

This illustrated the error of classification referred to earlier, for where such a grouping is used, it creates doubt as to the correct treatment of border-line incomes, e.g. £500, £750 and £1,000. Are they in the class with those figures as the upper limit, or in the next class where the figure is the lower limit? The lower and upper limits of successive classes in that year should have been written as £250 and under 500, £500 and under 750, etc., so that no uncertainty arose as to the treatment of incomes of £500, £750, etc. In Table 6 the Revenue statisticians have avoided the earlier mistake by defining precisely the actual limits of each class. Such a classification requires the individual incomes to be rounded to the nearest £.

A rather unusual method of defining the class interval is given in an official publication dealing with roads.[1] The frequency table classifies roads in Great Britain by their width in feet and gives the number of miles of each width of road. The classes are written as follows, width in feet: 8–10, 10–12, 12–14, etc., at first sight a faulty classification. Inspection of the table reveals a footnote to the title of the independent variable, explaining that '18–20ft., for example, includes roads of

[1] See Table 3 in Sample Survey of the Roads and Traffic of G.B. Road Research Technical Paper No. 62. H.M.S.O. 1962.

exactly 18ft. but not those of exactly 20ft.' Admittedly, the obvious mistake has been avoided, but this particular practice should be rejected. In such cases where the measurements of the independent variable are given so that the variable is continuous, then write the classes as '18ft. and under 20ft.', and avoid footnotes.

When interpreting the data in Table 6 it should be borne in mind that these are, in fact, *declared* personal incomes. To the extent that a return of income is not made by various individuals who wish to avoid their obligations to the Revenue, or where the income declared is incorrectly stated, the table fails to reflect the true state of affairs. Since the number of individuals in these two categories are relatively small in relation to the tax-paying population the error is probably not significant. Note too that the total is rounded to the nearest hundred thousand.

It is interesting to note that thirteen people in every hundred earned over £1,000 a year in 1960-61 and that over one-third the personal incomes were below the £500 a year level. As an exercise the student reader could calculate the proportion of incomes in each income group and their share of the total incomes returned. The results are interesting. For example, less than 2 per cent (335 out of 21,600) receive just 10 per cent of total incomes (£1,689 out of £16,690).

Table 7 is an example of a time series based upon more detailed information relating to consumer expenditure in the U.K. given in the National Income and Expenditure Blue book 1963. The latter publication, which appears annually, is an important source of economic statistics and their compilation is a necessary part of economic planning. The table analyses the main components of consumer expenditure for various years during the period 1952-62 and in order to facilitate comparison the money outlays in each of the selected years have been adjusted to 1958 prices.

The reader needs to treat such economic data with especial care for two reasons. First, by adjusting current year prices to a common base, *i.e.* 1958 prices, it is apparently possible to show the *real* changes in consumption. Over these years prices rose steadily so that movements in the figures for each year on a current cost basis would reflect the combined effect of rising prices and quantity changes. The use of a price index enables all the totals to be put on to a common base and the annual changes then reflect only quantity changes. Unfortunately, the calculation of such an index and the adjustment of prices is a highly complex operation and at best the adjusted figures provide only an indication of trends. Index numbers of prices and their construction are discussed in Chapter X. The second problem is the reliability of the annual totals for each category of expenditure,

TABLE 7

CONSUMERS' DOMESTIC EXPENDITURE. U.K. SELECTED YEARS 1952-62*

(£ million. Figures adjusted to 1958 prices)

	1952	1954	1957	1960	1962
Food ..	3,977	4,246	4,503	4,736	4,865
Alcoholic drink	835	845	916	1,023	1,088
Tobacco ..	915	949	1,012	1,087	1,055
Housing ..	1,216	1,273	1,347	1,431	1,477
Fuel and Light	595	625	643	742	829
Clothing ..	1,153	1,267	1,450	1,616	1,629
Durable and other household goods	982	1,349	1,477	1,948	1,954
Travel[1] ..	704	740	816	950	1,057
Other ..	2,037	2,376	2,442	2,747	2,806
Total ..	12,414	13,670	14,606	16,280	16,760

* Source: National Income and Expenditure Blue Book, 1963, Based on Table 19.
[1] Including running costs of vehicles.

both before and after conversion to 1958 prices. Some of the figures are highly reliable, e.g. tobacco and alcoholic drink, since these can be checked against the receipts of the Customs & Excise authorities in respect of these items. In contrast, that part of the 'Travel' figure which is ascribed to private motoring is subject to a much greater margin of error in view of the assumptions that must be made in deriving the estimate from various sources, e.g. licence duties, petrol tax receipts, garage costs, etc.

There is always the danger that, unless the limitations of such published data are appreciated, the figures will be extracted from official publications as if they were correct to the nearest £ million, as the table heading and the figures given imply. At best such data give some indication of the relative order of magnitude, e.g. that outlays on fuel and light are lower than on tobacco, as well as some indication of the longer run trend. The Blue Book containing these tables gives some fairly detailed explanation of the sources of the data which should provide the average reader with a clear warning of their limitations. Unfortunately, all too often such figures are extracted and quoted without first referring to such explanatory notes.

Table 8 is another illustration of a frequency distribution since it classifies the trades unions in the U.K. by their size, this being measured by their membership. Note the clear classification and the increasing size of the class-interval. The table gives three pieces of information. In the first half it shows the changing pattern of unions by reference to their size. It may be seen quite easily that nearly 600 of them had in 1951 a membership of less than 5,000. The second half of the table shows that while the smaller unions are more numerous, the large unions completely outweigh them by virtue of their massive individual membership. The third piece of information is given in the comparative figures in both sections of the table for 1961. These demonstrate the extent to which the pattern of union size and their membership have changed during the decade.

TABLE 8

Trade Unions in the U.K. Analysis by Membership in 1951 and 1961‡

Size (No. of Members)	Number of Unions		Membership (000's)	
	1951	1961	1951*	1961*
Under 100 members	145	120	7	6
100 and under 500	193	166	49	42
500 „ „ 1,000	79	58	55	40
1,000 „ „ 2,500	103	97	164	152
2,500 „ „ 5,000	77	66	273	220
5,000 „ „ 10,000	43	32	287	228
10,000 „ „ 15,000	26	20	317	244
15,000 „ „ 25,000	19	22	362	402
25,000 „ „ 50,000	18	16	650	527
50,000 „ „ 100,000	15	21	1,065	1,414
100,000 and over	17	17	6,305	6,609
TOTALS	735	635	9,535	9,883

* Totals do not agree due to rounding

‡ Source: Based on Table 150, Annual Abstract of Statistics No.100, 1963.

The same table and information therein could have been presented in slightly different form. For example, the two halves of the table could have been headed '1951' and '1961', and the two classifications by number (or frequency) and by membership set under the appropriate year. This would have facilitated the comparison over the decade between the size of the union in terms of its class, e.g. 'under 100 members' and their aggregate membership. For example, in 1961 the 120 unions with less than 100 members apiece had an aggregate membership of 6,000, i.e. an average membership of exactly 50. In contrast, in the same year, the 17 largest unions in the class '100,000 and over' comprised a total membership of 6,609,000. The actual form of the table depends on what aspect of the information contained therein one wishes to bring out most clearly. As set out in Table 8, the main comparison is the changing structure over the decade. If the alternative method had been used, then it would have brought out the dominance of the big unions. In fact, however, with a relatively small table like this, these differences are readily apparent from the data. However, as a general principle, when framing a table, especially if it contains a lot of information, always consider first its purpose, and then how that purpose may best be achieved. The next table is another compound of different facts which may be taken separately. Table 9 gives the regional distribution of the population in England and Wales, at three different dates; in addition it provides percentage figures of the distribution at each date. Any one of these pieces of information is by itself quite interesting; put together

TABLE 9

REGIONAL DISTRIBUTION OF THE POPULATION OF ENGLAND AND WALES IN 1911, 1931 AND 1962.†

Standard Region	1911[1]		1931[1]		1962[2]	
	Number 000's	Per cent.*	Number 000's	Per cent.*	Number 000's	Per cent.*
Northern	2,815	7·8	3,038	7·6	3,282	7·1
East and West Ridings	3,564	9·9	3,929	9·8	4,207	9·0
North Western ..	5,793	16·0	6,197	15·5	6,622	14·2
North Midland ..	2,623	7·3	2,939	7·4	3,696	7·9
Midland	3,277	9·0	3,743	9·4	4,845	10·4
Eastern	2,106	5·8	2,433	6·1	3,841	8·2
London and S. Eastern	9,100	25·2	10,330	25·9	11,149	23·8
Southern[3]	1,864	5·1	2,135	5·3	2,920	6·2
South-Western[3] ..	2,507	7·1	2,615	6·5	3,456	7·4
Wales	2,421	6·8	2,593	6·5	2,651	5·7
England and Wales..	36,071	100·0	39,952	100·0	46,669	100·0

† *Source: Annual Abstract of Statistics No.100, 1963. Based on Table 10.*
* Totals may not agree due to rounding.
[1] Census figure
[2] Mid-year estimate.
[3] 1962 figures affected by boundary changes in 1959.

in this table they provide a picture of what has happened to the population of England and Wales over the past five decades. From the total columns it is evident that between 1911 and 1962 it has grown from some 36 million to nearly 47 million. But what about the regions? It is obvious that their populations have grown too, but some have increased to a greater extent than others. This is where the percentage columns are especially useful. Thus, if the 1962 percentage were the same for a region as in 1931, then it means that the regional relative or proportionate change corresponds with the overall national change. If the percentage has fallen, then the region is declining relatively to other regions, although in absolute terms its population may still be increasing. This has happened, for example, in the case of

Wales. In contrast the population of the South-Western region has increased both absolutely and relatively over the past three decades. As in the previous table, it may be preferable to re-arrange the columns so that three totals come in adjacent columns, and the three percentages for each region are altogether. Once again, it depends on what the table is for.

Whenever information of this kind is extracted from an official publication, it is essential to check the basis of the figures. Several points need to be borne in mind with these data. To which population does the table refer? The Registrar-General gives three definitions: total, home and civilian.[1] To which of these do the figures in Table 9 relate? Furthermore, the *Annual Abstract of Statistics* from which the basic data were taken is a secondary source. These figures first appear in the Preliminary Reports of the Census of Population in 1931 and 1951, while the final column appears in the Registrar-General's annual White Paper on the population of administrative areas. Thus, the first two figures are on the census dates, *i.e.* in April, while the mid-year figure is based on 30th June. For this comparison such minor differences are not important, but they could be if more exact measurement of population changes were needed unless the secondary source indicates that such differences have been allowed for. Note, too, the footnote which warns the reader that the South-Western and Southern regions' boundaries were changed in 1959. Reference to the source shows that the change was quite modest and once again, for this simple illustrative table, it does not invalidate the overall picture. Whenever data are compiled from successive censuses, whether it be population, production or distribution, always check the areas and regions to which the data relate. There have been many changes in the past.

Data relating to a particular event or phenomena over time form the basis of a *time series*. A large proportion of the data given in economic publications, *e.g. Financial Statistics, Economic Trends*, and the *Monthly Digest of Statistics*, is of this type. Often several time series will be brought together in a single table since they are inter-related. Table 10 is an interesting example of such information reflecting several aspects of the coal-mining industry over the period 1952-62. Tables such as these are deceptively easy to interpret, primarily because the reader may believe that the basis for each column has remained unchanged throughout the period covered by the table. In this case, the figures for the earlier years have been adjusted in order, to quote the footnote, 'to achieve continuity'. Earlier versions of this table related to the entire coal-mining industry, whereas this relates to the National Coal Board. Admittedly, the

---

[1] These terms are explained in Chapter XVII,

TABLE 10
Output, Attendance and Productivity at National Coal Board Mines[1]

Year	Total output of saleable coal	Saturday output[2]	Average number of wage-earners on colliery books	Manshifts	Average number of shifts per week per wage-earner	Absence percentage			Output per manshift
	Thousand tons	Thousand tons	Thousands	Thousands		Voluntary	Involuntary	Total	cwts
1952	210,774	10,437	706·2	177,002	4·82	5·90	6·19	12·09	24·46
1953	209,898	11,654	707·5	174,017	4·73	4·99	7·42	12·41	24·86
1954	211,513	11,856	701·8	174,045	4·77	4·01	8·20	12·21	24·89
1955	207,879	11,812	698·7	172,104	4·74	4·12	8·42	12·54	24·72
1956	207,373	11,222	697·4	170,919	4·71	4·25	8·67	12·92	24·84
1957	207,441	10,902	703·8	171,358	4·68	6·12[3]	7·69[3]	13·81	24·86
1958	198,841	3,611	692·7	159,678	4·43	6·41	7·73	14·14	25·58
1959	192,558	—	658·2	146,581	4·28	5·95	8·74	14·69	26·93
1960	183,862	—	602·1	135,099	4·31	6·03	8·72	14·75	28·03
1961	179,661	181	570·5	127,764	4·31	6·39	9·01	15·40	28·94
1962	187,640	29	550·9	123,824	4·32	6·33	9·02	15·35	31·20

[1] Accounts for about 99% of total coal production.
[2] Included in total output of saleable coal.
[3] Since 1-6-1957 some involuntary absence has been reclassified as voluntary.
Note. Some figures have been adjusted to achieve continuity.
Source: Ministry of Power Statistical Digest, 1962.

latter accounts for 99 per cent of total output, but nevertheless unless this point is checked, then the comparison is invalidated. Note, for example, the sudden drop in the Saturday output figure for 1958. This is due to the fact that Saturday working was suspended during that year and has for all practical purposes not been resumed. Note, too, the line across the two columns under 'Absence percentage' after the 1956 figure. This is always done to warn the reader that a new basis of counting or classification has been introduced. Hence the earlier figures are no longer directly comparable with the later figures. In this case the comparability of these two series is affected by the new but unspecified definition of involuntary absence. The lesson to be learned from this particular table underlines sharply the point made in discussing Table 9; *i.e.* always check sources and definitions. If it were intended to use such data as are contained in Table 10 for policy decisions, then it would be advisable to get in touch with the National Coal Board in order to verify the basis of each series.

The Annual Reports of H.M. Commissioners of Inland Revenue are fruitful sources of statistical material and provide a variety of tabulations. Table 11 is an interesting example of compressing a lot of information into one table. As the title indicates, it is concerned with the distribution of earned income of surtax payers whose assessments were made up to 30th June 1963 and thus relates to the incomes of those taxpayers in the financial year 1961-62. Basically, it is no more than a collection of frequency distributions. For example, the range of total income in the first column is the independent variable. Note the classification with the varying class-interval determined by the progression of the surtax rates. The total column, the ordinary figures, form the total frequencies for each class of income. The italicised figures in the same column is the total earned income of that particular class of taxpayer. A similar frequency distribution can be made up from the horizontal top row which states, not the taxpayer's total income, but his earned income only. In this case the class interval has nothing to do with the tax rates, but is selected merely as a convenient and significant basis of classification. The frequencies corresponding to each of these classes are given by the figures in ordinary type in the bottom row.

Each of the foregoing frequency distributions could be prepared from this table for any single class of surtax payer, either by reference to his total income given in the first column, or by reference to his earned income, as set out in the top horizontal row of the table. As the table now stands, it is possible to ascertain how many taxpayers within a given class of income received a given amount of earned income. The balance between the total and earned income is therefore investment income. Thus, take the class in the first column with a

TABLE 11
Surtax, 1961–62: Analysis of earned incomes in each range of total income
(Assessments made up to 30th June 1963)

Amounts £ thousand

Range of total income			Nil	£1 -249	£250 -499	£500 -999	£1,000 -1,499	£1,500 -1,999	£2,000 -2,499	£2,500 -2,999	£3,000 -3,999	£4,000 -4,999	£5,000 -5,999	£6,000 -7,999	£8,000 -9,999	£10,000 -11,999	£12,000 -14,999	£15,000 -19,999	£20,000 & over	Total
Over £2,000	Not over £2,500	Number	7,371	3,610	1,488	1,669	840	193	19											15,190
		Total earned income		531	533	1,205	1,009	323	38											3,639
£2,500	3,000	Number	7,114	4,044	2,072	3,735	3,989	4,556	6,486	3,097										35,093
		Total earned income		613	753	2,745	4,964	7,985	14,441	8,145										39,646
£3,000	4,000	Number	8,561	4,744	2,320	4,203	4,904	5,782	9,510	13,109	10,467									63,600
		Total earned income		714	843	3,108	6,101	10,139	21,291	36,041	34,100									112,340
£4,000	5,000	Number	4,778	2,439	1,167	1,986	2,106	2,250	3,129	3,797	12,348	4,712								38,712
		Total earned income		358	420	1,447	2,621	3,932	6,969	10,446	43,487	20,035								89,715
£5,000	6,000	Number	2,970	1,260	650	1,089	1,132	1,084	1,440	1,533	4,425	6,451	5,998							28,032
		Total earned income		180	237	794	1,402	1,900	3,190	4,211	15,569	29,045	32,340							89,068
£6,000	8,000	Number	3,119	1,371	615	1,012	984	915	1,217	1,154	2,867	4,000	5,437	8,730						31,421
		Total earned income		195	220	734	1,212	1,593	2,680	3,172	10,013	18,022	29,895	58,359						126,095
£8,000	10,000	Number	1,605	629	299	446	376	367	441	391	1,053	1,248	1,479	3,245	3,027					14,606
		Total earned income		89	107	324	458	639	957	1,067	3,691	5,613	8,055	22,873	26,381					70,264
£10,000	12,000	Number	926	362	149	242	187	195	203	199	469	488	643	1,096	1,301	1,400				7,860
		Total earned income		48	53	173	227	338	444	541	1,645	2,290	3,494	7,670	11,782	14,967				43,572
£12,000	15,000	Number	799	262	117	203	187	119	178	161	308	372	448	652	603	789	1,017			6,215
		Total earned income		32	42	138	222	204	382	428	1,047	1,619	2,373	4,417	5,328	8,456	12,952			37,640
£15,000	20,000	Number	591	191	104	132	89	96	114	86	199	205	254	418	485	379	258	703		4,304
		Total earned income		24	38	93	107	164	247	232	685	918	1,577	2,897	5,174	4,043	3,467	11,757		31,423
£20,000	· ·	Number	662	169	94	149	120	95	99	86	180	167	179	312	108	122	484	345	873	4,248
		Total earned income		19	31	105	145	161	218	233	632	747	955	2,140	920	444	6,597	6,039	27,103	46,489
All ranges	· ·	Number	38,496	19,081	9,075	14,866	14,914	15,652	22,836	23,613	32,316	17,643	14,438	14,453	5,524	2,690	1,759	1,047	873	249,281
		Total earned income		2,803	3,277	10,866	18,471	27,378	50,867	64,516	110,869	78,189	78,689	98,356	49,793	28,902	23,016	17,796	27,103	689,891

The distribution of incomes shown in this table differs from that shown in previous reports because of the new reliefs for earned income.

Note: Totals of some columns may not agree due to rounding. Source: CMND 2283

TABLE 12

Disposals of Indigenous and Imported Coal for Consumption in the United Kingdom by Grade and Main Consumer in 1962

Thousand tons

Consumer	Large		Graded		Slacks and Smalls		Other Coal		Anthracite		TOTAL
	Tons	%	Tons	%	Tons	%	Tons	%	Tons	%	Tons
Electricity supply industry	10	0·0	338	0·6	55,873	94·7	2,255	3·8	532	0·9	59,008
Gas supply industry	872	4·0	16,564	76·1	536	2·5	3,782	17·4	1	0·0	21,755
Coke Ovens	702	2·8	995	4·0	21,516	86·4	1,702	6·8	—	—	24,915
Railways	5,443	96·4	107	1·9	90	1·6	—	—	6	0·1	5,646
Industry	1,102	3·7	10,771	36·4	16,795	56·8	423	1·5	476	1·6	29,567
Merchants[1]	24,386	83·2	4,012	13·7	30	0·1	14	0·0	867	3·0	29,309
Miscellaneous	2,221	23·3	3,516	36·9	2,970	31·1	209	2·2	619	6·5	9,535
TOTAL	34,736	19·3	36,303	20·2	97,810	54·4	8,385	4·7	2,501	1·4	179,735
Miners coal	3,221	67·9	992	20·9	13	0·3	438	9·2	83	1·7	4,747
Colliery consumption	181	4·3	260	6·2	2,468	58·8	1,265	30·2	23	0·5	4,197

[1] Mainly for domestic purposes.

Source: Ministry of Power Statistical Digest, 1962.

total income of £6,000 but not over £8,000 which contains 31,421 cases (read off in the final column). Of that number 2,867 have an earned income of between £3,000 and £3,999. In other words between 37½ and 50 per cent of the total income of this class of taxpayers is earned. The aggregate earned income of all taxpayers in this total income class £6,000 to £8,000 is £126,095,000, and of that sum £10,013,000 is received by those taxpayers in the above total income class whose earned income represents between 37½ and 50 per cent of their total annual income in 1961-62.

Table 12 illustrates the neatness with which a mass of information can be compressed into a single table which is immediately comprehensible. This table shows the main classes or grades of coal produced by the coal-mining industry in 1962 and how it was disposed of between the different customers and users. The main consumers are set down in the first column and their consumption of the various grades is read off horizontally. Thus railways bought 5,443,000 tons of large coal, 107,000 tons of graded and 90,000 tons of slacks and smalls, plus a handful of anthracite, making up a total demand of 5,646,000 tons for the year. The adjoining percentage figures which total 100 per cent horizontally (but not vertically) merely help to assess the relative importance of each class of coal for each consumer. If the reader has any doubt of the usefulness of tabulation for such purposes, let him try and write a report containing all the information in this quite small table bearing in mind, too, the need to make the report intelligible. This type of table can be adapted to any firm which produces a variety of products and different classes of customer, *e.g.* motor-vehicles with private and business buyers of various models.

The form of Table 13 will be more familiar to the accountant than to the statistician. It is included in this chapter primarily to show the dangers of trying to put in an excessive amount of information into a single table. Despite the sectional sub-headings and use of italicised print, the overall impression created by the table is forbidding. It would have been more helpful if it had been broken down, at least separating the main classes of revenue earning activities from the capital creating and maintenance work. Likewise, the figures could with advantage have been given to the nearest hundred thousand, or even millions, in the revenue section of the table.

A more unusual form of table is shown in Table 14 taken from the *Ministry of Labour Gazette* for August 1958. It shows an analysis of answers to questionnaires sent to nearly 7,800 people who entered Industrial Rehabilitation Centres in 1956 and completed their courses there. Six months after they had completed their courses they were asked by means of this questionnaire whether they were satisfactorily

TABLE 13 VALUE OF OUTPUT OF ELECTRICITY

	£ thousand					
	Generation and Main Transmission		Distribution		Total	
	1956	1961	1956	1961	1956	1961
Sales of electricity:						
To consumers (a) ..	8,931	16,364	438,566	687,829	447,497	704,193
Within the industry ..	302,279	493,601	302,279	493,601	—	—
Steam and hot water ..	781	1,865	—	—	781	1,865
Ashes	269	909	—	—	269	909
Scrap metal	267	320	1,508	1,253	1,775	1,573
Other out sold	9	50	—	—	9	50
Total	312,536	513,109	137,795	195,481	450,331	708,590
Rents: Meter rents ..			365	618	365	618
Hire of appliances			2,696	1,983	2,696	1,983
Total			3,061	2,601	3,061	2,601
Proceeds from sale of purchased appliances and sale of reconditioned appliances withdrawn from hire (b)			28,117	53,046	28,117	53,046
Less Cost of purchased appliances sold and value (before reconditioning) of appliances withdrawn from hire ..			20,997	} 38,008	20,997	} 38,008
Payment for renovation of appliances by other firms			88		88	
				15,038		15,038
Work charged for.						
Installation and maintenance of public lamps ..			3,052	4,220	3,052	4,220
Fitting and maintenance of wiring and appliances ..			13,520	19,049	13,520	19,049
Other work charged to consumers	83	183	5,519	7,592	5,602	7,775
Total	83	183	22,091	30,861	22,174	31,044
Other work done by industry's employees (value of materials and wages):						
On depots, workshops, offices and other buildings:						
New construction (including extensions)	1,070	490	1,280	2,746	2,350	3,236
Repairs and maintenance ..	1,379	1,978	3,239	4,198	4,618	6,176
On plant and machinery:						
New construction (including complete renewals) ..	2,402	1,628	11,321	21,417	13,723	23,045
Repairs and maintenance ..	14,334	20,555	10,524	12,117	24,858	32,672
On mains and services:						
New construction (including extensions and complete renewals) ..	538	969	37,165	48,101	37,703	49,070
Repairs and maintenance ..	541	1,081	7,336	10,434	7,877	11,515
Total new constructional work done (c)	4,067	3,153	50,402	73,825	54,469	76,978
Total repair and maintenance work done (c) ..	16,399	23,740	21,342	27,072	37,741	50,812
Total value of output and work done (excluding repair and maintenance work) ..	316,686	516,445	220,381	317,806	537,067	834,251
Less Cost of purchased material and fuel used (d)	187,038	273,818	44,712	58,314	231,750	332,132
Less Transport costs for carriage of goods outwards	46	260	133	—	179	260
Net output ..	129,602	242,367	175,536	259,492	305,138	501,859

Source: Ministry of Power Statistical Digest 1962, Table 79.

(a) The amount chargeable for the electricity actually supplied excluding rents shown separately.
(b) Includes sales on extended credit terms of £16,101 thousand and £33,508 thousand in 1956 and 1961 respectively.
(c) Includes value of work done between Divisions and Area Boards, details of class of work are not available.
(d) Includes carriage of goods inwards.

placed in employment. Those who were recorded as being 'in training' were in fact taking a further course under the Ministry's Vocational Training scheme. The table itself provides an example of the way in which almost any type of numerical data can be portrayed in a clear-cut fashion.

One important lesson to be learned from Table 14 is the need to enquire closely into the definition of categories in tables, particularly if they are of a qualitative nature, even though they may appear quite clear at first sight. Thus it appears from the table that the medical groups that had the greatest difficulty in retaining employment (recorded in the column 'Not in employment at date of enquiry but some work since course') were those suffering from mental or nervous disorders, with the notable exception of those in the 'able-bodied' group. However, on closer investigation, one finds that although these people were sent to the Rehabilitation Units as being 'able-bodied', upon medical examination after arrival nearly 90 per cent were found to have some disability which in half the cases was of a mental or nervous nature. Therefore the high proportion of able-bodied who had difficulty in retaining employment is not inconsistent with the general concept that it is the mental or nervous type of person who had the greatest difficulty in retaining employment. It may be of interest to consider the effects of other factors on people using these units, *e.g.* comparative youth of the cases of respiratory T.B., previous education and employment, unsettled compensation cases, etc.

The above selection of tables taken from official statistical publications serve to demonstrate the variety of information that can be presented in tabular form. Generally speaking, whenever more than a few simple numerical facts have to be presented in a report, a simple table is essential to rapid comprehension. Certainly it is preferable to a long recital of facts.

Some Simple Rules

The construction of a statistical table does not require profound thought or great skill. Provided attention is paid to the more obvious and simple points, it should not be difficult to produce an intelligible table. The following simple rules will help remind the student of the more prevalent pitfalls.

1. The table should have a title which should be short and self-explanatory. Try to avoid supplementing it with footnotes, although from time to time this is necessary.

2. Always give the source of the data in the table whether it is based upon your own collection of facts, or as is more often likely to be

TABLE 14
FOLLOW-UP OF PEOPLE WHO ENTERED INDUSTRIAL REHABILITATION UNITS IN 1956 AND COMPLETED THEIR COURSE

Medical Group	Number of replies to follow-up enquiry	Analysis of replies expressed as percentages of total replies in each medical group					
		In training	In employment and satisfied	In employment but not satisfied	Not in employment at date of enquiry but some work since course	No employment by date of reply	
						Unemployed	Sick
Amputations	154	19·5	44·2	3·2	14·3	14·2	4·6
Arthritis and Rheumatism	265	12·8	42·6	5·7	12·1	14·4	12·4
Diseases of:							
Digestive system	233	13·3	50·2	8·2	12·4	13·2	2·7
Heart or Circulatory system	486	11·7	45·7	6·8	15·0	15·4	5·4
Respiratory system (not T.B.)	402	8·7	48·3	7·0	14·9	11·2	9·9
Skin	71	16·9	45·1	8·4	11·3	15·5	2·8
Ear defects	57	12·3	40·4	10·5	19·3	17·5	—
Eye defects	86	16·3	48·8	8·1	10·5	13·9	2·4
Injuries and diseases of:							
Lower limbs	337	13·0	54·9	8·6	12·5	6·8	4·2
Upper limbs	267	13·1	56·2	7·5	10·9	9·8	2·5
Paraplegia	51	15·7	37·2	—	15·7	21·6	9·8
Other spinal disorders	291	16·8	47·8	5·2	12·0	13·4	4·8
Epilepsy	231	13·0	39·8	6·1	21·6	11·3	8·2
Other organic nervous diseases	357	10·1	41·7	4·5	14·3	18·8	10·6
Psychoneurosis	694	10·9	43·8	9·1	18·2	13·0	5·0
Mental deficiency	102	2·0	46·0	4·9	18·6	22·5	5·9
Psychosis	236	12·7	43·2	4·7	22·5	10·2	6·7
Respiratory Tuberculosis	1,656	28·3	45·3	5·8	10·1	8·1	2·4
Other Tuberculosis	171	23·4	44·4	5·3	9·9	13·5	3·5
All other disabilities	584	15·9	42·0	6·5	14·8	13·6	7·2
Able-bodied	41	12·2	43·9	2·4	24·4	17·1	—
All groups	6,772	16·8	45·6	6·5	13·8	12·2	5·1

Source: Ministry of Labour Gazette, August 1958.

3*

the case, where it is compiled from published sources, *e.g. Board of Trade Journal* or *Ministry of Labour Gazette*.

3. The units of measurement should be clearly shown and, if necessary, defined. What for example are 'metric' tons, as opposed to 'long' tons; or £ Australian compared with £? The column headings should also be clearly shown and explained.

4. Where the data are classified, *e.g.* in a frequency distribution as in Table 6, the classification must be quite clear, *i.e.* no overlapping of limits of the successive classes.

5. If the columns of figures are to be aggregated, consider whether their totals should be placed at the top of the column or at the foot. Likewise, if percentages are to be included; should they be shown alongside the actual figures, or kept separate with other percentages?

6. Where there is a considerable volume of data it is better to break them down into two or more tables, rather than try to incorporate them all into a single table. Remember that the primary purpose of a table is to convey information; the simpler the table, the easier it is to understand.

7. Use different thicknesses of column rulings to break up a large table; likewise, italicise some of the figures to make them stand apart from others.

Lastly, before drafting the table be quite clear what it is designed to reveal from the data. As with so many other things in life, practice in drawing up tables is the best means of learning the basic principles. Make a habit of scrutinising any published or printed table and asking whether or not it might have been better laid-out; whether there are any mistakes, *e.g.* in classification; or an excess of footnotes. For a start, look through the numerous tables which are reproduced in the chapters which follow[1] and consider how far they satisfy the criteria set out above.

[1] Especially Chapter XVIII on 'Economic Statistics'.

Questions

TABULATION

1. Draft a blank table to show annually 1955-60 and monthly from July 1959–December 1960, the following detail associated with gas production:

(a) fuel used for making gas, distinguishing coal and oil, in thousand tons;

(b) total gas available at gas works in terms of thousand million cubic feet and millions of therms;

(c) gas made at gasworks in million therms:

　　(i) coal gas;

　　(ii) water gas;

　　(iii) oil gas;

　　(iv) total;

(d) gas bought by gas works in million therms:

　　(i) from coke ovens;

　　(ii) from other sources.　　　　　　　　　　　　　　*I.M.T.A., 1962*

2. The numbers of vessels engaged in the foreign trade (a) entered, and (b) cleared, with cargo at United Kingdom ports quarterly during 1959 and during the first quarter of 1960 were:

(a) 11,945; 13,206; 13,815; 12,145; 12,696, and

(b) 　9,472; 10,248; 11,091; 10,551; 10,674.

The tonnage (thousand tons net) entered from (a) Commonwealth, and (b) foreign countries, was:

(a) 9,926; 11,186; 11,445; 10,753; 10,456, and

(b) 9,465; 10,692; 11,270; 10,043; 10,786.

The tonnage (thousand tons net) cleared for (a) Commonwealth, and (b) foreign countries, was:

(a) 7,635; 8,339; 8,816; 8,335; 8,151, and

(b) 4,568; 5,242; 5,757; 5,262; 5,297.

　　　　　　　(*Board of Trade Journal*, 17th June 1960, p. 1294.)

Express the above in the form of a table and include the total tonnage (thousand tons net) entered and cleared.　　　　　　　　*I.M.T.A. 1960*

3. In a recent survey 7,381 children were studied of whom 219 attended private schools, 78 % were the children of manual workers but only 40 of these latter attended private schools, 1 out of every 9 children were only children; among private school attenders the proportion was 20·1 % of whom 7 were the children of manual workers. Of the families with only children, 567 came from the manual working class.

　　Arrange these figures in a table calculating any secondary statistics you consider necessary and comment on the results.　　　　　*B.Sc. (Sociology) 1954*

4. In his annual statement, the Chairman of a group of three companies, *A*, *B*, *C*, gave the following analysis of the profit (in £'s sterling) from trading in various parts of the world. 'For Company *A* the total profit was £130,000; of this sum, £100,000 came from the United Kingdom and £10,000 from trade with the

Commonwealth, whereas profit in Europe amounted to £3,000 from the European Free Trade Area (E.F.T.A.), together with £15,000 from the European Economic Community (E.E.C.); profit from the U.S.A. was only £2,000. As for Company *B* this company made £67,000 in the United Kingdom but had no trade with the U.S.A.; profit from E.F.T.A. and E.E.C. countries was £1,500 and £5,000 respectively which, together with £2,500 from the Commonwealth, made a total of £76,000. Finally, Company *C* made the lowest total profit £52,800; of this, £40,000 was made in the United Kingdom, the Commonwealth profit being £5,700 compared with £2,100 from E.E.C. and £5,000 from E.F.T.A.'

Tabulate these data, adding any secondary statistics which may be helpful.

Institute of Statisticians 1962

5. The number of 78 r.p.m. gramophone records produced in 1955, 1956, and 1957 was 46,347; 47,508; and 51,359 respectively. In the period January-November, 1957, the output was 46,481 and in the corresponding period, 1958, 26,268. In the month of November 1957 it was 5,224 and in November 1958 1,881. The figures for 45 r.p.m. records (in the same order) were 4,587; 6,903; 13,161; 11,739; 23,604; 1,466; 3,463. For 33⅓ r.p.m. records the numbers produced were 8,989; 12,116; 13,766; 12,557; 14,132; 1,394; 1,644. (All figures are expressed in thousands.)

Present the above figures in a table, including a total column, and comment on them. *I.M.T.A. 1959*

6. 'In the five years, 1935-9, there were 1,775 cases of certain industrial diseases reported in Great Britain. This number was made up of 677 cases of lead poisoning, 111 of other poisoning, 144 of anthrax and 843 of gassing. The number of deaths reported was 10¼ per cent. of the cases reported for all four diseases taken together. The corresponding percentages were 10·9 per cent. for lead poisoning, 6·3 per cent. for other poisoning and 12·5 per cent. for anthrax.

'The total number of cases reported in the subsequent five years, 1940-4, was 2,807 higher. But lead poisoning cases reported fell by 351 and anthrax cases by 35. Other poisoning cases increased by 748 between the two periods. The number of deaths reported decreased by 45 for lead poisoning, but only by 2 for anthrax between 1935-9 and 1940-4. In the latter period, 52 deaths were reported from poisoning other than lead poisoning. The total number of deaths reported in 1940-4, including those from gassing, was 64 greater than in 1935-9.'

Construct a table from the above information, making whatever calculations are needed to complete the entries. Comment on the changes shown between the pre-war and the war periods. *B.Sc.(Econ.) 1959*

7. In a recent Census of Population a distinction was drawn between households containing only a primary family unit (P.F.U. households) and other households which were called composite households. There were in all 14,481,500 households containing 45,528,300 individuals, 37,025,000 individuals lived in the 12,501,200 P.F.U. households. The total number of married heads of households was 10,973,600 of whom 9,848,400 lived in P.F.U. households. 2,516,400 of the P.F.U. and 162,400 of the remaining households contained children under five years of age. The numbers of such children were 3,306,600 and 209,600 respectively.

Represent the above data in tabular form and comment on the results calculating any secondary statistics you think necessary.

B.Sc. (Sociology) 1962 – University of Ghana

8.   RURITANIA: ANNUAL FARM CENSUS
1954 VILLAGE No. 5

	Farm A	Farm B	Farm C	Farm D	Farm E	Farm F	Village Total
				Acres			
Wheat	20¼		5½	120			155¾
Barley	5		8¾	30			43¾
Oats	2		10½	10		25	45½
Mixed Corn	¼		4	4		12	20¼
Rye				4			4
Potatoes	10	1¼	5¼	55		19	93½
Sugar Beat	3			91	3		94
Other Crops	9½	18	7¼	29		3	66¾
Bare Fallow	2			9			11
Grass	58	5¾	17	288	6½	12	387¼
Total Area of Crops and Grass	110	25	56¼	640	9½	81	921¾

In calculating the row (village) totals for the above table, the census forms (which had been filled in by the farmers themselves) were laid side by side and the figures were then added across.

Describe carefully the procedure for checking this table for misrecordings and arithmetic errors. It may be assumed that the previous years' forms for the same farms are available for reference.

Institute of Statisticians 1955

CHAPTER V

GRAPHS AND DIAGRAMS

However informative and well designed a statistical table may be, as a medium for conveying to the reader an immediate and clear impression of its content, it is inferior to a good chart or graph. Many people are incapable of comprehending large masses of information presented in tabular form; the figures merely confuse them. Furthermore, many such people are unwilling to make the effort to grasp the meaning of such data. Graphs and charts come into their own as a means of conveying information in easily comprehensible form. It is for such reasons that the government has produced popular versions of important White Papers in the form of multi-coloured booklets full of charts and simple figures. Such pictorial representation admittedly reduces the amount of detail that can be put across to the reader, but very often it is not the detail which is important, but rather the overall picture. For example, few citizens can give figures of the extent of this country's post-war balance of payments position, but most of them have been made aware by publicity employing charts that an expansion of exports is still necessary to pay for our foodstuffs and raw materials.

Diagrammatic representation of statistical facts is not only popular with the lay public; it is also extremely useful to the statistician. For example, a few well designed but simple charts showing the trend of sales and costs will be infinitely more eloquent at a board meeting than a mass of detailed monthly figures. Even the statistician himself will employ diagrams to ascertain the pattern or distribution of his data because the character of the distribution will sometimes determine the type of statistical analysis he will employ. There is a large number of diagrammatic forms to choose from; some of the most popular types of chart are reproduced in this chapter. The variety does not arise because statisticians as a class are particularly artistic; the data will usually determine the type of chart used. While there are certain obvious rules regarding the construction of charts, the most important consideration is commonsense. A good policy to adopt is to consider the finished diagram and ask what conclusions can be drawn from it. If they differ substantially from the impressions derived from a brief study of the actual data, then the chart should be scrapped. Some loss of detail is inevitable, but the chart need not be misleading.

A good illustration of misleading design is given in Figures 1 and 2 below. Some years ago during a municipal election, one party anxious to impress upon the electorate its superior performance in house building put up a poster on the hoardings on the lines of the left hand part of Figure 1. By not drawing the base line upon which the vertical bars were drawn from zero, the relative performance of that party was greatly enhanced in the eyes of the casual observer. The fact that the correct figures were inserted in the chart probably did little to counter the first impression. The correct method of drawing this chart is given on the right hand side of the Figure 1. Some criticism

Figure 1

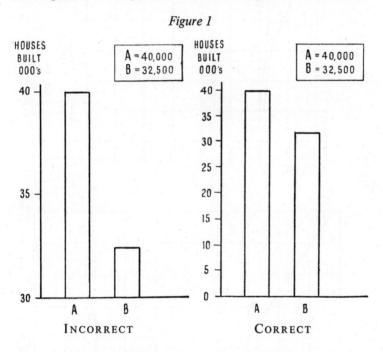

can also be directed against the left hand chart in Figure 2, which illustrates an advertisement used by one newspaper to demonstrate its popularity *vis-à-vis* its main rival. The vertical axis is clearly marked with the actual circulation figures, but once again, by omitting the base line and the entire lower part of the chart the performance of the advertiser's paper is greatly enhanced. It is undoubtedly true that the one paper had in the space of two years outstripped its rival, but by using a different scale and redesigning the chart, the picture can be made to look rather different. The reader should compare the right hand chart with the original on the left. This is also a bad chart, but

Figure 2

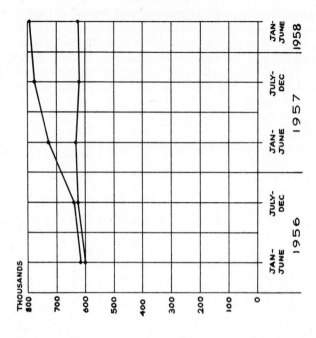

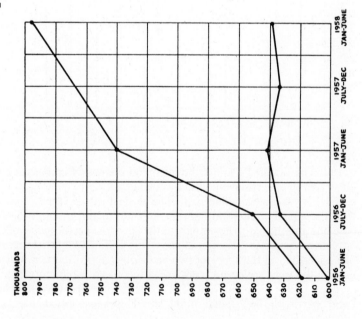

for different reasons. It is badly designed with the bulk of the space wasted. As an exercise the student should draw a new graph. The scale on the left-hand graph is sufficiently detailed to enable approximate figures of circulation to be extracted. In this new graph, the vertical axis should show the origin and at a point slightly above it show a distinctive break as in Figure 10 (page 73) up to the first figure of 600,000 from which point the scale can then be marked off.

Pictograms

Before discussing the various types of diagrams and their uses, a clear distinction should be drawn between the highly simplified and sometimes coloured pictorial diagrams, such as are employed by the Government Departments to explain the economic situation as well as by some leading companies to bring out the main features of their development in the past year to supplement the Chairman's speech and the graphs employed in statistical work proper. Within limits, the former type, 'pictograms', as they are sometimes called, are most useful. Their main advantage is the immediate visual impact on people who would not normally pay any attention to the more conventional line graph or column diagrams such as are portrayed in Figures 1 and 2. Pictograms are often printed in colour to heighten their impact. Figure 3 illustrates the conventional type of pictogram. Here the information to be conveyed to the reader is the increase in the consumption of light wine in this country between 1938 and 1958. The use of the small glass as the unit of measurement is effective, while the rising number of glasses from left to right indicates the growth of consumption. By placing a figure at the top of each column, a little more precision is given to the inter-year comparison although one can only guess at the precise content of each glass.

Figure 3

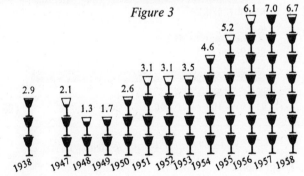

Light wine Glasses drunk per adult each year

Source: 'Barclays Bank Review', November 1962.

Sometimes, in place of rows or columns of such units, *i.e.* glasses or little men, cars, etc., the same facts, usually a startling increase over a period of time, are conveyed by drawing two similar figures but with the later one drawn several times larger than that for the earlier period. This certainly conveys the message, but such diagrams are considered inferior to the type illustrated in Figure 3, since the relationship between the two units is not always clear. Is it by height, or by area or even by volume? As in the above diagram, the appropriate figures are often inserted by the side or even within the unit itself in order to facilitate the comparison.

Figure 4

BAR DIAGRAM

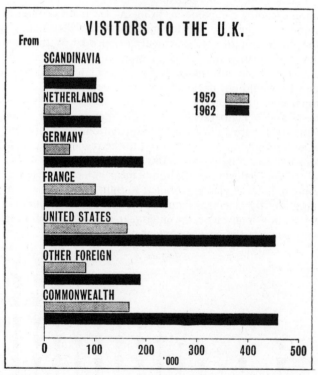

Source: Treasury Bulletin for Industry.

Simple Diagrams

The conventional method of depicting the information contained in Figure 3 would be to draw thick vertical lines or columns for each year, the height of each bar or column reflecting the relative change in the variable from year to year. A variant on this type of diagram is given in Figure 4. This illustrates very effectively both the country of origin of visitors to the U.K. and the increase in their numbers between the years 1952 and 1962. The growth during that decade of the number of Commonwealth and American visitors is immediately apparent. It will be realised that the bars in this diagram could equally well have been drawn vertically, with the names of the countries under the bar. On balance, however, the method adopted here is probably superior in its visual effect.

The same facts contained in Figure 4 might have been illustrated by a *pie chart* such as is depicted in Figure 5. The object of the pie chart is to show the composition of a particular aggregate. In the case of Figure 5 it shows the difference in the age structure of the population of England and Wales in 1841 compared with 1961. Had this diagram been used to depict the information contained in Figure 4, there would have been more sectors in the circle. They would then have been smaller and they would have to be labelled with the name of the country outside the sector, usually with an arrow from the name to the appropriate sector. This is the main weakness of the pie chart; it is

Figure 5

AGE DISTRIBUTION OF POPULATION IN ENGLAND AND WALES AT CENSUS
DATES 1841 AND 1961.

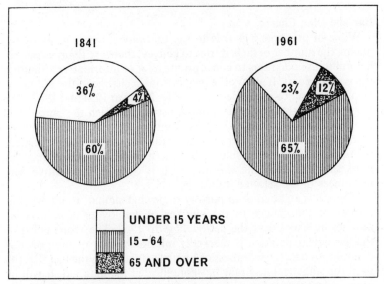

Source: Based on data from Registrar General's Annual Reviews.

really best suited to very simple comparisons such as that in **Figure 5** where there are only three age groups and the impact is immediate. To use this type of diagram where there are more than four sectors to be labelled usually results in a loss of the clear visual effect.

Figure 6 is a simple amalgamation of the merits of the two preceding figures. The use of the vertical column brings out the relative differences in the proportion of people who prefer one Summer month as against another for their holiday. The break-down of each bar on the lines of the pie chart indicates the reasons for the choice of that particular month. The column or bar diagram, with the column broken down into its component parts, is often used as an alternative to the pie chart. Its main advantage is that the difference in size between the two aggregates can be more effectively brought out by the columns. Where the two circles used in a pie chart represent different aggregates, for example the population of England and Wales in 1961 was about four and half times as large as that in 1841, one comes up against the problem of how the second circle should be drawn. Should the diameter be $4\frac{1}{2}$ times as large, or the area? If it is the former, it will exaggerate the difference; if the latter, then the precise difference in the sizes of the two circles is not so obvious. However, even when the bar diagram is used in preference to the pie chart, it is essential not to overload it with too many component sections. The effect of Figure 6 with its few basic facts is good, because it does not try to get too much information across.

Bar and Line Charts

While in principle it is true to say that any diagram gains by reducing the number of facts it tries to convey, there are times when it is helpful to use diagrams to convey quite a lot of information. Figure 7 is a case in point. It shows the proportionate change in the output of some of the major industries in the U.K. between 1954 and 1962. The bars on the right represent those industries in which the output has expanded; those on the left where output has fallen. If the exact figure of the changes in the output of individual industries were needed, then only a table would suffice. The merit of this particular diagram is its simplicity and the way in which it conveys the relative performances of the selected industries.

Figure 8 is an attempt to portray somewhat similar information in respect of the change in the number of employees in certain industries in Wales over the period 1948 to 1962. The visual effect of this particular diagram is markedly inferior to that of the previous diagram. In part the weakness of Figure 8 stems from the fact that the need for statistical accuracy has complicated the diagram. It will be noted that it covers two distinct periods, 1948-58 and 1959-62. This is

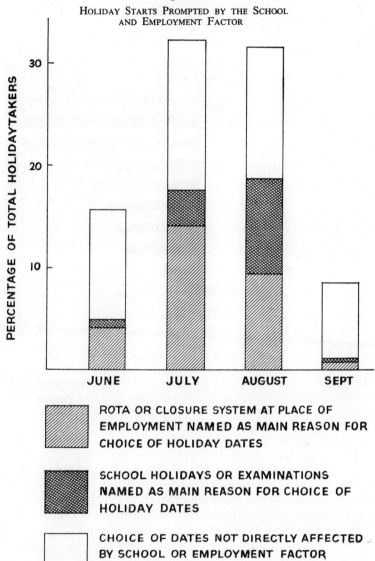

Figure 6

HOLIDAY STARTS PROMPTED BY THE SCHOOL
AND EMPLOYMENT FACTOR

Source: Social Survey Report S S 322.

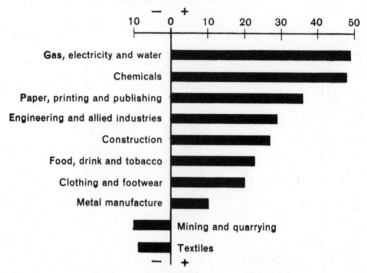

Figure 7

CHANGES IN OUTPUT OF MAJOR INDUSTRIES

% CHANGE—1962 on 1954

Source: Treasury Bulletin for Industry.

because the figures relating to employees in industrial employment in the two periods are not directly comparable owing to a change in the classification. The reader is told this in a footnote to the diagram. Another factor which weakens the effectiveness of the diagram is the attempt to put in so much information. It covers the experience of fifteen industries, whereas in Figure 7 only ten industries were included. The principle upon which the diagram is based is quite clear. It measures the proportionate change over the periods covered, and the steeper the slope of the line for any given industry, the greater is the relative change. In Figure 7 it was the length of the bar which reflected the degree of change.

Time Series Charts

The use of graphical methods for depicting time series is familiar to most people. Even the cartoonist drawing the business tycoon's office will show a number of charts relating to sales on the walls. Typical of this type of diagram are the two charts combined in Figure 9. These

Figure 8

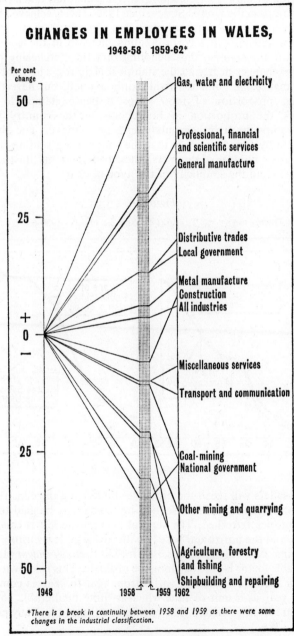

CHANGES IN EMPLOYEES IN WALES,
1948-58 1959-62*

Per cent change

50

25

+

0

−

25

50

1948   1958   1959 1962

Gas, water and electricity

Professional, financial and scientific services

General manufacture

Distributive trades
Local government

Metal manufacture
Construction
All industries

Miscellaneous services

Transport and communication

Coal-mining
National government

Other mining and quarrying

Agriculture, forestry and fishing

Shipbuilding and repairing

*There is a break in continuity between 1958 and 1959 as there were some changes in the industrial classification.

Source: Treasury Bulletin for Industry, June 1963.

are good illustrations of how effective a simple diagram can be. They are designed to show the tremendous increase that has taken place in trading stamps in the United States since 1950. This is achieved by the left-hand diagram which shows a steep and rising curve of stamp sales to retailers, linked with a corresponding increase in the goods exchanged for the stamps, thus denoting that the purchasing public uses, *i.e.* exchanges nearly all the stamps it acquires. The right-hand diagram makes the point that while only a small, but nevertheless increasing, proportion of food shops dispense stamps to their customers, the proportion of households in the country which acquire stamps through their purchases is much higher. The obvious point made by this diagram is that the shops using trading stamps attract a larger number of shoppers. These two diagrams illustrate the effectiveness and the simplicity of this type of chart.

Figure 9

DEVELOPMENT OF TRADING STAMPS IN U.S.A. 1950-63

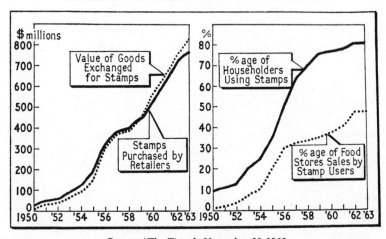

Source: 'The Times', November 30 1963.

Most readers will recall the principles underlying the construction of a graph as taught them at school, but some may be glad to have their memories refreshed. The vertical axis on the left is called the ordinate, and the horizontal base line the abscissa. It is customary to plot the time factor along the abscissa, for it is the *independent* variable. The variable factor is plotted along the ordinate. This is known as the *dependent* variable because it varies from year to year as compared with the year as a unit of time which is quite unalterable and 'independent' of other factors. The position on the graph of any point is located by the co-ordinates of that point. Thus, if we use Figure 10 for illustration, it is clear that in June 1952 the gold and dollar reserves were some £600 million. This point is located on the graph by the intersection of a vertical line drawn from the abscissa where it is marked mid-1952, and a horizontal line from the point on the ordinate marked £600 million.

Figure 10

U.K. GOLD AND DOLLAR RESERVES AND OVERSEAS COUNTRIES' STERLING HOLDINGS 1950-60.

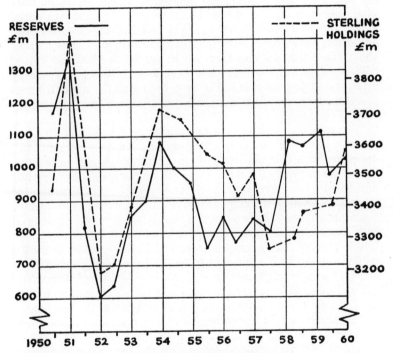

Source: Economic Survey 1960.

The main interest in Figure 10 as an illustration of graphical methods lies in the use of separate scales for the two series. The ordinate on the left is marked off in units of £100 million and the curve representing the reserves is read off against this scale. Because the sterling holdings of other countries greatly exceed the United Kingdom's reserves, if the curve representing fluctuations in their size had to be plotted against the same ordinate, then the left-hand ordinate would have to be extended up to £3,800 million. The sterling holdings curve would then be stuck up at the top of the graph and there would be a great empty space between the curves. This could be reduced by breaking the left-hand ordinate so that it consisted of two separate scales, the lower one reading as at present £600 – 1,300 million and above it, broken as shown in Figure 10 at the bottom of each of the ordinates a scale reading £3,200 – 3,800 million. This would eliminate a large part of the 'empty space' created by using a continuous scale from £600 – 3,800 million without any break. But it would still take up a lot of space which, in the average sized graph, would entail reducing the scale used. Instead of each space on the ordinate representing £100 million, it would represent (say) £500 million. Such a scale would then virtually obliterate the minor movements in the two values. By using both the vertical scales these differences are avoided. Note that the two scales both show a break above the origins so that they can be started at £600 and £3,100 million respectively. Unless such a break were shown it would be necessary, as was the case in Figure 2, to plot all the values from £0 upwards along the ordinates, again wasting space and reducing the scale of the graphs.

The method of plotting the two curves used in Figure 10 is extensively employed for all the above reasons. It entails the use of two ordinates on a common scale so that like movements in the two figures are shown by the same rise or fall on the graph. By using two ordinates the two curves can be almost superimposed upon one another thereby facilitating the comparison of their respective movements. Note that although the two scales are equal, *i.e.* the same distance on each ordinate equals £100 million, this is not essential to such a graph. If the Sterling figures ranged from say £2,600 million to £3,800 million then to fit this range of values into the same graph the distance representing £100 million would be reduced by half. This would not greatly matter, the graph could still be drawn in that way. It would mean, however, that more care in interpreting the fluctuations of the curves would be needed since the distance on one curve represents twice as much as on the other.

Had the two curves intersected at more than one point, they would have been more difficult to interpret. In such cases, the right-hand scale can be lifted bodily up the graph an inch or two so that frequent intersection of the curves is avoided. The actual construction, however, of Figure 10 may be criticised on one point, namely the two scales are equal; *i.e.* the same vertical distance on each equals a change of £100 million. Such a change is, however, of very different *relative* significance to the two totals. For example, the drop of over £700 million in both figures in 1952 represents almost half the reserves, but barely twenty per cent. of the sterling holdings, although the con-

clusion drawn from the graph as drawn is that it is an equivalent fall –
as it is, in *absolute* terms. On the other hand, since the object of the
chart is to emphasise the inter-dependence of the reserves and overseas liabilities, it serves its purpose.

Both the diagrams illustrated in Figures 9 and 10 are what are
known as line charts. This is the most usual way of depicting a time
series diagrammatically. Sometimes, however, one comes across a
time series which is represented in the form of a bar of column
diagram. For example, the pictogram in Figure 3 is really a form of
bar diagram. When is a bar diagram more appropriate than a line
chart and vice versa? Strictly speaking, the bar diagram is most
suitable for depicting a time series in which the dependent variable is
measured at particular points of time. For example, if we had the total
advances made by the London clearing banks for the decade 1954-64
and the figure for each year was the end-year balance sheet total, then
one could represent these data in a series of vertical bars or columns,
one for each year, and the variation in the total of advances from year
to year would be indicated by the varying heights of the successive
bars.

If, however, the data for advances were monthly, then a line chart
would be more suitable. To start with, a column diagram would be
too tedious to draw; each year would require twelve separate bars.
Also the main advantage of the line chart is that the movement from
month to month and year to year is clearly brought out. With month
by month data the gradual movement in a given total is best brought
out by the line chart. Such a chart is less appropriate to yearly data,
which are often best shown by bars, since the nature of the fluctuations within the successive years is not known. For example, in the
case of Figure 9 the monthly sales of trading stamps, probably even
the weekly sales, would be known and can be depicted in the form of
a continuous line chart. In the case of Figure 10, however, the data
are given half-yearly, *i.e.* the reserves and liabilities as at 30th June
and 31st December of each year. A bar chart similar to that of
Figure 4 would have served for these data, but in that case the
adjacent bars for each point of time would have been one for the
reserves and the other for the sterling liabilities. The primary purpose
of Figure 10, however, is to bring out the relationship between the
two series and to illustrate more effectively the sharp fluctuations in
the size of the reserves and liabilities.[1] Thus, as with the construction
of statistical tables, the best diagram is the one which meets the
particular requirements most adequately, *e.g.* is it the amplitude of
the fluctuations or as in Figure 10, the relationship between the

---

[1] The apparent absence of related movement in the two series in 1959 and 1960 is due to special
factors involving extra sales of gold.

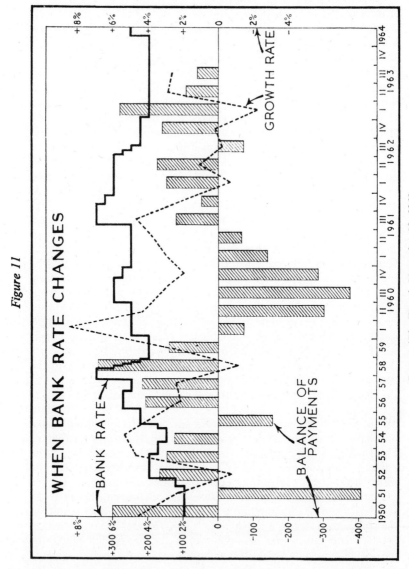

Figure 11

Source: 'The Times', November 28, 1963.

series; the relative size of the quantities or values at particular points of time? Commonsense, not rule of thumb methods, is the best guide.

Diagrams are sometimes used, not merely to convey several pieces of information such as several time series on one chart, but also to provide visual evidence of relationships between the series. Figure 11 is an interesting example of the use of diagrammatic methods to illustrate the relationship between three important economic variables. First, it shows the changes in Bank Rate over recent years; these changes are shown as a 'discrete' series of steps and not a smooth

Figure 12

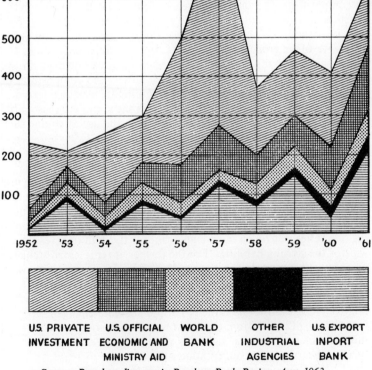

£ MILLION   MAIN SOURCES OF AID AND INVESTMENT IN LATIN AMERICA

| U.S. PRIVATE INVESTMENT | U.S. OFFICIAL ECONOMIC AND MINISTRY AID | WORLD BANK | OTHER INDUSTRIAL AGENCIES | U.S. EXPORT INPORT BANK |

Source: Based on diagram in Barclays Bank Review, Aug. 1963.

curve since the change is not gradual, but is made in steps of half or one point. Changes in Bank Rate exercise an indeterminate effect on the level of economic activity and on the balance of payments. The reader will recognize the vertical bars, which reflect the recorded changes in the quarterly balance of payments figures, as a variant on the diagram illustrated in Figure 7. The bars below the horizontal line are read off against the left hand ordinate as deficits, the bars above the line represent surpluses. Lastly, the dotted line which oscillates above the horizontal, and is there described as the 'growth rate', is presumably based upon the changes in the quarterly figures of the national product. To increase the clarity of the diagram, each of the series is clearly marked. The reader may judge for himself whether it achieves the object of conveying the essentials of the relationships between the three variables.

Figure 12 is a useful diagram where the composition of an aggregate changes over time. Here the aggregate is the amount of funds lent for development and investment to the Latin American countries. The aggregate changes widely from year to year; likewise its composition. The largest source of funds, and the most volatile constituent in the total, is U.S. private sector investment. This type of diagram can be used to illustrate the make-up of any large aggregate of heterogeneous elements. For example, changes in the composition of bank assets over a period of years; changes in the pattern of bank advances classified by industry; the composition of the labour force by age and sex. Such diagrams are usually referred to as *strata* charts. Note, however, that unless the strata are reasonably wide and relatively few in number, even with the use of colour such diagrams may be rather confusing to read.

Logarithmic Scale Graphs

In the discussion of Figure 10 depicting the fluctuations in the U.K. gold reserves and the sterling liabilities, the point was made that a change of £100 million in either figure occupied the same amount of space measured against the ordinate. Yet, such a change in relation to the two series was much more significant for the gold reserves than it was for the sterling liabilities. For the former, it represented a change of about 10 per cent; for the sterling liabilities it was just about 3 per cent.

Sometimes it is more useful to use a graph which brings out the relative changes in totals rather than the absolute change. Suppose, for the sake of illustration, the gold reserves were to increase by £100 million each year. At first sight this would suggest a constant rate of increase year by year, but although the absolute increase was constant, the relative increase would fall, *i.e.* the rate of increase would decline. This may happen with an expanding business. Turnover and profits may be rising each year, but the rate of expansion may be declining. Comparisons between differing totals are facilitated by measuring the relative instead of absolute changes. Assume, for example, two companies whose net profits over a period of four years rise in both cases by £80,000. The following table compares their performances.

Year		1	2	3	4
Profits £000's	A	20	40	80	100
	B	140	160	200	220
Absolute increase £000's ..	A	—	20	40	20
	B	—	20	40	20
Percentage of relative increase each year on previous year	A	—	100	100	25
	B	—	14	25	10

It is quite clear which is the better investment for growth. While each year both companies experience the same absolute increase in profits, the relative increase, *i.e.* the rate, is much better in Company A than Company B. Where the primary consideration is the rate of change in a series, rather than the absolute figures, or where comparisons with series involving figures of rather different magnitude as in the above simple illustration, a logarithmic scale graph is most useful.

At first sight a logarithmic scale chart or graph as illustrated in Figure 13 is the same as the line charts illustrated in Figures 9 and 10. The years are marked off along the abscissa, while the dependent

TABLE 15

Number of Private Cars With Current Licences 1950-62[1]

Year	Number (000's)	Logarithm	Year	Number (000's)	Logarithm
1950	2,258	3·3537	1957	4,187	3·6219
1951	2,380	3·3766	1958	4·549	3·6579
1952	2,508	3·3993	1959	4·966	3·6960
1953	2,762	3·4412	1960	5,526	3·7424
1954	3,100	3·4914	1961	5,979	3·7766
1955	3,526	3·5473	1962	6,556	3·8166
1956	3,888	3·5897			

[1] Licences current any time during the September quarter.

Source: Annual Abstract of Statistics, Nos. 96 and 100.

variable, *i.e.* the number of registrations in the September quarter of each year, is marked off against the ordinate. Closer examination of the ordinate will reveal that the distances between successive points tend to diminish. For example, the distance between 2,250 and 3,000 is the same as that between 3,000 and 4,000. They both represent an increase of one-third on the year. The interval between 4,000 and 5,000 is larger than that between 5,000 and 6,000 because the former represents an increase of 25 per cent and the latter one of

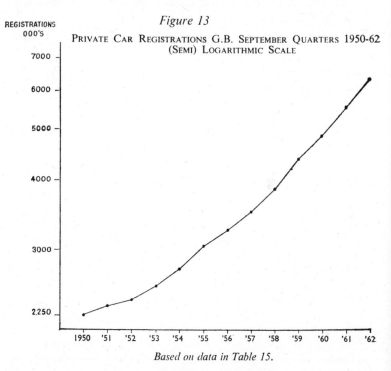

Figure 13

PRIVATE CAR REGISTRATIONS G.B. SEPTEMBER QUARTERS 1950-62 (SEMI) LOGARITHMIC SCALE

REGISTRATIONS 000'S

Based on data in Table 15.

only 20 per cent. This is the essential characteristic of a logarithmic scale. Any given increase, regardless of its absolute size, is related to a given base quantity. Thus, a perfectly straight line on such a graph denotes a constant percentage rate of increase, and not a constant absolute increase.

It is the slope of the line or curve which is significant in such a graph. The steeper the slope, whether it be downwards or upwards, the more marked is the *rate* of change. Note, for example, the slope of the curve between 1952 and 1953 when the actual increase was 254,000 from 2,508,000, *i.e.* approximately 10 per cent. Between 1961 and 1962 the actual increase was 577,000, more than twice as much as the earlier quoted increase, but the slope of the curve between those points is about the same as the earlier section because the relative increase in that year is just over 10 per cent on the total at the beginning of the year. Had these values been plotted on a normal scale then the 1961-62 increase would have occupied more than twice the distance of the 1952-53 increase along the ordinate.

This type of graph possesses a number of advantages. It is possible to graph a number of series of widely differing magnitudes on a single chart and bring out any relationship between their movements. However wide the amplitude of the fluctuations in the series, a logarithmic scale reduces them to manageable size on a single sheet of graph paper, whereas, on a normal scale, it might prove impossible to get the larger fluctuations on to a single chart, except by so reducing the scale that all the other smaller movements in the series are almost obliterated.

Two further points should be noted by the student reader. First, although this graph is referred to as logarithmic, it will be noted that in the heading of the graph the word 'semi-' has been inserted in brackets. This is because only one of the two scales on the graph is logarithmic and, to distinguish this type of graph from one which has both scales on a logarithmic scale, it is sometimes referred to as 'semi-logarithmic'. The other type of graph is of limited application and most references to log-scale charts relate to the type depicted in Figure 13. Second, unlike the conventional chart where all the points are measured from the origin, *i.e.* the intersection of the ordinate and base line, with the log-scale chart there is no such origin. The log scale along the ordinate starts at any value which is appropriate. In the case of Figure 13, the lowest point marked on the ordinate was 2,250, a value slightly below the actual number of registrations for 1950.

There are various ways of drawing this graph. Semi-log. scale paper can be purchased, but usually the statistician can make his own scale. The simplest method is to take the log. scale of a slide rule and set it against the ordinate and then mark off on the graph the actual values given in the table. Alternatively, look up the logarithms of the actual numbers as has been done for Table 15, mark off the ordinate between the lowest and highest logarithm on a normal scale, *e.g.* the distance on the ordinate between the logarithms 3·3500 and 3·5500 will be the same as between 3·7500 and 3·9500, and plot the logarithms of the values. Just to convince himself that this method works, the student can plot the logarithms in Table 15 and compare the resultant graph with that in Figure 13.

4

The Histogram

The data given in Table 3, *i.e.* the outputs of factory operatives, can be plotted, as depicted in Figure 14. This diagram, comprising what appears to be a series of contiguous rectangles, is known as a *histogram*. This particular diagram is commonly used to depict data given in the form of grouped frequency distributions. For this type of diagram the independent variable is plotted along the base and the frequencies against the vertical. In this example the outputs per class are measured along the horizontal axis and the total frequencies in each class are read off the vertical axis.

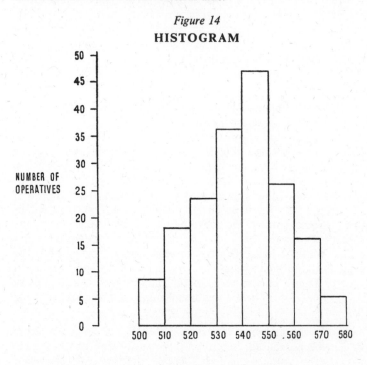

Figure 14

HISTOGRAM

OUTPUT IN UNITS

Source: Data in Table 3.

The histogram is most easily constructed by drawing vertical lines from the mid-points of each class interval to the height representing the frequencies in successive classes measured off against the ordinate. If flat lines representing the width of each class interval are drawn across the top of each vertical, which then forms the mid-point of each flat bar, and the limits of the horizontal bars joined by perpendiculars to the base, a series of contiguous rectangles is obtained. The aggregate of the frequencies is represented by the total area of the histogram, while the areas of the different rectangles are proportional to the frequencies in the respective classes.

This means that if the class interval is constant, *i.e.* the same throughout the frequency distribution, the bars will be equal width and the frequencies within each class are then indicated by the height of the bar. If, however, as is sometimes the case the class-interval varies, then the height of the bar must be adjusted. For example, the final class or two often contain relatively few cases so that they are often merged, the class interval of the final group then being twice that of all the other classes. When this class is plotted in a histogram, the length along the base will be twice that of all other classes, *i.e.* the bar will be twice as wide because the class-interval is twice as large as the others. But it will then have its height reduced by half of the total frequencies in that class so that the area of the bar is proportional to all the others. For the time being the histogram or block diagram, as it is sometimes called, may be regarded as no more than another method of diagrammatic representation. When the principles of sampling are discussed, it will be seen that the tendency for certain types of frequency distribution to conform to this rather peaked and symmetrical distribution depicted in Figure 14 is of great value to the statistician.

Continuous and Discrete Variables

Suppose we are classifying households in a given town by reference to the number of children of school age in the household: then we should get a frequency distribution in which the independent variable would take values of 0, 1, 2, 3, 4, etc., according to the number of children in the household. Such measurements are exact, we cannot have less or more than a whole child. The unit is clearly and unequivocally defined for us. Such a variable is termed *discrete*, as is any other variable which can take only certain restricted values, *e.g.* the distribution of theatre tickets sold during the week classified according to their individual price, *e.g.* 3s 6d, 6s, 8s 6d, etc.; the number of living rooms in a house, and so on. On other occasions, however, the limitations of our measuring rod tend to give approximate values. For example, if we take the temperature of a furnace at one minute intervals, the best readings we get will be rounded to the nearest degree centigrade. If we measure the height or weight of schoolchildren, the recorded heights and weights will be expressed to the nearest unit practicable, *e.g.* $\frac{1}{2}''$ or 1 pound. Two children may be identical in height and weight, yet the records may show one to be $\frac{1}{2}''$

shorter than the other but one pound heavier. The difference is partly explained by the human error in taking these measurements, but it is partly the fault of our height measures and scales which will give only an approximate result, *e.g.* to the nearest inch or pound. In practice, however, the approximation is quite adequate for most purposes. Where a variable can take any value within the range of its observed minimum and maximum values, it is referred to as a *continuous* variable.

The importance of this distinction between these types of variable, *i.e.* discrete and continuous, is discussed further in the chapter on averages (page 98). It is also relevant in discussing the appropriate form of graph for depicting certain data. Take, for example, the distribution of households by the number of children per household. A line graph with the frequencies plotted against the ordinate and the values 0, 1, 2, etc., along the base would not make sense. It could be interpreted from such a chart that there were (say) 549 households with 1·6 children apiece. A continuous line implies that a reading can be taken from the line at any point. Thus in Figure 10 even if the actual data plotted were half-yearly totals of gold and dollar reserves, a reading between any two dates will still give a sensible if not precise result. For example, if at 30th June the reserves were £800 million and by 31st December they totalled £1,000 million, it would not be silly to suggest that half-way through that period the size of the reserves was approximately £900 million.

As a general rule, line charts and histograms in which the bars are contiguous should be kept solely for plotting the course or distribution of given values of a continuous variable. If the variable is discrete, then column diagrams, such as Figure 6, with a gap between each bar or column are conventionally used. But in, some cases the distribution is such that although the variable is discrete it can be regarded for practical purposes as continuous. The reason for doing this is that if we can assume a variable to be continuous, the statistician can use certain satistical techniques which are extremely useful but which lie beyond our purview. The point can be illustrated by noting the histogram in Figure 14 which relates, strictly speaking, to a discrete variable. After all, a single article produced in a factory can be no more nor less than one unit. But since the differences between successive values of this variable are so small in relation to the value of any single worker's output, *e.g.* 580 units, we are in the same position as with temperature recordings given to the nearest unit which we stated represented a continuous variable. The same is true of frequency distributions of money sums where the difference between successive recorded values of the variable are small in relation to the individual values themselves.

Lorenz Curves

The periodic Census of Production, discussed in detail in Chapter XVIII, provides for each industry a variety of information, in particular the number of firms in the industry classified by reference to their size, as is shown in Table 65. For each class of firms so classified the census statisticians also estimate the net output. This is also given in Table 65. One of the features of modern British industry is what is termed the 'concentration' of production in the larger units. In many British industries today the bulk of the output is produced by a handful of very large firms, while the numerous small firms produce a very much smaller part of total output.

Working from such data as are given in Table 65, in particular columns 1, 2 and 5, the relative role, of different classes of firm, classified by size as in Column 1, can be compared with the relative

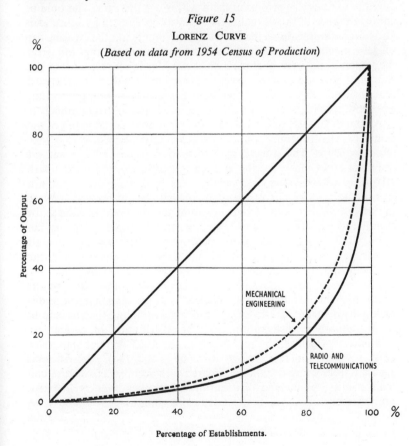

Figure 15

LORENZ CURVE

(*Based on data from 1954 Census of Production*)

Percentage of Establishments.

contribution they make to total output. For example, the thirty-six firms employing between twenty-five and forty-nine workers represent over 15 per cent of all firms in the brewing and malting industry. Their contribution to the aggregate net output of the industry is only £2,078,000, or less than 1½ per cent. In contrast, the three largest firms representing less than 2 per cent of the industry by number, account for over 15 per cent of the net output of the entire industry.

A measure of the degree of concentration in an industry can be provided by graphical methods. Figure 15 depicts a Lorenz curve illustrating the relative concentration as evidenced by the 1954 Census of Production in two industries, mechanical engineering and radio and telecommunications. The flatter the curve, the less is the degree of concentration, *i.e.* the more diffused is the total output of the industry over the various size groups of firms within the industry. The Lorenz curve in Figure 15 shows that there is more concentration in the radio and telecommunications industry than the other; in other words, the bigger firms contribute a larger part of the output. In that industry, the upper 20 per cent of firms by size account for 80 per cent of the aggregate net output of the industry. If the industry were so organised that each size class of firms produced the corresponding proportion of output, then the curve would be replaced by the diagonal line.

An important use of the Lorenz curve is in the measurement of the extent to which income is unevenly distributed between the various income groups. The student may turn to Table 6 on page 44 and draw a Lorenz curve for the distribution of incomes shown in that table. For each income class calculate the percentage of all incomes and the percentage of total income enjoyed by that class. Cumulate the successive percentage in each column and then plot the one set of cumulative percentages against one axis, and the other set against the other.[1] In this case the cumulative percentages for the incomes in each successive class could be plotted against the vertical axis, while the cumulative percentages of the total income received by each successive income class could be plotted against the other axis. To the extent that the resultant curve diverges from the diagonal, income is unequally distributed. Obviously, such a curve should then be related to a comparable distribution, *e.g.* the distribution of capital, an estimate of which is now published in the 1963 Annual Report of H.M. Commissioners of Inland Revenue (Table 144). Alternatively, such a Lorenz curve could be compared with a similar curve based on data relating to income distribution in another country, *e.g.* the United States. As with most statistical techniques, the object is to

---

[1] The cumulation of successive percentages is done in the same way as the cumulation of frequencies for the derivation of the Median. See p. 106.

facilitate comparisons and interpretation of the data. The Lorenz curve is especially valuable for the former purpose.

Summary

The type of diagram to be used for depicting given data does not usually pose serious problems. As with tabulation, diagrammatic representation of statistical data is largely a matter of commonsense coupled with a few rather obvious rules. Since experience shows that these rules tend, all too often, to be overlooked, it may be helpful to set them out below. The charts and graphs in the previous pages and elsewhere in the book can then be assessed in the light of these rules.

1. The title should be brief but self-explanatory.
2. The source of the data plotted should be given so that the reader may consult them for himself.
3. The axes of the graph or chart must be clearly labelled so that the quantities and values to which they refer are immediately apparent.
4. Except for logarithmic graphs, always show the origin. Whenever the values are such that the points plotted will normally lie a considerable distance from the origin and lead to a compression of the scale, then the graph may be 'broken' or 'torn' across the vertical axis and the relevant values marked off along virtually the entire length of the ordinate.
5. Always bear in mind that graphs are meant to simplify the picture; a detailed or overloaded chart or graph defeats its own object.

Finally, it should not be forgotten that the addition of a diagram or graph among the published data adds nothing to what is already available. Its sole *raison d'être* is that it drives home more effectively than the tabulated data the main findings of the enquiry. If the diagram does not meet this requirement, then it is better omitted. Generally speaking, diagrams are useful aids to comprehension. The reader interested in charts will find a varied collection in any issue of *Economic Trends* published monthly by H.M. Stationery Office.

Questions

GRAPHS AND DIAGRAMS

1. The Utilisation of Milk used for Manufacture.

THE UTILISATION OF MILK USED FOR MANUFACTURE

| Year to 31st March | Creamery Butter | Million gallons | | |
		Cheese	Condensed Milk	All Other
1955	56	126	73	56
1956	54	126	77	65
1957	99	196	102	74
1958	153	213	95	77
1959	81	161	85	87
1960	70	175	83	94

(*27th Annual Report of the Milk Marketing Board*)

Show these figures in the form of a diagram. *I.M.T.A. 1961*

2.	1959 Index of Employment (1954 = 100)	1959 Index of Industrial Production (1954 = 100)
Mining and Quarrying	95	92
Chemical and Allied Trades	109	131
Metal manufacture	101	106
Engineering, Shipbuilding and Electrical Goods	107	116
Vehicles	106	131
Metal Goods not Specified	103	114
Textiles	85	92
Leather, Leather Goods and Fur	90	91
Clothing and Footwear	93	112
Food, Drink and Tobacco	106	114
Paper and Printing	109	117
Gas, Electricity and Water	101	123

(*Ann. Abs. Statistics 1960*, Tables 132 and 155)

Construct a diagram illustrating these figures. Does it suggest any strong relationship between the change in industrial production and the change in output per head between 1954 and 1959 in the specified industries? *I.M.T.A. 1961*

3.				Average number Unemployed ('000)	Average number of Jobs Vacant ('000)
1954	1st Quarter	..	..	367·4	268·8
	2nd ,,	..	..	281·8	337·0
	3rd ,,	..	..	231·9	370·8
	4th ,,	..	..	257·8	336·7
1955	1st Quarter	..	..	283·1	253·5
	2nd ,,	..	..	231·4	428·4
	3rd ,,	..	..	195·6	448·4
	4th ,,	..	..	218·9	390·7

1956	1st Quarter	..	..	268·3	373·9
	2nd ,,	..	..	237·2	389·1
	3rd ,,	..	..	250·7	364·4
	4th ,,	..	..	271·9	299·3
1957	1st Quarter	..	..	375·2	248·3
	2nd ,,	..	..	306·8	292·1
	3rd ,,	..	..	259·1	308·5
	4th ,,	..	..	309·1	254·2
1958	1st Quarter	..	..	417·7	215·4
	2nd ,,	..	..	440·2	212·7
	3rd ,,	..	..	444·5	198·9
	4th	..	..	527·2	162·9
1959	1st Quarter	..	..	593·3	163·6
	2nd ,,	..	..	474·9	218·1
	3rd ,,	..	..	409·0	264·7
	4th ,,	..	..	423·5	247·5
1960	1st Quarter	..	..	441·3	256·7
	2nd ,,	..	..	345·8	331·3
	3rd ,,	..	..	306·2	357·6
	4th	..	..	348·5	309·3

(Adapted from Ministry of Labour Gazette)

Present these figures in the form of a graph and briefly comment on the result.

I.M.T.A. 1962

4. Suggest a suitable type of chart or diagram to show:

 (i) progress of work being made in the completion of a plan;
 (ii) the circulation of the *Morning Star* in each county;
(iii) the Deposits of the London Clearing Banks;
(iv) the relation between the Bank Rate and Gold and Dollar Reserves.

Institute of Statisticians 1959

5. What are the principles underlying the construction of a histogram? Explain the relations between the histogram, the frequency polygon, and the frequency curve. Do you consider that the frequency polygon has any particular advantages over the other two forms of representation? Illustrate with a histogram the distribution of males which is given below, and use a frequency polygon to illustrate the distribution of females.

Students Admitted to Courses for a First Degree

Age on admission (years)	Males	Females
17½—	15	27
18 —	67	73
18½—	67	56
19 —	41	49
19½—	22	20
20 —	11	8
21 —	5	5
21 —	5	5
22 —	12	2

Intermediate D.M.A. 1963

6. In many cases arithmetically-scaled graphs are used where only semi-logarithmic graphs can give the desired information. Give three examples of cases where a semi-logarithmic graph is necessary and state in each case why arithmetically-scaled graphs, if used, would give misleading conclusions.

Institute of Statisticians 1957

7· The following information is available from British Railways Headquarters concerning traffic receipts in 1961 and 1962:

		Period No.9 – 1961		Period No.9 – 1962	
Source of Revenue		*4 weeks to* *Sept. 10, 1961*	*Aggregate* *for 36 weeks*	*4 weeks to* *Sept. 9, 1962*	*Aggregate* *for 36 weeks*
Passengers	..	14,203	111,973	14,456	114,760
Merchandise	..	7,066	69,505	6,694	65,189
Minerals ..	..	2,909	30,613	2,484	25,612
Coal and Coke	..	7,335	71,105	6,817	69,694
Parcels ..	..	4,410	38,603	4,465	39,471
Total	..	35,923	321,799	34,916	314,726

All Data in £'000's

Prepare a diagram designed to show the changes in revenue from the various sources for the 'annual holiday' period (4 weeks to 9th or 10th September), and for the first 36 weeks of a year, for 1961 and 1962.

Institute of Statisticians 1963

8. The following table shows the distribution of income after tax in two fictitious societies, A and B. Plot the cumulative frequency distribution curves of these two societies so as to exhibit the similarities and differences between them. Comment on these.

Society A			*Society B*		
Income after Tax		*No. of*	*Income after Tax*		*No. of*
Not under £	*Under* £	*incomes* *(Thousands)*	*Not under* £	*Under* £	*incomes* *(Thousands)*
50	250	5,170	50	250	5,220
250	500	6,930	250	500	1,820
500	750	6,070	500	750	1,590
750	1,000	4,970	750	1,000	1,220
1,000	2,000	3,387	1,000	2,000	2,830
2,000	4,000	308	2,000	3,000	1,470
4,000	6,000	50	3,000	4,000	270
6,000 and over		15	4,000 and over		80
	Total	26,900		Total	14,500

Institute of Statisticians 1963

9. Give examples of the application to social statistics of the following types of diagram; histogram, logarithmic graphs, bar chart, 'pie' chart, pictogram.

B.A. Makerere University 1962

10. The following figures come from the report on the Census of Production for 1958: Textile Machinery and Accessories:

Establishments Nos.	Net Output £000
48	1,406
42	2,263
38	3,699
21	2,836
26	3,152
16	5,032
23	20,385
214	38,773

Analyse this table by means of a Lorenz Curve and explain what the curve shows.   *I.C.W.A., 1962*

11. The following table gives the age distribution in 1958 of male teachers in maintained grammar schools in England and Wales who have degrees in science and mathematics. Compare the advantages of two different diagrams (other than cumulative frequency diagrams) that might be used to illustrate these data. Draw the diagram that in your opinion provides the better illustration.

Age Group	1st Class Honours	2nd Class Honours	Other Degrees	Total
20-24	23	119	77	219
25-29	68	380	533	981
30-34	34	206	356	596
35-39	57	165	203	425
40-44	104	335	250	689
45-49	160	314	336	810
50-54	147	282	369	798
55-59	72	155	260	487
60-64	41	72	153	266
65 and over	11	13	29	53

Intermediate D.M.A. 1961

12. The following table is reproduced from the *Monthly Digest of Statistics:*

Permanent House Building: Construction begun. (U.K.)

	All houses	For local housing authorities	For private owners	For other authorities
1958	263,249	119,675	139,076	4,498
1959	324,976	147,721	172,336	4,919
1960	316,741	123,405	186,061	7,275
1961	320,054	120,651	192,950	6,453

Show how this information can be presented in the form of: (*a*) pie charts; (*b*) bar (or column) diagrams.

Comment briefly on the advantages and disadvantages of the two types of presentation.   *I.C.W.A. 1963*

CHAPTER VI

AVERAGES: TYPES AND FUNCTIONS

The Function of Averages

Few people can assimilate a mass of detailed information expressed in numerical form, even when it has been substantially reduced by tabulation. It is helpful, therefore, if instead of merely tabulating the information derived from a specific enquiry and depicting it in graphs or diagrams, it can be expressed in more abbreviated numerical form, yet in such a way that the salient features of the tables are clearly brought out.

For instance, in the case of the firm owning the plant with 180 employees whose outputs were given in Table 1 it may be assumed that this firm controls several such plants of varying size in different parts of the country. The management would be anxious to compare the outputs of the operatives in the various plants. If conditions in each plant are similar, the results should closely correspond. If there are serious discrepancies in their respective production levels, then some explanation must be found. It would be a tedious process comparing all the individual outputs in every plant, finding out the lowest output, the highest, and the most frequent, by such tables and graphs as we have so far employed for illustrative purposes. If all the significant features of the data relating to each plant can be brought out by one or two figures, their comparison is a far simpler task than the detailed scrutiny of the data suggested above. These 'summary figures' may for the moment be described simply as 'averages', illustrated by the following three examples:

1. If, for example, it is stated that the average weekly output of an operative in Plant 1 is 539 units and in Plant 2 with identical working conditions it is 519, such information warrants investigation.

2. Further, if more operatives each produce 539 units per week than any other output in Plant 1, and the corresponding most frequent output in Plant 2 is 515, the apparent conclusion is that the operatives in Plant 1 are for some reason generally more productive.

3. If the individual outputs of the operatives in both plants are ranged separately in two arrays, i.e., in order of magnitude as in Table 1, it may appear that the middle worker in the array for

Plant 1, *i.e.* the 91st out of 180, has a weekly output of 540 units, while in Plant 2, with an equal number of operatives, 120 have a smaller weekly output than this.[1] The management confronted with this information would invariably seek an explanation.

Given the facts above, together with the ranges of the two distributions, it would be possible for anyone conversant with statistical methods to estimate with reasonable accuracy the general level of productivity in each plant, and even depict distribution of outputs as a frequency curve sufficiently accurately to bring out the same essential features as a graph of the complete data. It is because 'averages' summarise the salient features of most data so usefully that they are so widely employed in statistics. In fact, statistics has been described as 'the science of averages', although this is a little misleading in so far as averages form only a·part of the techniques employed, particularly in the later stages of a statistical enquiry.

The three specific comparisons made above have now to be considered separately and in detail.

The Arithmetic Average or Mean

'*The "average" output of the operatives in Plant 1 was 539 units.*' Most people are acquainted with the use of the term 'average' in this context. Thus in cricket, when a batsman has 'averaged' 50·0 runs per innings, no one assumes that exactly 50 runs have been scored in each innings or possibly in any innings; but if the total runs scored are divided by the number of completed innings the result will represent the batting 'average'. Assuming the above batsman completed 30 innings it is clear that his aggregate is 1,500 runs.

Using the data in Table 1, by aggregating the individual outputs and dividing the total by 180, an average output per operative of 539 units is obtained. Reference to the detailed array reveals that only five workers actually produced this output, and with only this figure as a guide, a somewhat limited picture of the situation would be obtained. This fundamental weakness in this type of average, or arithmetic 'mean' as it is known in statistics, is even more clearly demonstrated by the following example. A prospective investor is informed that three companies, X, Y and Z, have during the past six years each averaged a net profit of £6,250 each. From this information alone he would conclude that three investments are equally attractive. An examination of the profits record over the last six years suggests otherwise. The actual figures for each company over the period are as follows. It is assumed for the purpose of this example that the annual profit figures have been adjusted to a common basis in order

---

[1] To be exact, the *middle* operative in a series of 180 'lies between the 90th and 91st'. This point is discussed later in the chapter, it does not affect the present argument.

to eliminate non-recurring or capital items and are therefore comparable.

TABLE 16

COMPARATIVE PROFITS OF COMPANIES X, Y AND Z, 1958-63.

Year to Dec. 31st	X Co. Ltd.	Y Co. Ltd.	Z Co. Ltd.
	£	£	£
1958	1,500	8,800	12,000
1959	4,000	7,200	4,000
1960	6,000	6,600	Loss 2,000
1961	7,500	6,000	8,000
1962	8,500	5,900	15,000
1963	10,000	3,000	500
	6)37,500	6)37,500	6)37,500
Average Annual Profit ..	6,250	6,250	6,250

It will be self-evident that confronted with this more detailed information the investor would promptly forget all about the 'average' profits and completely revise his first impression based on the original statement of equal average profits that the investments are equally attractive. In the case of X Company the profits of the past six years show a rising trend; in the Y Company the reverse is true and the two companies are thus poles apart in terms of their attraction to a prospective investor. The last company Z is clearly a highly speculative prospect for the investor, although if the capital were highly geared, *i.e.* a small equity capital and a very large fixed interest loan, so that sharp increases in annual profits would mean disproportionate rises in equity dividends, the investor might be attracted. However, the main point is clear. The 'mean' by itself can give quite a misleading impression of any series or distribution as it provides no indication of the variation between the actual values within the distribution.

Calculation of the Mean

Most people know how to calculate an average, *e.g.* the batsman's average, or the average annual profits of the three companies illustrated above. In practice, however, the Arithmetic Mean is not always quite so simple to compute, as is illustrated by the following example. In a certain works, the works staff comprises 100 skilled men, 200 semi-skilled operatives and 50 unskilled men, all of whom are paid on a time basis at £20, £16 and £12 per week respectively. The 'average' or 'mean' wage paid in the works is *not* £16 per week computed as follows: $\dfrac{£20 + 16 + 12}{3} = \dfrac{£48}{3}$, *i.e.* £16 per week. The inaccuracy of this result may be easily proved. The total sum required to pay the weekly wages of the factory is £5,800, *i.e.* (£20 × 100, £16 × 200, £12 × 50). The total yielded by multiplying the first mean of £16 by 350 workers is £5,600, or £200 short. The correct mean wage is £5,800 divided by 350, *i.e.* £16 11s. 5d. to the nearest penny.

This type of average is sometimes described as a 'weighted' mean, *i.e.* the separate values or items within the series are each multiplied by the frequency with which each item or value appears. In the preceding example, the weights were the number of employees within each group, 100, 200 and 50 respectively. Such a computation is required when a compound made up of several constituents has to be priced for the purpose of cost accounts or final stock valuation. Thus, if A, B, C and D are four chemicals costing £15, £12, £8 and £5 per cwt. respectively, and are contained in a given compound in the ratio of 1, 2, 3 and 4 parts respectively, the resultant compound must be priced out at £ $(1 \times 15) + (2 \times 12) + (3 \times 8) + (4 \times 5)$ divided by 10, equalling £8 6s. 0d. per cwt. In the correct statistical sense of the term these figures (numbers of workers or cwt.) are *not* weights – they are simply the frequencies of each value of the independent variable. Unfortunately, the distinction is not always clearly made and all too often the frequency of a single value or group of values is termed its 'weight'.[1]

The calculation of the Mean from a *grouped* frequency distribution is different from that employed for the simple series or frequency distribution above. Where the data have been grouped, the exact frequency with which each value of the independent variable occurs in the distribution is unknown. Our knowledge is limited to the fact that, within successive class limits of the independent variable, a certain number of frequencies occur. The procedure for calculating the Mean in such cases is illustrated in Table 17. Here it will be seen that for the purpose of 'averaging', the mid-point of the class-interval is selected. This is the same as saying that the average, or 'mean', of the values lying between the limits of each class corresponds with the mid-point of the class-interval. Thus, in Table 17 there are 34 operatives whose individual outputs contain between 36 and 40 rejects. How many there are with 36, 37, 38, 39 or 40, is not known. To make any progress at all, however, it is therefore assumed that the mean number of rejects, *i.e.* average, is 38, *i.e.* the mid-point of

[1] The main use of weighting is discussed in the Chapter on Index Numbers.

the class 36-40. Note that in this example the variable is discrete. Hence the mid-point is derived by summing the two limits and dividing by two. This arbitrary procedure is justified on the score that if the number of frequencies is large, the frequencies within each class will probably be spread evenly over the range of the class-interval, *i.e.* there will be as many items below the mid-point as above it.[1]

TABLE 17

REJECTS PER OPERATIVE IN PLANT 4 DURING 4-WEEK PERIOD

(1) No. of Rejects		(2) Mid-point	(3) No. of Operatives	(4) Products of cols. (2) × (3)
21-25	..	23	6	138
26-30	..	28	17	476
31-35	..	33	22	726
36-40	..	38	34	1,292
41-45	..	43	20	860
46-50	..	48	12	576
51-55	..	53	5	265
			116	4,333

Average rejects per operative $= \dfrac{4,333}{116} = 37$ to nearest unit.

It should be noted that the procedure of multiplying the mid-points of the classes within a grouped frequency distribution by the number of items within the respective classes does not provide the Mean as such. It produces the Mean of all the mid-points of the classes 'weighted' by the frequencies within each class of the distribution. Since, as stated above, the assumption is made that the mean of all the values within each class is equal to the mid-point of that class, the use of the mid-points in order to obtain the Mean of the distribution is permissible. Generally, the smaller the class-interval and the larger the number of frequencies in each class, the more likely it is that the 'mid-point' average will correspond to the average calculated exactly,

---

[1] The same arithmetic result will, of course, be obtained if the majority of the frequencies were concentrated on the mid-point of the group as in (2), and the remainder of the frequencies spread equally over the other four values in the group, *i.e.* two on each side of the mid-point. Using hypothetical figures we get:

(1)	Value in units	f	f.x.	(2)	Value in units	f	f.x.
	1	16	16		1	5	5
	2	16	32		2	10	20
	3	16	48		3	50	150
	4	16	64		4	10	40
	5	16	80		5	5	25
		80	240			80	240
	Average = 3				Average = 3		

The student should note that these results arise because the frequencies are distributed symmetrically about the mean.

i.e. if the data given as an ungrouped frequency distribution in Table 2 were employed.

When preparing the grouped frequency distribution it will be seen whether all the items are dispersed evenly throughout the range of the independent variable. If this is the case, the classes may be taken at the most convenient intervals, *e.g.* multiples of 5 or 10 units as with the data in Table 1 (p. 39). But where, as is frequently the case, there are irregular concentrations at intervals throughout the range of the independent variable, the obvious class-limits may not be suitable. It will then be necessary, as was explained on p. 42, to revise the class intervals in such a way that within each class the mean value will be found around the mid-point. Clearly this is an ideal seldom attainable in practice, but it is important to remember it when a simple frequency distribution such as is given in Table 2 is converted into the grouped distribution shown in Table 3. At some later date someone else may wish to use the data.

Determination of the Mid-point

Apart from the questions of selecting a suitable number of classes for the grouping of any frequency distribution, and the size of the class interval, care must be taken to ensure that no uncertainty can arise in allocating any particular value to its appropriate class. The most common slip is to state the class interval as follows: £10-20, £20-30, £30-40 and so on throughout the range of the independent variable. If, after its compilation, a grouped frequency distribution in this form were to be examined, three alternatives concerning the disposition of those units which are multiples of £10, *i.e.* £20, £40, etc., spring to mind. The person responsible for the classification may have had no system at all, sometimes the item was put in the lower class, sometimes in the upper class; *e.g.* £30 could have been put in £20-30 or £30-40. The second course would have been to place them systematically in the upper class, *i.e.* assume that the upper limit of the preceding class was read as 'under £30'. The third alternative would be to assume that the lower limit of the class £30-40 meant all items *over* £30 to be included. The value of any calculations performed on such dubious tables would be problematical to say the least.

In the chapter on Tabulation, the need for accurate classification was stressed, and the above example should emphasise the reasons.

The classes quoted above can be written in several ways, and although the differences are not important one method may be more suitable than another for a particular distribution. Thus:

(1)	(2)	(3)	(4)
Under £10	£0— Signifying	—£10 Signifying	£0 — £9
£10 and under £20	£10— up to but	—£20 up to and	£10 — £19
£20 „ „ £30	£20— not includ-	—£30 including	£20 — £29
	ing £10,	£10, or	
	£20 etc.	£20	

The first is clear enough and can be used for values quoted to the nearest penny. A frequently used alternative to the classification in (1) is given in (2); they are the same. The third example differs from (2) since a value of exactly £10 will fall in the second class in the first example but in the first class in (3). The last grouping is based on the assumption that all the items are given to the nearest pound. The conventional methods for deriving the mid-points of the classes in a distribution are as follows. If the variable is *discrete*, and the classi-fication written as follows: 1 — 5, 6 — 10, 11 — 15, and so on, the mid-points are clearly 3, 8 and 13 respectively. The method may be summarised by stating that the limits of each class are aggregated and their sum halved, *e.g.* $\frac{6+10}{2} = 8$.

Continuous variables are slightly more difficult, since much de-pends on the correct demarcation of the class limits. If the classes are written as in example (1) above, 'Under £10', and so on, the class limits will depend on the way in which the individual values in the distribution have been expressed. For example, if all are expressed to the nearest penny, the upper limit of the first class is £9 19s. 11d.; the limits for the second class are £10 and £19 19s. 11d., and so on. Strictly speaking, money values should be treated as a discrete variable, but as for example in the above illustration, the smallest unit of one penny is so very small in relation to the individual values, the error introduced by treating the values as continuous may be ignored. Normal practice with *continuous* variables is to derive the mid-point by halving the sum of the lower limits of two successive classes. Applying this rule to example (3), the mid-point for the second class would be £15. If the values in a continuous distribution are written to the nearest decimal place, the classes could then be written — 10·0, — 20·0, — 30·0, — 40·0, etc., with mid-points of 5·0, 15·0 and 25·0.

When in doubt as to the limits of the class interval the student should consider firstly the unit of measurement and secondly how the individual values have been defined. For example, if no payment is less than a multiple of a pound then the series is discrete, since the difference between the upper limit of a class interval and the lower limit of the next must be one pound. Such a classification would then read as in example (4). If, however, the values have been rounded to the nearest pound, then clearly a value of £10 in the distribution could represent any value ranging from £9 10s. 0d. to £10 9s. 11d. Such a distribution should be treated as a continuous series and the classification should be similar to that given in any of the first three examples. It should be remembered that the grouping of a distribution and the use of mid-points for calculating averages by themselves give rise to possible error in the average. The choice of the class-interval should be as accurate as is compatible with such considerations.

The Short-cut Method

When using the mid-points of successive classes in a frequency distribution to compute the Mean, the volume of arithmetic calculation can be reduced in the following ways:

TABLE 18

CALCULATION OF MEAN FROM
INDIVIDUAL OUTPUTS OF 180 FEMALE OPERATIVES AS PER TABLE 1

Output in Units (1)	Mid-points (2)	Mid-points less 504·5 (3)	No. of Operatives (4)	Products (3) × (4)
500 to 509 ..	504·5	Nil	8	Nil
510 to 519 ..	514·5	10	18	180
520 to 529 ..	524·5	20	23	460
530 to 539 ..	534·5	30	37	1,110
540 to 549 ..	544·5	40	47	1,880
550 to 559 ..	554·5	50	26	1,300
560 to 569 ..	564·5	60	16	960
570 to 579 ..	574·5	70	5	350
			180	6,240

Mean output per operative = 504·5 + 34·7 = 539 to nearest unit.

In the example in Table 18, by using the mid-points and subtracting the figure of 504·5 which is common to every mid-point, the arithmetic involved is reduced to very simple proportions.

Such a simple example involving easily manageable figures does not arise very frequently, and a more usual method of computing the arithmetic mean is given in Table 19 below. This method is based upon the simple rule of algebra that the sum of the individual differences between a series of numbers and their mean is always equal to zero. Take for example the following distribution: 4, 7, 9, 10, 15,

17 and 22. Their aggregate is 84 and the average of the seven figures comprising that total is therefore 12. From this figure, *i.e.* their mean, subtract each of the figures in the distribution in turn. The following result is obtained: —8, —5, —3, —2, 3, 5, 10. When aggregated the differences are equal to zero. The reader may check the rule by testing any selection of values he cares to make. Given this rule the accuracy of an estimated mean may be tested quite simply. Suppose that for the above series we had guessed that the true mean equalled 10. The differences, or as they are usually termed, the *deviations* from the mean, would then be: —6, —3, —1, 0, 5, 7, 12 and their total is 14. Clearly then, if the rule is valid the estimate of the mean is wrong. But if the 'error' is apportioned, *i.e.* 14 units, between the seven constituent numbers, the 'average error' is 2 and if this is then added to the estimated value of the mean, *i.e.* 10, we arrive at the correct value of the mean of the distribution, *i.e.* 12. The student should amuse himself setting out short series of figures and proving to himself the validity of the rule.

The foregoing principle is illustrated in Table 19. The successive stages in the calculation are as follows:

1. Select as the assumed mean the mid-point of the class which contains a high proportion of the units and is nearly central. Since wages are paid to the nearest penny, the true limits of the first class are £4 10s. —£4 14s. 11d. and the variable is, strictly speaking, discrete. But, for all practical purposes, this variable may be treated as continuous since one penny is so small a unit and the mid-points are derived by adding together the upper limits of successive classes and halving them.

2. In the column headed 'deviations from assumed mean in class-intervals', enter 0 against the class whose mid-point is to be used as the assumed mean, *i.e.* £5 10s. 0d. —£5 15s. 0d. Against the mid-points on either side of this latter class enter 1, against the next above and below those mid-points enter 2 and so forth. Where the mid-point is smaller than the selected mid-point representing the assumed mean the deviation will be negative; thus the upper part of Col. (4) before the mid-point marked 0 will contain all the negative deviations. The reverse applies to the lower part where the mid-points are greater in magnitude than the assumed mean, *i.e.* the deviations are positive. Before inserting the figures, the student should note whether all the class-intervals are of equal size; *i.e.* as in Table 19.

3. Each deviation is multiplied by its respective frequency, *i.e.* the frequencies in each class; the negative quantities being kept apart from the positive products to avoid confusion. Both the negative

and positive products are then aggregated separately and the balance obtained, *i.e.* + 1,216 —1,461   —245.

TABLE 19

EARNINGS OF 1,783 PART-TIME FEMALE EMPLOYEES OF THE
XYZ MANUFACTURING CO., LTD. IN THE WEEK ENDING 8TH NOVEMBER 1963

Weekly Earnings (1)		Fre- quencies (2)	Mid- points (3)	Deviations from assumed Mean in Class Intervals (4)	Products of cols. 2×4	
					Negative	Positive
					(5)	
£ s. d.	£ s. d.					
4 10 0 but under	4 15 0	64	4·12·6	— 4	— 256	
4 15 0 ,, ,,	5 0 0	126	4·17·6	— 3	— 378	
5 0 0 ,, ,,	5 5 0	224	5· 2·6	— 2	— 448	
5 5 0 ,, ,,	5 10 0	379	5· 7·6	— 1	— 379	
5 10 0 ,, ,,	5 15 0	474	5·12·6	0		
					— 1,461	
5 15 0 ,, ,,	6 0 0	227	5·17·6	1		227
6 0 0 ,, ,,	6 5 0	108	6· 2·6	2		216
6 5 0 ,, ,,	6 10 0	74	6· 7·6	3		222
6 10 0 ,, ,,	6 15 0	31	6·12·6	4		124
6 15 0 ,, ,,	7 0 0	43	6·17·6	5		215
7 0 0 ,, ,,	7 5 0	19	7· 2·6	6		114
7 5 0 ,, ,,	7 10 0	14	7· 7·6	7		98
		1,783				1,216

Assumed mean = mid-point of the class £5 10s. 0d. but under £5 15s. 0d. = £5 12s. 6d.

Sum of deviations from assumed mean = — 245 i.e., — 1,461 + 1,216

Average deviation in class intervals = $\dfrac{-245}{1,783}$ = — ·137

Correct arithmetic mean = £5 12s. 6d. — ·137 of the class-interval
= £5 12s. 6d. — ·137 × 5s. 0d.
= £5 12s. 6d. — 8d. to nearest 1d.
Arithmetic mean = £5 11s. 10d. to nearest 1d.

4. The balance, in this case negative, is divided by the sum of the frequencies, *i.e.* 1,783. The result in this example is a *fraction of the class-interval*, not of a single unit. In other words the result is expressed in 5s. units. Reference to columns (3) and (4) will reveal that the deviations are measured in units of 5s.; thus 2 deviations equal 10s. as is apparent by subtracting the mid-point of the class £5 10s. 0d. —£5 15s. 0d. from that of the class £5 0s. 0d. —

£5 5s. 0d. It is important therefore that the quotient of —·137 is converted into shillings before it is subtracted from the assumed mean, from which the deviations were measured. The result gives the true arithmetic mean of the frequency distribution.

Working With Class Interval Units

The figures entered under the heading of 'deviations from the assumed mean', might have been multiples of 1s. instead of 5s., or, for that matter, of any unit always provided the difference between the negative and positive totals expressed in terms of those units is finally converted to the original unit of measurement. If the deviations had been measured in actual shillings instead of multiples of 5s., then the figures in column (4) would have read 0, 5, 10, 15 and so on instead of 0, 1, 2, 3, etc. Equally, the '0', *i.e.* the mid-point assumed to be the mean from which all the deviations are measured, may be placed anywhere in the series, but it simplifies the calculation if put against the class limits between which the largest number of frequencies occur. To exercise his arithmetic and prove this point to his own satisfaction, the reader should re-work the data in Table 19 taking another mid-point as the assumed mean and measuring the deviations in shillings.

It will be realised that there is no need to work in class-intervals, although this is usually the most convenient method when all the class-intervals are equal in size. If, however, they vary, the mid-point method of deriving the Mean may still be used. With varying class-limits the differences will be in multiples (sometimes fractions) of the class-interval 'unit'. If, for example, there had been another class at the upper end of Table 19, say £7 10s. 0d. to £8 5s. 0d. then the mid-point of this class interval is £7 17s. 6d. Had this been the case the deviation from the mean allowed for that class would be 9 and not 8, since the difference between the mid-point of this class and that immediately preceding it is equal to 10s.; twice as much as the unit of 5s. in which previous deviations have been measured.

Where 'open-end' classes are involved, *e.g.* '£7 10s. and over', the difficulty is still greater. If the open-end was necessary there is every reason to assume that one or more of the frequencies did not fall within the limits of any normal class, *i.e.* there is (or are) extreme item(s) which would affect markedly the value of the Mean. The usual assumption in the absence of further information is to assume the limits of that class are identical with the others and select a mid-point accordingly. It is probable that the use of this arbitrary mid-point will tend to under-estimate the true Mean of the distribution. Since extreme or unrepresentative items distort the Mean this compromise avoids that danger, but the method is still unsatisfactory.

Sometimes in the case of an open-ended class at the upper end of the distribution, because this indicates some values which do not fall within the range of a single class interval, a compromise is made. This is to assume that the open-ended class has an interval of twice the interval of the other classes in the body of the table. A similar difficulty arises with open-ended classes at the lower end of a distribution, *e.g.* under £4 10s. Can we assume that the range of this class is 5s., *i.e.* no value under £4 5s., or should it be a double interval on the assumption that all the values fall within the limits of £4 and £4 10s.?

The only guide is knowledge of the data being handled. For example, if the distribution is concerned with the age of women at marriage, with the first class in the distribution reading 'under 20 years', and the next class '20 but under 25', then knowledge of the law provides the answer regarding the lower limit since no girl under 16 may marry. In practice, the decision about the lower or upper limit of such open-ended classes often depends on the number of frequencies in that class. For example, suppose the top class reads '£7 10s. and over'. If only one or two cases are recorded in this class, compared with a marked concentration of frequencies in the body of the table, then it hardly matters whether one takes £8 10s. or £10 as the upper limit. This is because it will be swamped in the calculation by the large number of other values in the rest of the distribution. If, however, a large proportion of the total frequencies are in that class, then clearly the choice of the upper limit and the consequent mid-point does matter a great deal. Any error in the choice, *e.g.* a mid-point rather higher than the true mean of that class, will tend to raise the Mean of the distribution disproportionately. In such cases the assumptions used in fixing the lower limit of the bottom class and the upper limit of the highest class should be clearly stated next to the calculation. In examination questions such distributions are often used to test the candidate's understanding of the process and, if he is doubt, then it is a good practice to write a short footnote explaining the choice of limit made. A worked example illustrating the problem of open-end classes and varying class-intervals in the same distribution is given at the end of the next chapter.

Apart from its relative simplicity of computation, the arithmetic Mean has other advantages. The statistician considers it a useful measure since it is itself the result of an arithmetic process and therefore lends itself to further mathematical treatment. In so far as the Mean takes into account every item in the series or distribution, it generally provides a reasonably accurate summary of the data, hence its popularity with the lay public who use the term 'average man' to refer to the representative of the majority of the male members of the community. On the other hand this advantage also lies at the root of its outstanding weakness; by including every item in the series the presence of extreme, or even a single non-representative, items may, especially in a short series, so seriously distort the Mean that it no longer provides an accurate indication of the nature of the data. Thus, if seven directors receive annual fees £100, £200, £200, £250, £250, £300 and £1,500 respectively, the Mean is £400. In fact, this particular amount is not received by any director, is in excess of what six out of seven receive and provides no indication whatsoever of the nature of the series on account of the extreme item of £1,500. This weakness occasionally provides one of the reasons for needing other measures which will amplify and even replace the Mean. Nevertheless, the latter remains one of the most frequently employed measures in statistical work.

Formulae for the Mean

Reference to the majority of books on statistics will reveal a formidable array of mathematical symbols dealing with the calculation of the Mean, which convey little to the non-mathematical reader, and may even confuse the issue for him. Nor is the beginner helped by the fact that the notation has not yet been completely standardised and statisticians may use slight variations of symbols which are comprehensible to the informed reader, but confusing to the student.

Most of the symbols employed are merely 'shorthand' or abbreviations of simple procedures which would be cumbersome if expressed in simple English. The following are typical of the various notations and symbols used.

Thus:

1. The Arithmetic Mean of a *simple* series, *e.g.*, 2, 6, 9, 12, 15, is written $\bar{x} = \dfrac{\Sigma x}{N \text{ (or } n)}$ to represent

$$\text{A.M.} = \frac{(2 + 6 + 9 + 12 + 15)}{5}$$

$\bar{x}$ = the arithmetic mean.
where Σ (termed large sigma) = the sum of.
x = the individual items.
N or n = total frequencies.

By using Σ the need for the following notation is avoided:

$$\frac{x_1 + x_2 + x_3 + x_4 + \ldots x_n}{n}$$

where x_1, x_2, $x_3 \ldots x_4$ refer to the individual values of the variable and $n = $ the total frequency.

2. For frequency distributions, such as the wages or chemical compound examples on page 95, the formula may be written:

$$\frac{f_1x_1 + f_2x_2 + f_3x_3 \ldots + f_nx_n}{f_1 + f_2 + f_3 \ldots + f_n} \; i.e., \; \frac{(15 \times 1) + (12 \times 2) + (8 \times 3) + (5 \times 4)}{(1 + 2 + 3 + 4)}$$

which is normally abbreviated to $\frac{\Sigma fx}{\Sigma f}$, the letter f representing the frequencies. Σf, it should be noted, is the same as N or n.

3. The A.M. of a *grouped* frequency distribution is written $\bar{x} = \frac{\Sigma fx}{\Sigma f}$, where $f = $ the number of observations in each class of the distribution. Note that x in this case, *i.e.* in fx, represents the midpoints of the classes.

4. When the A.M. is derived from a frequency distribution by using the *deviations from an assumed Mean*, the formula is written $\bar{x} = x' + \frac{\Sigma fd'}{\Sigma f}$, where $d' = $ the deviations from the assumed Mean written as x'. (Note: if the deviations are expressed in class-intervals, they must be converted into the original unit values. Thus $\frac{\Sigma fd}{\Sigma f} \times i$ where i represents the class-interval). This formula applies to the example in Table 19. The student reader may commit these formulae to memory in case they should appear on an examination paper, or more probably, in another text. For practical purposes they are unnecessary at this stage, it is the method, not the formula, which should be learnt.

The Median

The nature of the third average employed in describing statistical data was indicated in the passage on page 92. 'If the individual outputs of the operatives are ranged in order of magnitude, *i.e.* as an array, the central figure has a value of 540 units'. The central value, known as the Median, divides the distribution into two equal parts. In other words, it is the value which divides a distribution so that an equal number of values lie on either side of it. In this example the one half contains the better operatives and the other the less productive.

Median of Ungrouped Data

In contrast to the Mean the task of finding the Median is sometimes extremely simple. All that is necessary is to arrange the individual items in order of magnitude; the middle item is then the Median.

Thus in the following series, 2, 3, 4, 5, 6, 7, 8, 9 and 10 the Median is 6, *i.e.* the fifth figure, with four figures on either side of it. Such is the procedure when the data is given in an array, more usually it is described as *ungrouped* data. It may be located by the formula $\dfrac{N+1}{2}$ where $N =$ the number of items in the series. When the number of items, *i.e. N*, is odd, then the Median is an actual value with the remainder of the series in two equal parts on either side of it. If N is even, then the Median is a derived figure, usually half the sum of the two middle values. If these are the same, as they often are, then the Median of an even series will also be an actual value.

Median of Grouped Data

More frequently, however, it is necessary to select the Median from grouped data. In this case the ranging of the data has already been effected since the classes will clearly be in order. The normal method is to add the class frequencies together cumulatively, as has been done in the example below and divide the total frequencies into

TABLE 20

DERIVATION OF MEDIAN FOR DATA AS IN TABLE 19

Earnings	No. earning Wages shown opposite	Cumulative Total	
£ s. d. £ s. d.			
4 10 0 and under 4 15 0.. ..	64	64	
4 15 0 ,, ,, 5 0 0.. ..	126	190	(64 + 126)
5 0 0 ,, ,, 5 5 0.. ..	224	414	(190 + 224)
5 5 0 ,, ,, 5 10 0.. ..	379	793	(414 + 379)
5 10 0 ,, ,, 5 15 0.. ..	474	1,267	(793 + 474)
5 15 0 ,, ,, 6 0 0.. ..	227	1,494	etc.
6 0 0 ,, ,, 6 5 0.. ..	108	1,602	
6 5 0 ,, ,, 6 10 0.. ..	74	1,676	
6 10 0 ,, ,, 6 15 0.. ..	31	1,707	
6 15 0 ,, ,, 7 0 0.. ..	43	1,750	
7 0 0 ,, ,, 7 5 0.. ..	19	1,769	
7 5 0 ,, ,, 7 10 0.. ..	14	1,783	
	1,783		

two halves. The formula for deriving the Median from *grouped* data, *i.e.* as in a grouped frequency distribution, is $\dfrac{N}{2}$. Thus, in the distribution given in Table 20, the Median value is located as follows:

Median $= \dfrac{N}{2} = \dfrac{1,783}{2} = 891\frac{1}{2}$.

The Median is located between 793 and 1,267, *i.e.*, in the class with 474 individuals receiving at least £5 10s. 0d. but under £5 15s. 0d. Therefore the median wage is greater than £5 10s. 0d. but below £5 15s. 0d.

$891\frac{1}{2} - 793 = 98\frac{1}{2}$, thus the Median is the $98\frac{1}{2}$th of the 474 items ranged in order of size, these 474 values ranging from £5 10s. 0d. to £5 14s. 11d.

£5 10s. 0d. $+ \left(\dfrac{98\frac{1}{2}}{474} \times 5\text{s.} \right) = $ £5 10s. 0d. $+$ 1s. 0·3d. or 1s. 0d. to nearest penny, *i.e.*, £5 11s. 0d. = Median wage.

Alternatively the calculation may be carried out by assuming that £5 15s. 0d. is greater than the median wage, and 375 (474 — 99) employees in that class earn more than the median wage.

Median wage = £5 15s. 0d. $- \left(\dfrac{375}{474} \times 5\text{s.} \right)$

calculated to nearest penny = £5 11s. 0d.

It will be noticed that the same assumption has been made in the computation of the Median as was made in determining the Arithmetic Mean, *i.e.* that the values falling within any given class are ranged evenly throughout and, as before, the validity of this assumption will determine the accuracy of the result.[1] This is justified with a continuous series with a large number of classes. If the variable is discrete and the class-interval large, the Median may be little better than an approximation, and the result is best given to a round number.

The outstanding advantage of the Median resides in the fact that it is not affected by extreme items, as is the Mean. Thus if seven salesmen take £700, £750, £780, £800, £830, £870 and £1,600 respectively, the Median value is £800, which gives a fair indication of the typical salesman's results; the Mean on the other hand is over £900 and quite unrepresentative. The Median value often corresponds to a definite item in the distribution; the Mean seldom. A further important advantage of the Median is that it can be located just as easily in a grouped distribution in which the first and last classes are open-ended and the lower and upper units are not available. In contrast it may be virtually impossible to compute the Mean of such a distribution with any degree of accuracy.

Median by Interpolation

The Median can also be interpolated approximately from the ogive, *i.e.* the cumulative frequency distribution plotted on a graph as shown in Figure 16. This is true regardless of whether the ogive is

---

[1] This assumption need only be valid for the class containing the Median, since the Median is unaffected by the values in any other class.

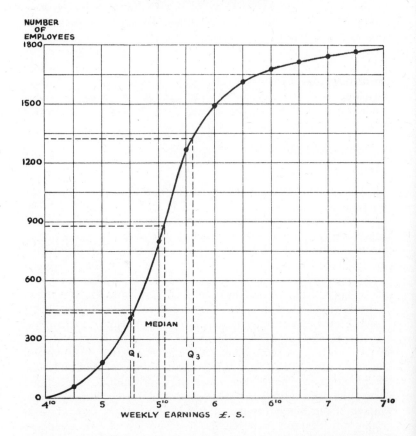

Figure 16

DERIVATION OF MEDIAN AND QUARTILES FROM A
CUMULATIVE FREQUENCY CURVE.
(DATA FROM TABLE 20)

drawn by cumulating the series upwards or downwards as was explained on p. 41. The reader should cumulate the distribution from
the highest value and sketch the corresponding curve. Care is required when drawing either curve that the data are correctly plotted.
Thus the curve shown starts at zero frequencies and rises continuously. The first point plotted against the vertical axis, in this case
64 (see Table 20, p. 106) will be above the upper limit of the class
£4^{10} —£4^{15}, the next figure 126 against the upper limit of the next
class, *i.e.* £4^{15} —5. Thus, when the frequencies of any given output are
read from the curve, they will be interpreted as ' 190 employees below

£5, 64 below £4^{15}, etc. The student should not plot the frequencies over the mid-points of the class-intervals as is done with the histogram, otherwise he will read off the wrong results, *i.e.* the values of the independent variable will be too low.

Having drawn the ogive, find the mid-point on the scale representing the frequencies, which in Fig. 16 is the vertical axis. From that point draw a line parallel to the horizontal axis until it intersects the ogive. The value of the Median will then be read off against the scale along the horizontal axis directly below the point of intersection. The reader can compare the values derived from the graph with those calculated from the data in the example on p. 107. Since both methods of deriving the Median are approximate it is not surprising that there are often slight differences between the results.

Quartiles and Deciles

In precisely the same way as is used to obtain the Median, it is possible to locate the *quartiles* and *deciles*, the values of which are useful in describing a distribution. As the name of the former suggests, the *quartiles* divide the series into four equal parts, *i.e.* they perform for each equal part of the distribution on either side of the Median what the Median has done for the whole series. The *deciles*, less frequently employed in practical work than the quartiles, divide the series into 10 equal parts. The method of computing the quartiles of *grouped* data is the same as with the Median, except that instead of $\frac{N}{2}$, the denominator is 4, *i.e.* $\frac{N}{4}$. The two quartiles in any distribution are known as the *lower* and *upper* quartiles, the former indicating the smaller value and obtained by $\frac{N}{4}$, and the latter, the higher value in the position $\frac{3N}{4}$. The lower quartile is usually written as Q_1, the upper quartile as Q_3. The calculations for the deciles are similar, the denominator being 10, thus the fourth decile is the observation which is $\frac{4N}{10}$ from the lower end of the range.[1]

Working on the data given in Table 22, the following results are obtained for the two quartiles:

$$Q_1 = \frac{1,783}{4} = 446 \text{ to nearest unit, } i.e.. Q_1 = \text{ value of 446th item.}$$

The 446th item lies in the class £5 5s. 0d. — £5 10s. 0d. containing 379 items, which are assumed to be spread evenly over the class-interval of 5s. 0d.

[1] When the series is ungrouped, the formula for Q_1 is $\frac{N+1}{4}$, and for Q_3 $\frac{3(N+1)}{4}$. if the division yields an odd quarter in the quotient, the answer may be expressed to the nearest unit.

$$\therefore Q_1 = £5 \text{ 5s. 0d.} + \left(\frac{446 - 414}{379}\right) \times \text{5s. 0d.}$$

$$= £5 \text{ 5s. 0d.} + \frac{32}{379} \times \text{5s. 0d.}$$

Lower Quartile wage = £5 5s. 0d. + 5·7d. = £5 5s. 6d. to nearest 1d.

$$Q_3 = \frac{3 \times 1,783}{4} = 1,337 \text{ to nearest unit, } i.e., \; Q_3 = \text{value of 1,337th}$$

item.

1,337th item lies in group £5 15s. 0d. – £6 0s. 0d. containing 227 items;

$$Q_3 = £5 \text{ 15s. 0d.} + \left(\frac{1,337 - 1,267}{227}\right) \times \text{5s. 0d.}$$

$$= £5 \text{ 15s. 0d.} + \frac{70}{227} \times \text{5s. 0d.}$$

Upper Quartile wage = £5 15s. 0d. + 18·5d. = £5 16s. 6d. to nearest 1d.

Apart from their value in providing a description of any distribution, the quartiles and deciles are especially useful for comparison of two distributions, *i.e.* contrasting the values in each series at the lower quartile position, upper quartile and so on. As explained above, the values of the quartiles or any of the deciles can be estimated from the ogive, as was the Median in Fig. 16. Note that neither the quartiles nor deciles are averages, they are measures of dispersion and as such are discussed in the next chapter. They are discussed at this stage simply because they are derived by the same methods as those employed in calculating the Median.

The Mode

Statements such as 'the average man prefers this brand of cigarettes', or that 'the average woman uses cosmetics', are frequently read and overheard. Used in this context, the term 'average' means the majority and not the Arithmetic Mean discussed earlier in this chapter. The fact that the Mean does not always provide an accurate reflection of the data due to the presence of extreme items has already been stated; similarly, the Median may prove to be quite unrepresentative of the data owing to an uneven distribution of the series. For example, the values in the lower half of a distribution range from, say, £20 to £100, while the same number of items in the upper half of the series range from £100 to £5,000 with most of them nearer the higher limit. In such a distribution the Median value of £100 will provide little indication of the true nature of the data.

Both these shortcomings may be overcome by the use of the Mode, which refers to the value which occurs most frequently within a distribution. This particular 'average' is the easiest of all to find in some distributions, since it is the value corresponding to the largest frequency. Thus in the following distribution which is discrete:

No. of Rooms ..	1	2	3	4	5	6	7	8	9	10	11
Frequencies ..	4	9	15	19	24	38	26	18	13	7	1

the modal value or mode is '6', since it appears more times in the series than any other value. The Mode is a particulary useful average for discrete series, *e.g.* number of people wearing a given size shoe, or number of children per household, etc.

The Mode by Interpolation

Ascertaining the Mode is not always quite so easy as the above simple example suggests, although it is seldom necessary to find it exactly. When, as is frequently the case, it has to be located in a grouped frequency distribution, the Mode lies within a given class, *i.e.* within the limits of the maximum and minimum values of that class. The simplest course is to select the mid-point of that particular class; this is no more arbitrary than computing the Mean from a grouped frequency distribution by multiplying the frequencies by the mid-points of the corresponding classes. As was pointed out, if the distribution were evenly dispersed throughout its range of values, the result from calculating the Mean by this arbitrary method should correspond with the Mean derived from a detailed computation.

TABLE 21

Commission Payments for January 1964			No. of Salesmen
£10 and under £15	..	..	6
£15 ,, ,, £20	..	..	12
£20 ,, ,, £25	..	..	30
£25 ,, ,, £30	..	..	53
£30 ,, ,, £35	..	..	77
£35 ,, ,, £40	..	..	96
£40 ,, ,, £45	..	..	54
£45 ,, ,, £50	..	..	37
£50 ,, ,, £55	..	..	19
£55 ,, ,, £60	..	..	8

The assumption that the frequencies in a given class are spread evenly over all the values within the limits of that class is arbitrary but, as stated above, provides in many cases a fair enough approximation to the truth. In some distributions there are more items below the modal value than above it, *e.g.* in the classes below the modal group in value, the number of frequencies may be far greater than the number of frequencies contained in the classes in the upper regions of the table. Such a case is illustrated in Table 21, where the modal class, *i.e.* the largest with 96 cases, lies within the class limits £35-40.

Here, it is found that more salesmen (77) were in the class below the modal class than in the one above, containing 54. Because of this, it is likely that the concentration of the salesmen within the modal class (£35 – 40) is more marked between, say, £35 to £37 10s. 0d. than between £37 10s. 0d. and £40. In other words, had a different class interval been selected for this frequency distribution, it is possible that instead of most frequencies being within the class £35 – 40, yielding an arbitrary Mode of £37 10s. 0d., *i.e.* the mid-point; the Mode would have been located in a new group, say, £35 – £37 10s. 0d., yielding a modal value of £36 5s. 0d. Such a breakdown of the distribution into smaller or different groups is not possible unless the full data are given elsewhere, and in passing it may be noted that the Mode can be markedly affected by the classification adopted in compiling the grouped frequency distributions. The above theory underlies the following formula for estimating the Mode which involves a simple exercise in proportions.

$$\text{Mode} = L + \left[\frac{fa}{fa + fb} \right] \times C.I.$$

Where:

L = lower limit of modal group (£35).
fa = frequencies in group following the modal group (54).
fb = frequencies in group preceding the modal group (77).
$C.I.$ = class interval (£5).

Thus:

$$\pounds35 + \left[\frac{54}{54 + 77} \right] \times \pounds5$$

$$= \pounds35 + \left(\frac{54}{131} \right) \times \pounds5$$

$$= \pounds35 + 2 \text{ to nearest } \pounds = \pounds37.$$

In this example, the use of the formula does not result in the same modal value as may be obtained by simply taking the mid-point of the modal class, *i.e.* £37 10s. 0d. This arises because the relative sizes of the two classes adjacent to the modal class, *i.e.* 77 and 54, are unequal. The principle of the theory may be tested by substituting more closely similar figures, *e.g.* 70 and 68 in place of 77 and 54 respectively. The Mode derived by using the above formula is then £37 10s. 0d. to the nearest 5s. 0d., the same as the mid-point.

The above formula for estimating the Mode is one of several based upon the same principle of proportions. The more complicated formula sometimes encountered represents an attempt to obtain a better estimate. In practice, however, the Mode is quite difficult to calculate if some degree of precision is required. The only satisfactory method is to fit a curve to the distribution and determine the highest point of the curve in relation to the independent variable. Even this method can yield only an approximate result; so much depends on the shape of the distribution and the size of the class interval. Hence, in elementary statistical work the Mode is not often used except where it is fairly self-evident, or where it is especially appropriate, *e.g.* the modal size in men's shoes is 8½.

Which Average?

While the actual calculation of any of the three averages described above is fairly simple, the choice between them for purposes of describing a distribution is not always so clear-cut. By examining the advantages and limitations of each of the three averages, it is possible to draw up some general rules. What can be said of each of them?

Arithmetic Mean:
   i Its main characteristic and virtue is that in its calculation every value in the distribution is used. To this extent the arithmetic mean may be regarded as more representative than the other two.
  ii The foregoing advantage also provides its main defect. If in a distribution there are few very large, or alternatively very small values, these may distort the mean. Thus it may be rather lower than the bulk of the values, or it may be rather higher, and as such becomes unrepresentative.
 iii It is basically the simplest of all the averages. Most people understand this type of average. Since it is the result of arithmetic processes it can be used for further calculation. For example, knowing the mean and the total frequency of the distribution, their product gives the aggregate of the entire distribution.
 iv Since the mean is often calculated from grouped frequency distributions, the presence of open-ended classes at the extreme ends of the distribution has the result that the mean can only be estimated on the basis of assumptions regarding the size of the class-interval of the open-ended classes. If such classes contain a large proportion of the values, then the mean may be subject to substantial error.

Commonsense suggests that the defects and limitations of the mean provide the *raison-d'être* of the other two averages. These too have their merits and shortcomings:

Median:

i It is a useful average in so far as it divides a distribution into two equal parts conveying the information that the same proportion of values lie above the median value as lie below it.

ii It is especially valuable for distributions with extreme values or with open-ended classes since it remains unaffected by these values.

iii To the extent that it does not reflect the distribution in the way that the mean does, *i.e.* by including all values, it needs to be supplemented by other statistics.

Mode:

i This can be both the easiest average to obtain and also, in the case of a grouped frequency distribution, the most difficult since, strictly speaking, unless it can be easily picked out from the distribution it should be derived by algebraic methods.

ii When it does not need to be calculated it is an actual value in the distribution and as such forms an important part of the distribution.

iii It enjoys with the median the advantage that it is not affected by the remainder of the distribution, *e.g.* opened-ended classes or extreme values.

From the above summary it is evident that no single average is preferable at all times to the others. As a rough guide, however, the arithmetic mean and mode are preferable where the frequency curve of the distribution is hump-backed and there are no extreme values in the distribution since, it will be remembered, such values may distort the arithmetic mean. The Median is generally the best choice with open-ended grouped distributions especially where, if plotted as a frequency curve, instead of a hump-backed curve one gets a J or reverse J curve. This signifies, in the case of the former, that there are relatively few frequencies at the lower values of the independent variable and many at the upper values. In the latter case, *i.e.* reverse J, the frequencies are concentrated at the lower values of the independent variable and their number declines continuously as one approaches the upper values.

Clearly then, the choice of the average in any given case must be determined by the nature of the data and the purpose to be served by the average. If it is not forgotten that a single average is designed to replace the detail, yet at the same time to provide the outline of that detail, then the selection of the average will be seen to depend on which measure fulfils this requirement most adequately. Since the three 'averages' comprise rather different concepts, the data may be such as to warrant the use of all three, and as will be shown in the next chapter, the relationship between the three measures may be significant. In any case, the chief use of averages is to compare those of one series with the same averages of another but comparable series. In practice, the Mean is a firm favourite in so far as it is so readily computed and understood; generally speaking, it should be used instead of the others. But either the Median or even the Mode will be preferable if the generalisation concerning mid-points in the calculation of the Mean is unjustified, or the Mean is seriously affected by extreme items.

Questions

AVERAGES: TYPES AND FUNCTIONS

1.

Weekly Income (£)	No. of Incomes
Under 10	34
10 and under 12	58
12 and under 14	69
14 and under 16	103
16 and under 18	95
18 and under 20	70
20 and under 24	34
24 and under 30	13
	476

For the above distribution calculate the arithmetic mean and the standard deviation.   *Rating and Valuation Association 1963*

2. Define: (a) Median; (b) Mode; (c) Arithmetic Mean.
Use the following eleven values as a frequency distribution to illustrate your definitions: 8, 10, 6, 5, 2, 13, 6, 11, 16, 4, 7.   *Institute of Statisticians 1959*

3. The following table shows the age distribution of the estimated population of U.K. at 30th June, 1960:

Age	Numbers (ten thousands)
0– 9	795
10–19	782
20–29	670
30–39	720
40–49	707
50–59	692
60–69	494
70–79	292
80+	100

Calculate: (a) the mean and the standard deviation; (b) the median and the quartile deviation.
Why does the mean differ from the median?   *I.C.W.A., 1962*

4.   Dwellings Occupied by Private Households 1951
                            England and Wales

No. of Rooms per Dwelling	No. of Dwellings Occupied ('000)	No. of Dwellings Occupied by one Household ('000)
1	94·6	94·6
2	456·6	454·9
3	1,346·0	1,336·5
4	3,399·4	3,321·5
5	4,270·0	4,079·0
6	1,548·3	1,307·8
7	486·9	378·5
8	245·7	165·2
9	110·4	69·3
10	54·3	34·0
11	25·2	14·5
12	17·0	10·1
13	7·4	4·0
14	5·8	3·6

(*Ann. Abs. Statistics 1960, Table 64*)

To what extent are those dwellings occupied by one household smaller than
those occupied by more than one household.   *I.M.T.A. 1961*

5.   Distribution of Earnings of Dock Workers
                        in the Week Ended 19th November 1960

Range of Earnings				Numbers
Under £8				521
£8 but less than £10	..	..	..	1,860
£10 ,, ,, ,, £12	..	..	..	3,095
£12 ,, ,, ,, £14	..	..	..	5,381
£14 ,, ,, ,, £16	..	..	..	5,129
£16 ,, ,, ,, £18	..	..	..	5,182
£18 ,, ,, ,, £20	..	..	..	4,344
£20 ,, ,, ,, £22	..	..	..	3,871
£22 ,, ,, ,, £24	..	..	..	2,900
£24 ,, ,, ,, £26	..	..	..	2,112
£26 ,, ,, ,, £28	..	..	..	1,383
£28 ,, ,, ,, £30	..	..	..	1,058
£30 ,, ,, ,, £35	..	..	..	1,058
£35 ,, ,, ,, £50	..	..	..	497
£50 and over	..	..	..	17
				38,408

(*Min. Lab. Gazette, April 1961*)

What are the earnings corresponding to the mean, median, and quartiles in
this distribution?   *I.M.T.A., November 1961*

6.   Annual Receipts and Year-End Stocks
              in Grocery and Provision Businesses 1959

Businesses with receipts of				Receipts £000	Stocks at end of year £000
Under £10,000	..	..	..	..	534	98
£10,000 and under £50,000	..	..	..	8,257	613
£50,000 and under £200,000	..	..	..	86,906	5,716
£200,000 and under £1,000,000		..	..	308,796	17,031
£1,000,000 and under £5,000,000		..	..	260,445	13,601
£5,000,000 and over	..	..	..	..	916,121	42,836

(Board of Trade Jnl., 7 April 1961, p. 803)

Calculate the ratio of year-end stocks to annual receipts for the total of all businesses and estimate the size of business in which this ratio would occur. What is the probable relationship for a business with annual receipts of £50,000?

I.M.T.A., November 1961

7. From the table given below calculate any useful averages; then present them in such a way as to show their significance. Can you draw any conclusion from your table and what investigations, if any, would you make in the light of the figures shown therein?

Works	Production (tons)	Man-hours worked	Total Manufacturing Cost (£)
Westworth ..	31,171	124,735	93,497
Northend ..	15,823	59,316	39,821
Southlands ..	14,119	42,887	31,604
Hilltop ..	23,882	97,992	71,386
Riverside ..	47,915	203,818	143,698
Eastleigh ..	43,112	191,828	137,789
Beaumont ..	8,171	34,612	25,322

The hourly wage-rates of operatives and the price of raw materials are the same for all works.   *Institute of Statisticians 1958*

8.

Weekly income of household					Average weekly housing payment s. d.	Average total household expenditure s. d.
£50 or more	..	..	..	..	65 6	975 5
£30 but under £50	..	..	..	46 3	559 11
£20 „ „ £30	..	..	..	29 9	415 10
£14 „ „ £20	..	..	..	25 3	308 10
£10 „ „ £14	..	..	..	21 2	237 6
£8 „ „ £10	..	..	..	18 4	195 8
£6 „ „ £8	..	..	..	15 5	159 11
£3 „ „ £6	..	..	..	14 10	111 9
Under £3	..	..	..	11 7	67 9

(Local Government Finance, April 1958, p. 92)

What would you expect the average total household expenditure to be if the average weekly housing payment were (a) 40s., and (b) 60s.?   *I.M.T.A. 1959*

9. In October 1958 there were 136,000 working days lost in disputes by the 54,200 men involved. In the 10 months January-October 1958 the corresponding figures were 3,339,000 and 476,100. In Industry X in October 4,700 men were responsible for 23,000 lost working days and during the 10 months 95,600 men lost a total of 2,113,000 working days. From these figures prepare a table to show, for the periods January to September 1958 and October 1958, for (a) Industry X, (b) all other industry, and (c) total, the number of men involved and the average number of days lost per man involved.   *I.M.T.A. 1959*

10. The following table gives the total number of persons dying by homicide in 1960, classified by age last birthday.
  (i) Plot these figures in the form of a histogram.
 (ii) Calculate the mean age at death.
(iii) If the last class had been specified as '70 and over', what would you consider to be the appropriate measure of central tendency?

Age last birthday		No. of
Not under	Under	Homicides
—	1	28
1	5	18
5	10	13
10	20	22
20	35	59
35	50	46
50	70	35
70	85	10

Institute of Statisticians, June 1963

11.   *Students admitted to Courses for a First Degree*

Age on admission (years)	Males	Females
$17\frac{1}{2}$ –	15	27
18 –	67	73
$18\frac{1}{2}$ –	67	56
19 –	41	49
$19\frac{1}{2}$ –	22	20
20 –	11	8
21 –	5	5
22 –	12	2

Calculate from the above data the arithmetic mean age of admission for males and females separately. State precisely any assumptions that you make. Show how the two means may be combined to give the overall mean age for all students admitted.   *Inter D.M.A. 1963*

12. Combine the data of question 11 above into one distribution irrespective of sex. Illustrate this distribution with a cumulative frequency graph. Show how the graph may be used to estimate the following quantities:
(a) the median age of admission;
(b) the semi-interquartile range;
(c) the minimum proportion of students who will attain the age of 22 years before graduation, assuming that no student will graduate in less than $2\frac{3}{4}$ years from admission.   *Inter D.M.A. 1963*

CHAPTER VII

MEASURES OF VARIATION

Introduction

A brief recapitulation of the content of the preceding chapters may assist the reader in understanding the purpose of the measures to be discussed in this chapter. It may be assumed that the data have been assembled in tabular form so that the initial semblance of order, so necessary to further progress, has been achieved. In the chapter on Tabulation the various forms of presenting the data in full or, more usually, in abbreviated form were discussed. The conclusion was drawn that, helpful as tabulation undoubtedly is in providing some indication of the nature of the data, it is still inadequate to permit rapid comparison with comparable data drawn from other but similar sources. Graphical representation, it was found, was particularly valuable in conveying rapidly, and often very effectively, an impression of the nature of the data. They greatly facilitated comparison, although in such diagrams much of the detail had to be sacrificed.

The next stage was to summarise the data from the state of tables and frequency distributions into simple figures which would indicate the characteristics of the series. To this end three averages were discussed in the last chapter, each with its particular advantages and shortcomings. The Mean and Mode are sometimes referred to as measures of central tendency. The reason will be apparent from the examples already given, since the major part of many distributions appears to concentrate around a central value with the remaining items distributed on either side of that value. It is only because of this tendency, to which further reference will be made below, that the Mode and, sometimes the Mean, have value as representative items. If all the items in a distribution are widely dispersed and there is no tendency to concentrate around any one value, then clearly no average can adequately summarise the distribution.

These averages nevertheless provide only rather incomplete summaries of any frequency distribution, and important as, say, the central section of any distribution may be, it is also essential to know what form the rest of the distribution takes. (Thus, if the mean age of a group of six people was 25, many varieties of combinations of ages would yield this Mean. Thus, 10, 16, 20, 22, 31 and 51 years yield a Mean of 25, as do the following: 22, 23, 24, 25, 27 and 29). The position is improved if the *range*, *i.e.* the difference between the maximum and minimum values in the distribution is known. In some frequency distributions the range cannot be given, since the extreme values are unknown. Such an example is given by many income distributions in which, for example, the lower extreme is some unknown quantity 'below £500', while the upper limit is concealed in the group 'over £2,000'. When the range is given, this together with the averages, provides a good deal of information about the frequency distribution. But since the existence of a single extreme value in a distribution will greatly distort the range, its value in describing the distribution is limited.

For all distributions it is necessary to know how typical, *i.e.* representative, of the distribution the average is; whether most of the values are concentrated around that average or widely dispersed through the range. Clearly, if the intermediate values throughout the range and their distribution can be described in some numerical form, a whole series can be summarised for comparative purposes in a few simple figures. The methods used to this end produce measures of *dispersion* and *skewness*.

The Meaning of Dispersion and Skewness

To illustrate these measures, three frequency curves are shown in Figure 17, the independent variable being plotted along the X axis and the frequencies against the ordinate on Y axis. In (1) the apex or peak of the curve lies to the left of centre of the X axis; in (3), the apex is to the right of centre. The former frequency curve indicates that the majority of the frequencies are to be found around the lower values of the independent variable; in the latter, that the modal value is in the higher range of the independent variable. Figure 17 (2) shows two frequency curves superimposed, the continuous line is taller and narrower at the base, the dotted flatter and broader. The apex of each curve lies at the centre of the range of the independent variable, but whereas the smooth curve depicts a distribution most of whose values lie very close to the modal value (*i.e.* given by the apex), the dotted curve depicts a distribution in which the frequencies are dispersed fairly evenly over the entire range of the independent variable. For both these curves the distribution (given by the shape of the curve) is identical on either side of the apex, so that both the distributions which they portray will have the same Mean, but the range for the 'dotted' distribution will be greater than that of the other. Such curves as these, in Figure 17 (2), are described as *symmetrical*, those in Figures (1) and (3) are *skewed*, or asymmetrical.

Further inspection of the two central curves will reveal that any distribution which, when plotted, forms a symmetrical frequency curve (*i.e.* the curve is of the same shape on either side of the apex),

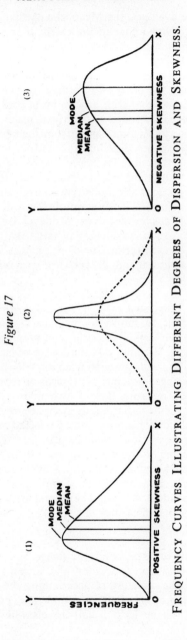

Figure 17

Frequency Curves Illustrating Different Degrees of Dispersion and Skewness.

5*

will have all three averages, the Mean, the Median and Mode equal to each other. It follows that to the extent that a frequency curve diverges from symmetry, *i.e.* it is skewed or asymmetrical, the three averages will differ from each other. As long, however, as the distribution is reasonably symmetrical, the Mode can be ascertained approximately from the other two averages by yet another method. Experience has shown that with any hump-backed distribution, the Median lies between the Mode and the Mean, usually one-third of the difference between the two measured from the Mean. As shown in Figure 17 (1) and (3) in any distribution the Mode is at the apex of the frequency curve. If the curve is skewed, the Mean is pulled towards the longer 'tail', while the Median which divides the area under the curve into equal parts, is also pulled away from the Mode, and lies nearer the Mean. From this observation the following formula to ascertain the Mode has been evolved: Mode = Mean — 3 (Mean — Median), but the reader should bear in mind that as with so many tools, it can only be used for its particular purpose with a reasonably symmetrical distribution and *at best will generally yield only an approximate result.*[1]

The Measures of Dispersion

It is now possible to return to the measures of variation and skewness which are to serve the purpose of amplifying the generally imperfect summary of any distribution provided by the three averages. It will be readily appreciated that as the dispersion increases, so the averages become less typical or representative of the distribution. These measures are of two main types. The first is designed to measure the variation, or more accurately, the deviation of each value in the distribution from the selected measure of central tendency, usually the Mean or Median. The second group provides a measure of the degree of asymmetry in the distribution. The first are called measures of *dispersion, i.e.* they measure the extent to which the individual items in a series are dispersed or distributed over the whole range of the independent variable. The second are known as measures of *skewness* rather than measures of symmetry or asymmetry.

The Range

The first measure of dispersion is the *Range*. This is usually defined as the difference between the smallest and largest values of a distribution or series. The difficulty of ascertaining the range where the classes at the extremes of a frequency table are 'open', has already been mentioned. Where, as is usual, the class-limits are given, by

---

[1] The reader who has not forgotten his elementary algebra will realise that the equation will also serve to give either the Mean or Median, provided both the other averages are known.

convention the range is taken as the difference between the mid-point of the first class in order of magnitude and the mid-point of the last class. Thus, in Table 24, on page 127, the range of marks awarded is 45, *i.e.* from 13 to 58. This is quite arbitrary, since as a result of the grouping of the data the actual value of the smallest and largest items cannot be ascertained. The weakness of the range is almost self-apparent. It requires only one extreme value at either end of the series to render it virtually useless. In brief, dependence on the two extreme items renders the Range most unreliable as a guide to the dispersion of the values within a distribution. Its chief merit lies in its simplicity.

The Quartile Deviation

The weakness of the Range can be partially overcome by using a measure of dispersion which covers only a restricted range of items so that any extreme values are effectively excluded. The majority of the frequencies of a frequency distribution are often found in the central part, hence, it is natural that the *Quartile Deviation* should have been evolved. This measures the dispersion of that part of any distribution lying between the two quartiles, *i.e.* upper and lower quartiles.

The formula for the Quartile Deviation is written $\dfrac{Q_3 - Q_1}{2}$

The smaller the result given by this formula, the less is the dispersion of the middle half of the distribution about the Median. This is the average normally used with this dispersion measure.

The Quartile Deviation is an absolute measure which is affected by the values of the observations in the distribution. Thus, the Q.D. of one distribution may be much greater than that of another, although the dispersion of frequencies is in fact smaller. This may occur when the actual values in the latter distribution are much smaller than those in the former. For example, with a distribution of weights of adult males the quartile values might be 161 and 199 pounds with a median value of 180 pounds. A similar distribution of adult women's weights might give quartile values of 110 and 130 pounds with a median value of 120 pounds. It cannot be inferred from these data that male weight is more widely dispersed than is that for females. The Quartile Deviation for men of 19 lbs as against 10 lbs for women is due simply to the fact that men are on the whole heavier than women. This illustration will serve as a reminder of the basic rule of statistics, compare like with like.

This measure depends finally, like the Range, on only two derived limits, in consequence it is sometimes described as the '*semi-inter-quartile* range'. Unfortunately, it provides no indication of the degree of dispersion or grouping of the other half of the distribution lying beyond the limits of the two quartiles. Consequently, some further measure may be required which will indicate the dispersion of all the items throughout their range.

The Mean Deviation

The Mean Deviation measures the *average* or mean of the sum of all the deviations of every item in the distribution from a central value (either the Mean or Median). The Mean Deviation, therefore, provides a useful method of comparing the relative tendency of the values in comparable distributions to cluster around a central value or to disperse themselves throughout the range.

It will be recalled that in Table 19 on page 101 the calculation of the mean was carried out by use of deviations from an assumed mean. It will be recalled that if the true mean were selected, then the sum of the deviations would equal zero; the difference between the assumed mean and the true mean is given by the average of the sum of the deviations. The same notion is applied in the calculation of the Mean Deviation. As before, the deviations from the true mean (they may also be taken from an assumed mean, but then a further correction is necessary) are taken and aggregated. To avoid the situation of a zero total, the signs in front of the deviations are ignored. The reason for this is that we are only concerned with measuring the extent of the variation about the mean; not whether more values are above or below it. Having derived the sum of the deviations, it is divided by the total frequences to give the Mean Deviation. The principle is illustrated in Table 22, for both the Arithmetic Mean and the Mean Deviation.

TABLE 22

(1) Values	(2) Frequency	(3) Deviations in Class Interval Units from AM = 45	(4) F × CI Units	
			with signs	ignoring signs
10 and under 20 ..	2	— 3	— 6	6
20 ,, ,, 30 ..	4	— 2	— 8	8
30 ,, ,, 40 ..	4	— 1	— 4	4
40 ,, ,, 50 ..	8	0	0	0
50 ,, ,, 60 ..	6	1	6	6
60 ,, ,, 70 ..	3	2	6	6
70 ,, ,, 80 ..	2	3	6	6
	29		0	36

$$\text{A.M} = 45 + \frac{0}{45} \times 10 = 45.$$

$$\text{M.D. (ignoring signs)} = \frac{36}{29} \times \text{C.I.} = 1 \cdot 24 \times 10 = 12 \cdot 4.$$

The computation of the Mean Deviation is much more complicated where the Mean proves to be an awkward number entailing tedious arithmetical calculation in deriving the products of the deviations and their respective frequencies. In such cases an arbitrary origin or assumed mean is selected and the adjustment is made at the end on lines similar to the correction made when calculating the Mean from an arbitrary origin as on p. 101. This difficulty usually arises with grouped distributions in which the Mean proves to be an awkward fraction for further calculations. Often the mid-point of the middle group is arbitrarily selected as the value,[1] from which the deviations may be computed. These points will be illustrated in the examples showing the computation of the next measure of dispersion. The rather complex method of calculating the M.D. from an arbitrary origin is not dealt with here, since this particular measure of dispersion is little used. The Mean Deviation is nowadays only of academic interest, and in practice has been replaced by the Standard Deviation, which enters into so many statistical formulae.

The Standard Deviation

From the point of view of the mathematician, the practice of ignoring the signs before the deviations when computing the Mean Deviation is quite unjustifiable, and in consequence the Mean Deviation is unsuitable for use in further calculation. On the other hand, to leave the signs in, will, as has already been pointed out, reduce the Mean Deviation to zero. The Standard Deviation which is the most important measure of dispersion overcomes this problem by 'squaring' the deviations.

Thus, $(-2)^2$ is 4, just as the square of $+2$ is 4. As with the Mean Deviation, the sum of the products is divided by the total frequencies. The mean of the sum of the squared deviations is known as the *variance*, but before it can be related to any other statistic, *e.g.* the mean, the square root of the variance must be obtained. This is known as the Standard Deviation. Table 23 illustrates the principle; the reader should compare the result with the Mean Deviation computed from the same data above in Table 22. He will note that the Standard Deviation is larger, owing to the fact that the process of squaring gives relatively greater emphasis to the extreme values in the distribution.

Short Method of calculating the Standard Deviation

The deviations are usually measured from the Mean of the distribution, but if, as is often the case, the Mean is not a round number

Termed the 'arbitrary origin' or the 'assumed' mean.

TABLE 23
CALCULATION OF THE STANDARD DEVIATION

Classes	f	d Deviation from A.M. (45)	$f \times d$	$(fd \times d)$ $= f\,d^2$
10 and under 20 ..	2	− 30	− 60	1,800
20 ,, ,, 30 ..	4	− 20	− 80	1,600
30 ,, ,, 40 ..	4	− 10	− 40	400
40 ,, ,, 50 ..	8	0	0	0
50 ,, ,, 60 ..	6	10	60	600
60 ,, ,, 70 ..	3	20	60	1,200
70 ,, ,, 80 ..	2	30	60	1,800
	29		− 180 + 180	7,400

$$\text{S.D.} = \sqrt{\frac{\text{Sum of frequencies} \times \text{deviations squared}}{\text{Sum of frequencies}}} = \sqrt{\frac{\Sigma f\,d^2}{\Sigma f}}$$

$$\text{S.D.} = \sqrt{\frac{7,400}{29}}$$

$$\text{S.D.} = \sqrt{255 \cdot 172} = 16 \text{ (answer rounded to nearest unit since the}$$
original data do not justify greater accuracy).

coinciding with say the value of the mid-point of any class, then the calculations could be extremely tedious. Consequently, a method has been evolved to avoid this, which in its essence is the same as the short method for calculating the Mean itself.[1] It will be remembered that an assumed Mean was chosen and from it the differences for each class worked out in terms of class intervals and the residual or net difference converted and deducted from or added to the value of the assumed mean. To calculate the Standard Deviation the same procedure is employed with the additional step of multiplying the *products* of the frequencies and deviations, by the deviations. This is the 'squaring' of the deviations required for the Standard Deviation. Then, as for the A.M., a correction is introduced to derive the Standard Deviation of the distribution from the true mean. Table 24 provides a simple example; the successive steps in the calculations are detailed below.

1. The selection of the mid-points of the classes is relatively simple. The series is discrete, since the individual values are determinate amounts, *i.e.* one unit is the minimum variation. The mid-points are derived by halving the sum of the limits of the individual class.

---

[1] If the student reader has forgotten the process he should refresh his memory by reference to pp. 100-101.

TABLE 24
EXAMINATION MARKS AWARDED TO 392 CANDIDATES

Classes	Mid-points (2)	Frequencies (3)	Deviations (4)	$f \times d'$ $= fd'$ (5)	$fd' \times d'$ $= fd'^2$ (6)
11 .. 15	13	6	— 5	— 30	150
16 .. 20	18	12	— 4	— 48	192
21 .. 25	23	30	— 3	— 90	270
26 .. 30	28	53	— 2	— 106	212
31 .. 35	33	77	— 1	— 77	77
36 .. 40	38	96	0	0	0
41 .. 45	43	54	1	54	54
46 .. 50	48	37	2	74	148
51 .. 55	53	19	3	57	171
56 .. 60	58	8	4	32	128
		392		— 134	1,402

2. The deviations are measured from an assumed origin (38, the mid-point of 36-40 group), instead of the true Mean, which is unknown, and a correction will have to be introduced later to offset the discrepancy this method will introduce into the calculation.

3. The deviations from the assumed origin are expressed, as can be seen in Col. 4, in terms of class intervals (multiples of 5) and the result will be expressed in these terms.

4. Col. 5 gives the products of the frequencies and the deviations (Col. 3 × Col. 4). This column is important since from it is derived the correction to adjust the error introduced by calculating the Standard Deviation from an arbitrary origin. The net products will also provide the fraction for computing the Mean, thus the *true* Mean equals $38 + \left(- \dfrac{134}{392}\right) \times 5 = 38 - 1 \cdot 7$, or 36 to the nearest unit.

5. Col. 6 gives the results of multiplying Col 5 (*i.e.* products of frequencies and deviations) by the deviations. This can be verified by squaring all the deviations in Col. 4 and multiplying the squared products by their respective frequencies in Col. 3. The results will be the same as given by the method shown in Col. 6.

The Standard Deviation from the assumed mean can now be calculated by dividing the total of Col. 6 by the total frequencies and taking from this the correction for the use of an assumed Mean. The student should note that the correction fraction $\left(- \dfrac{134}{392}\right)$ is the same for the S.D. as it is for the A.M. except that for the former calculation it is squared. Then the square root of the balance is calculated. The calculations are as follows: the usual symbol for the Standard Deviation is σ the Greek letter (little sigma), but S.D. will serve equally well.

$$\sigma^2 = \left(\frac{1,402}{392}\right) - \left(-\frac{134}{392}\right)^2$$

$$\sigma = \sqrt{\left(\frac{1,402}{392}\right) - \left(-\frac{134}{392}\right)^2}$$

It should be noted in passing, that the correction fraction should be squared and subtracted from the other fraction, before the square root is calculated.

$$\sigma = \sqrt{3\cdot5765 - (\cdot3418)^2} = \sqrt{3\cdot5765 - \cdot1169}$$
$$= \sqrt{3\cdot4596} = 1\cdot86$$

But σ is still in class-interval units, to convert to the original units 1·86 is multiplied by 5; then σ in original units = 9·3 or 9 to nearest whole unit.

The Standard Deviation is the square root of the average of the squared deviations measured from the Mean. It is sometimes described as the root-mean-square deviation, although the only advantage to be derived from this name is that it indicates the method of calculation.

Characteristics of Dispersion Measures

The major characteristics of the measures of dispersion may be summarised as follows:

Range
  i The simplest to derive and the easiest to comprehend.
  ii Its value as an indication of the variation in the data may be virtually nullified by the existence of one exceptionally large or small value.
  iii It provides no indication as to the distribution of the frequencies between the limits of the range.

Quartile Deviation
  i This measure is not difficult to calculate, but covers only half the items within the distribution. It does eliminate, however, the risk of extreme items which seriously distort the Range.
  ii As with the Range, its value is based on the values of the two limits, *i.e.* Q_1 and Q_3 with all the attendant disadvantages arising from this fact.
  iii It bears no relationship to any fixed point in the distribution as do the M.D. and S.D.; nor is it affected by the distribution of the individual values lying between the quartiles.

Mean Deviation
  i Unlike the Q.D. it is affected by every value in the distribution.

ii It indicates the extent of the deviation of all values from a given value, in this case the Median or Mean of the distribution.

iii For the purpose of further mathematical treatment the M.D. is unsatisfactory.

Standard Deviation

i Like the M.D., it includes every value of the distribution.

ii It is itself the result of correct mathematical processes and thus further calculations may be based upon it.

iii It is the best measure of dispersion and, as will be seen in the later chapters, is of very great importance for sampling theory.

Coefficient of Variation

It cannot be sufficiently emphasised that all measures of dispersion are in terms of the units in which the original values are expressed. Thus, the S.D. of men's heights, weight of cotton bales and salesmen's salaries will be expressed in inches, hundredweights and pounds sterling, respectively. These measures of absolute variation cannot be compared with each other, if expressed in differing units, or if the average values of two distributions in a comparable field are widely dissimilar. Thus, in the case of differing units, if the A.M. and S.D. of one distribution are expressed in centimetres, and for the other in feet, the units must either be converted to a common base, *e.g.* both series in feet, or a standard measure devised which ignores the original units of measurement. Similarly, where the means of the distributions are widely dissimilar, *e.g.* the average levels of remuneration received by the administrative and labouring sections of a large organisation, respectively, then the dispersions within the two groups can only be compared by relating them to some 'equalising' factor.

This is done by turning the absolute measure of dispersion, *i.e.* the S.D., into a relative measure. More precisely, the S.D. is related to some other measure directly connected with the same distribution, *e.g.* it is frequently expressed as a *percentage* of the Mean of that distribution. This new measure is termed the *Coefficient of Variation* normally written as $CV = \dfrac{100\sigma}{\bar{x}}$ Where the series being compared are expressed in the same unit of measurement and the Means are similar, no advantage is to be gained by calculating the coefficient of variation. The S.D. is then quite sufficient.

Measures of Skewness

So far, only the first group of measures of variation have been covered, those which indicate the *dispersion* of the frequencies throughout the range of the independent variable.

The second group was described as measuring the degree of

symmetry of any distribution plotted as a frequency curve. Any curve which is not 'symmetrical' may be described as 'asymmetrical', or *'skewed'*. The latter term is generally employed and the statistician refers to measures of skewness. Most 'hump-backed' or uni-modal frequency distributions are skewed (*i.e.* not symmetrical); and to that extent the characteristic of a symmetrical distribution, *i.e.* identity of values of A.M., Median and Mode, is absent.

It is a logical step, therefore, to develop some measure which shows the degree to which these three measures of central tendency diverge. The difference between them provides the first measure of skewness. This difference, however, could be unsatisfactory on two counts.

1. It would be expressed in the unit of value of the distribution and could therefore not be compared with another comparable series expressed in different units.

2. Distributions vary greatly and the difference between, say, the Mean and Mode in absolute terms might be considerable in one series and small in another, although the frequency curves of the two distributions were similarly skewed.

If the absolute differences were expressed in relation to some measure of the spread of the values in their respective distributions, the measures would then be *relative* and not *absolute* and therefore directly comparable.

The above considerations form the basis of Professor Karl Pearson's formula for deriving what is known as a coefficient of skewness:

$$sk = \frac{\text{Mean} - \text{Mode}}{\text{SD}}$$

The Mode was criticised in an earlier chapter as being particularly difficult to determine precisely for many frequency distributions. Consequently a variation of the above formula is used:

$$sk = \frac{3 \ (\text{Mean} - \text{Median})}{\text{SD}}$$

An alternative measure of skewness has been proposed by the late Professor Bowley. This is based on the relative positions of the Median and the Quartiles. If the distribution were symmetrical then Q_1 and Q_3 would be at equal distances from the Median. Then it follows that $(Q_3 - Me) - (Me - Q_1) = O$. The more skewed the distribution, the larger will be the diference between these two quantities, which can be re-arranged as follows: $(Q_3 - Me) - (Me - Q_1) = Q_3 + Q_1 - 2Me$. This measure of skewness is expressed in absolute terms so that if there were two distributions, one highly skewed and the other much less, comprising very different-sized variables, *e.g.* pounds and ounces, then despite the fact that the distribution with the large values was less markedly skewed than the other, the above measure of skewness would yield a larger result. To overcome this weakness, the absolute measure is converted into a relative measure (as in the Pearson coefficient above) by relating it to the Quartile Deviation, which is itself a reflection of the absolute variation of the independent variable. Thus, we get:

$$sk = \frac{Q_3 + Q_1 - 2Me}{\dfrac{Q_3 - Q_1}{2}} = \frac{2(Q_3 + Q_1 - 2Me)}{Q_3 - Q_1}$$

Beyond the fact that symmetry (*i.e.* complete absence of skewness) is indicated by O (zero) in both the above formulae, *i.e.* Bowley's and Pearson's, the coefficient of skewness derived by the two measures are *not* comparable.

Formulae for Measures of Dispersion

The mathematical notation of the measures of dispersion like those of the Arithmetic Mean, given in the preceding chapter, is fundamentally simple.

Only two of the four measures of dispersion are so expressed, the M.D. and the S.D.

The Mean Deviation calculated from the true Mean of an ungrouped series:

$$\text{M.D.} = \frac{\Sigma |d|}{N} \quad i.e., \quad \frac{\text{Sum of the deviations (ignoring signs)}}{\text{number of items}}$$

When the data are in the form of a frequency distribution,

$$= \frac{\Sigma f |d|}{\Sigma f} \quad i.e., \quad \frac{\text{Sum of frequencies} \times \text{deviations (ignoring signs)}}{\text{Sum of frequencies}}$$

The *Standard Deviation* is usually represented by the sign σ (small Greek letter 'sigma'), but S.D. is often used. For an ungrouped distribution with the deviations taken from the true mean: $\sigma = \sqrt{\dfrac{\Sigma d^2}{N}}$

The S.D. of a grouped frequency distribution computed from the true Mean:

$$\sigma = \sqrt{\frac{\Sigma f d^2}{N}} \quad \text{Equally one can write: } \sigma = \sqrt{\frac{\Sigma f d^2}{\Sigma f}}$$

since Σf and N both refer to the total frequency.

If the S.D. is computed from an assumed Mean or arbitrary origin, then:

$$\sigma = \sqrt{\frac{\Sigma f (d')^2}{\Sigma f} - \left(\frac{\Sigma f d'}{\Sigma f}\right)^2} \quad \text{Where the symbol } d \text{ denotes deviations from arbitrary origin or assumed mean.}$$

If the calculation has been performed with the deviations expressed in class intervals:

$$\sigma = \sqrt{\frac{\Sigma f d'^2}{\Sigma f} - \left(\frac{\Sigma f d'}{\Sigma f}\right)^2} \times i$$

where i = class interval.

The same comment applies here as was made in connection with the formulae for the various averages. If these processes are really understood, these formulae can always be constructed. Nevertheless, most students prefer, sometimes unwisely, to rely on their memories.

Conclusions

The student has now reached the end of what may be termed 'descriptive statistics'. After collection of the data, all the processes described so far have been in the nature of summarizing with the object of making simple comparisons. Such comparisons can be made as between two or more distributions by means of simple tables, or by use of diagrams as was shown in Chapter V. The next stage was to summarise the data still further by means of 'averages' and at the beginning of this chapter (p. 119) it was explained that such averages may describe a distribution very imperfectly. In fact, two widely disparate distributions can have identical means, but whereas in the one case the range of the values is very slight, in the other it is considerable. Hence, it is essential to calculate statistics which will measure not merely averages, but also describe the degree of dispersion of the values within the distribution. In most statistical work, it is not only the tendency for observations or values to conform to an average that is of vital importance. Any tendency for values to diverge from the norm may be of even greater interest to the statistician. Hence, just as we think of Tweedledum and Tweedledee as inseparable, so each measure of central tendency (*i.e.* average) should be quoted with its appropriate measure of dispersion. Thus, the arithmetic Mean should always be given with the standard deviation, while the Median is linked with the Quartile Deviation. The student need not worry himself about the calculation of the Mean Deviation, but he should know what is meant by the Range. Likewise, the measures of skewness are of less importance for elementary statistical work; it is the concept which needs to be remembered.

Some Worked Examples

The only satisfactory way of learning how to calculate the measures of central tendency and of dispersion is by practice. Each frequency distribution can pose its own particular problems, *e.g.* uneven class-intervals, or open-ended classes. To provide such practice and to ensure that the reader has grasped the principles involved, the next few pages contain worked examples of the mean and median, quartile and standard deviations. These are the standard fare of all examination papers The student would be well advised to check that he understands the methods by calculating these statistics from the following frequency distributions and comparing his working with the answers given.

The first example is really two frequency distributions which offer extra practice. It is suggested that the student works on the data for Great Britain first and compares his working with those overleaf. If he has worked the example correctly, he can pass on to the third example. If, however, he has slipped up and made mistakes, he should work out the same statistics from the second frequency distribution relating to London and South East England.

TABLE 27

ANALYSIS BY AGE OF EMPLOYED MALES IN GREAT BRITAIN, MAY 1957 (THOUSANDS)

Age	Great Britain	London and S.E. England
under 20 years	1,094	211
20–25 years.. ..	1,237	286
25–30 „	1,512	359
30–35 „	1,594	382
35–40 „	1,546	369
40–45 „	1,515	373
45–50 „	1,560	398
50–55 „	1,473	374
55–60 „	1,192	291
60–65 „	880	212
65 years and over	597	164
TOTAL	14,200	3,419

Source: Based on data in Ministry of Labour Gazette, June, 1958.

The third example is more complicated. The distribution in Table 28 contains both irregular class-intervals and open-ended classes. In this particular illustration the student would be well advised to have an extra column in his working to show the mid-points of each class and this will avoid the risk of calculating the deviations from the assumed mean incorrectly.

Note that whatever the assumed mean chosen for purposes of the calculation, the result should be the same. Thus, if in the case of Table 28 one student chose £12 as the assumed mean and calculated all the deviations from that figure, and another reader worked from £25, the true means will still be the same. Thus, if the reader finds on comparing his working with those given below that different assumed means have been selected, this does not mean that his method is wrong. If his working is correct, the final results by both methods should agree.

TABLE 28

SAMPLE OF HOUSEHOLDS IN U.K. CLASSIFIED BY INCOMES OF HEAD OF HOUSEHOLD

Gross Weekly Income			Number of Households
Under £4			350
£4 but under	£6		303
6 ,, ,,	10		426
10 ,, ,,	14		896
14 ,, ,,	20		932
20 ,, ,,	25		288
25 ,, ,,	30		109
30 ,, ,,	40		96
40 and more			86
			3,486

Source: Family Expenditure Survey 1960 and 1961.

Note, however, the answers will probably disagree if different mid-points have been used in the first and last open-ended classes. For example, if instead of assuming for the final class in Table 27 that the upper limit is 70, it is taken as 75, so that the mid-point is 70 instead of $67\frac{1}{2}$, then this will tend to increase the value of the mean. Likewise, in the first class where a lower limit must be assumed, the lower the mid-point of that class the smaller will the mean prove to be. Thus, it is important for the reader, when he compares his workings, if he has used different mid-points or assumed means, to follow through the effects of these differences and judge to what extent they have affected the answer. When answering examination questions, it is good practice to indicate for any open-ended class, what limit has been chosen.

In this distribution there are two open-ended classes, the first and the last. Of the first, the question is what is the lower limit? This is easy. Since no boy may leave school and start work before his 15th birthday, the lower limit is clearly 15. In the case of the last class, the choice is more difficult. Quite a few men work after their 70th year, but most of them retire soon after reaching 65. Bearing in mind that the mid-point is merely the best estimate of the mean of all the members of that class and the majority of them are likely to be nearer 65 than 70, much less 75, it seems reasonable to assume that the mid-point should be $67\frac{1}{2}$, *i.e.* the upper limit is taken as 70. Note that the relative frequency of that class compared with the other class frequencies is small. This means that even if a slight error has been made in determining the mid-point, it will not affect the answer very much.

Given the classification, it may be assumed to be a continuous series, although for this purpose the opening class will have to be read as '15 and under 20 years'. The mid-point for a continuous series is derived by adding the lower limits of the successive classes and dividing by 2, *e.g.* $(20 + 25) \div 2$.

Illustration (1)

CALCULATION OF MEAN, S.D. AND MEDIAN AND Q.D.
Data for Great Britain from Table 27

Age	No. of employed males	Mid-point	$d' \div$ CI	$fd' \div$ CI	$f(d')^2 \div$ CI	Cum.f.
under 20 years	1,094	$17\frac{1}{2}$	−5	— 5,470	27,350	1,094
20 – 25 years	1,237	$22\frac{1}{2}$	−4	— 4,948	19,792	2,331
25 – 30 ,, ...	1,512	$27\frac{1}{2}$	−3	— 4,536	13,608	3,843
30 – 35 ,, ...	1,594	$32\frac{1}{2}$	−2	— 3,188	6,376	5,437
35 – 40 ,, ...	1,546	$37\frac{1}{2}$	−1	— 1,546	1,546	6,983
40 – 45 ,, ...	1,515	$42\frac{1}{2}$	0	— 19,688	0	8,498
45 – 50 ,, ...	1,560	$47\frac{1}{2}$	+1	+ 1,560	1,560	10,058
50 – 55 ,, ...	1,473	$52\frac{1}{2}$	+2	+ 2,946	5,892	11,531
55 – 60 ,, ...	1,192	$57\frac{1}{2}$	+3	+ 3,576	10,728	12,723
60 – 65 ,, ...	880	$62\frac{1}{2}$	+4	+ 3,520	14,080	13,603
65 years and over	597	$67\frac{1}{2}$	+5	+ 2,985	14,925	14,200
TOTAL ..	14,200			(+ 14,587 − 19,688)	115,857	
				— 5,101		

Working on the figures for Great Britain, the first calculation provides the Arithmetic Mean. It is simplest to use the short-cut method of working in deviations from the assumed mean ($42\frac{1}{2}$ years, *i.e.* the mid-point of the class 40 and under 45 years) measured in group intervals:

$$AM = 42\frac{1}{2} + \left(\frac{-5,101}{14,200}\right) \times 5 \text{ years}$$

$$= 42\frac{1}{2} + 5\,(-0\cdot36) \text{ years}$$

$$= 42\frac{1}{2} - 1\cdot8 \text{ years}$$

$$= 41 \text{ years (to nearest year)}$$

The Standard Deviation is derived from the formula:

$$\sqrt{\frac{\Sigma f d'^2}{\Sigma f} - \left(\frac{\Sigma f d'}{\Sigma f}\right)^2} \times i$$

i.e., working in class intervals from the assumed Mean.

Substituting the true values for the symbols in the above formula

$$S.D. = \sqrt{\frac{115,857}{14,200} - \left(\frac{-5,101}{14,200}\right)^2} \times 5$$

$$= \sqrt{8 \cdot 1589 - (0 \cdot 36)^2} \times 5$$

$$= \sqrt{8 \cdot 1589 - 0 \cdot 1296} \times 5$$

$$= \sqrt{8 \cdot 0293} \times 5$$

$$= 5 (2 \cdot 834) = 14 \cdot 17 = 14 \text{ years to nearest year.}$$

The Median from a grouped frequency distribution is derived by the formula $\frac{N}{2}$. Thus:

$$\frac{14,200}{2} = 7,100\text{th item which lies in the group 40–44 years.}$$

Thus the value of the Median by interpolation

$$= 40 + \left(\frac{7,100 - 6,983}{1515}\right) \times 5 \text{ years}$$

$$= 40 + \left(\frac{117}{1,515}\right) \times 5$$

$$= 40 + 5 (0 \cdot 077) = 40 \text{ years 5 months.}$$

$$= 40 \text{ years to nearest year.}$$

The values at the Quartiles are derived in the same way

$$Q_1 = \frac{14,200}{4} = 3,550\text{th item}$$

$$= 25 + \left(\frac{3,550 - 2,331}{1,512}\right) \times 5$$

$$= 25 + \left(\frac{1,219}{1,512}\right) \times 5$$

$$= 25 + 5 (0 \cdot 805) = 29 \text{ years to nearest year.}$$

$$Q_3 = \frac{3(14,200)}{4} = 10,650\text{th item}$$

$$= 50 + 5 \left(\frac{10,650 - 10,058}{1,473}\right)$$

$$= 50 + 5 \left(\frac{592}{1,473}\right) = 50 + 5 \,(0.402) = 52 \text{ years to nearest year.}$$

The Quartile Deviation is easily obtained by using the above results in

the formula $QD = \dfrac{Q_3 - Q_1}{2}$

$$= \frac{52 - 29}{2} \text{ years} = \frac{23}{2} = 11\tfrac{1}{2} \text{ years}$$

The Coefficient of Variation is derived from $\dfrac{100 \times \text{S.D.}}{AM}$

$$= \frac{100 \times 14 \cdot 2}{40 \cdot 75} = \frac{1,420}{40 \cdot 75} = 35\%$$

If the reader has performed the above calculations correctly he may omit the next illustration and attempt the third illustration, on page 139.

Illustration (2)

CALCULATION OF MEAN, S.D. AND MEDIAN AND Q.D.
Data for London and South East England from Table 27

Age	No. of employed males	Mid-point	d ÷ CI	fd ÷ CI	f(d)² ÷ CI	Cum.f.
under 20 years	211	17½	—5	— 1,055	5,275	211
20 – 25 years ..	286	22½	—4	— 1,144	4,576	497
25 – 30	359	27½	—3	— 1,077	3,231	856
30 – 35	382	32½	—2	— 764	1,528	1,238
35 – 40	369	37½	—1	— 369	369	1,607
40 – 45	373	42½	—0	— 4,409	0	1,980
45 – 50	398	47½	+1	+ 398	398	2,378
50 – 55	374	52½	+2	+ 748	1,496	2,752
55 – 60	291	57½	+3	+ 873	2,619	3,043
60 – 65	212	62½	+4	+ 848	3,392	3,255
65 years and over	164	67½	+5	+ 820	4,100	3,419
TOTAL ..	3,419			(+ 3,687 — 4,409)	26,984	

The calculations follow the same pattern as for the first distribution and no comment is required.

$$AM = 42\tfrac{1}{2} + \left(\frac{-722}{3,419}\right) \times 5 = 42\tfrac{1}{2} + (-0.21)\,5$$

$$= 42\tfrac{1}{2} - 1.05 = 41 \text{ years to nearest year}$$

$$SD = \sqrt{\frac{\Sigma f d'^2}{\Sigma f} - \left(\frac{\Sigma f d'}{\Sigma f}\right)^2} \times i$$

$$= \sqrt{\frac{26,984}{3,419} - \left(\frac{-722}{3,419}\right)^2} \times 5$$

$$= \sqrt{7 \cdot 8924 - (-0 \cdot 21)^2} \times 5$$

$$= \sqrt{7 \cdot 8924 - 0 \cdot 0441} \times 5$$

$$= \sqrt{7 \cdot 8483} \times 5$$

$$= 5 \, (2 \cdot 801) = 14 \cdot 005 = 14 \text{ years to nearest year.}$$

$$\text{Median} = 1,709 \cdot 5 = 40 + 5 \left(\frac{1,709 \cdot 5 - 1,607}{373}\right) = 40 + 5 \left(\frac{102 \cdot 5}{373}\right)$$

$$= 40 + 5 \, (0 \cdot 275) = 41 \text{ years.}$$

Quartile Deviation:

$$\text{Position of } Q_3 = \frac{3(3,419)}{4} = \frac{10,257}{4} = 2,564\tfrac{1}{4} \text{ rank}$$

$$\text{Value of } \quad Q_3 = 50 + \frac{5(2,564\tfrac{1}{4} - 2,378)}{374}$$

$$= 50 + \frac{5(186\tfrac{1}{4})}{374} = 50 + \frac{931\tfrac{1}{4}}{374} = 50 + 2 \cdot 489$$

$$= 52 \text{ years to nearest year.}$$

$$\text{Position of } Q_1 = \frac{3,419}{4} = 854\tfrac{3}{4} \text{ rank}$$

$$\text{Value of } \quad Q_1 = 25 + \frac{5(854\tfrac{3}{4} - 497)}{359}$$

$$= 25 + \frac{5(357\tfrac{3}{4})}{359} = 25 + \frac{1,788\tfrac{3}{4}}{359} = 25 + 4 \cdot 98 \text{ years}$$

$$= 30 \text{ years}$$

$$\text{Quartile Deviation} = \frac{Q_3 - Q_1}{2} = \frac{52\tfrac{1}{2} - 30}{2} = \frac{22\tfrac{1}{2}}{2}$$

$$= 11\tfrac{1}{4} \text{ years}$$

Coefficient of Variation

$$\frac{\text{S.D.} \times 100}{AM} = \frac{14 \cdot 005 \times 100}{41 \cdot 5} = \frac{1400 \cdot 5}{41 \cdot 5} = 33 \cdot 74\%$$

It will be noted that several places of decimals have been used in the calculations above. However, in the comparison of results below all the statistics have been rounded to the nearest year as this approximation is adequate for the data involved and any greater degree of accuracy is unobtainable from data classified as in the original tables.

	Great Britain	London and S.E. England
Number in thousands	14,200	3,419
Arithmetic Mean	41 years	41 years
Median value	40 ,,	41 ,,
Lower Quartile Q_1 value	29 ,,	30 ,,
Upper Quartile Q_3 value	52 ,,	52 ,,
Quartile Deviation	12 ,,	11 ,,
Standard Deviation	14 ,.	14 ,,
Coefficient of Variation	35%	34%

Illustration (3)

The data in Table 28 are derived from a government report on family expenditure and show the distribution of incomes of heads of households. As in the preceding illustrations the reader is required to calculate the arithmetic mean, the standard deviation, the median and the quartiles. In this case the insertion of an extra column for the mid-points of each class is essential for accurate working. As the distribution is continuous, the mid-points are derived by halving the sum of the lower limits of the successive class-intervals, *e.g.* £4 + £6 divided by 2. The next problem is to determine the mid-points of the first and last class-intervals which are open-ended. In the first, obviously no one has a zero income so the mid-point would not be £2. Even the State pension is over £3 and those who do not have as much as that will usually be supplementing their incomes from the National Assistance Board. We can choose either £3 or £3 10s. as the mid-point and, since it simplifies the arithmetic, the former will serve. In the final class 'over £40' it can be assumed that the majority of income in this class are nearer the lower limit than over, say, £50. It is also a very small group so whatever mid-point is chosen, within reason, it is not likely to affect the final answer greatly. In the event it is assumed that the class interval is £40 to £50 and the mid-point is then £45.

CALCULATION OF MEAN, S.D., MEDIAN AND QUARTILES. DATA FROM TABLE 28

Gross Weekly Income	No. of House-holders	Mid-Points £	d'	fd'	fd'²	Cum. f.
Under £4	350	3	— 14	— 4,900	68,600	350
£4 but under 6	303	5	— 12	— 3,636	43,632	653
6 ,, ,, 10	426	8	— 9	— 3,834	34,506	1,079
10 ,, ,, 14	896	12	— 5	— 4,480	22,400	1,975
14 ,, ,, 20	932	17	—	—	—	2,907
20 ,, ,, 25	288	22½	+ 5½	+ 1,584	8,712	3,195
25 ,, ,, 30	109	27½	+ 10½	+ 1,144½	12,017	3,304
30 ,, ,. 40	96	35	+ 18	+ 1,728	31,104	3,400
40 and more	86	45	+ 28	+ 2,408	67,424	3,486
TOTAL	3,486			— 16,850		
				+ 6,864½	288,395	
				— 9,985½		

$$\bar{x} = £17 + \left(-\frac{9,985\frac{1}{2}}{3,486}\right) = -2·9 \text{ to one dec. place.}$$

$$\bar{x} = £14 \text{ .o nearest } £$$

$$\sigma = \sqrt{\frac{288,395}{3,486} - \left(-\frac{9,985\frac{1}{2}}{3,486}\right)^2} = \sqrt{82·7 \quad -(2·9)^2}$$

$$= \sqrt{82·7 \; -8·4} = \sqrt{74·3} = £8·6 \text{ approx.}$$

$$Me = \frac{3\,486}{2} = 1,743 = £10 + \left(\frac{1,743 - 1,079}{896}\right)4 = £14 \text{ to nearest } £$$

$$Q_1 = \frac{3,486}{4} = 871\frac{1}{2} = £6 + \left(\frac{871\frac{1}{2} - 653}{426}\right)4 = £8 \text{ to nearest } £$$

$$Q_3 = \left(\frac{3,486}{4}\right)3 = 2,614\frac{1}{2} = £14 + \left(\frac{2,614\frac{1}{2} - 1\,975}{932}\right)6 = £18 \text{ to nearest } £$$

Since the class interval is irregularit is not possible to work as in the two previous illustrations, *i.e.* in deviations in class-interval units. The actual differences of each mid-point from the assumed mean are used. The arithmetic is in consequence rather laborious, but there is no alternative unless we are prepared to approximate extensively. Note that in the final calculations there is a small element of approximation; products and quotients have not been worked to several decimal places, etc., and in the case of the mean, median and quartiles, all have been approximated to the nearest £. Any attempt to express these statistics more precisely, *e.g.* to the nearest shilling, would be rather silly in view of the nature of the data, in particular the large class intervals and the assumptions involved in using mid-points.

Questions

MEASURES OF VARIATION

1. Explain, with simple illustrations, how you would calculate the arithmetic mean, median, quartile deviation and standard deviation. What purpose do such statistics serve and when would you use the mean in preference to the median?
I.M.T.A. 1958

2. What are the uses and limitations of different measures of dispersion
B.A. Makerere University 1962

3. Pneumonia notifications, 1945:

Age group	Male	Female
0—	4,593	3,734
5—	2,773	2,090
15—	4,533	3,682
45—	4,700	2,761
65—	2,160	2,102

From the above data estimate the median age and the quartile deviation of the age distributions for both male and female patients. Comment on the accuracy of your estimates. Would it have been more useful to have calculated the arithmetic mean and standard deviation? For the male group *only* estimate by graphical methods the median and two quartiles.   *I.M.T.A. 1962*

4. Income in 1955-56 of General Medical Practitioners and General Dental Practitioners (*registered in 1950 or later and under the age of 30 at the time of registration*)

Income (£)	General Medical Practitioners	General Dental Practitioners
0-	2	1
200-	5	1
400-	26	2
600-	25	2
800-	80	1
1,000-	88	8
1,200-	94	5
1,400-	41	4
1,600-	63	10
1,800-	29	4
2,000-	28	13
2,200-	10	18
2,400-	3	7
2,600-	8	4
2,800-	—	3
3,000-	—	21
4,000- over	—	9
All incomes	502	113

Source: Report of the Royal Commission on Doctors' and Dentists' Remuneration (1960)

Compare the above distributions in terms of their medians and quartile deviations, and comment on your results. *Final D.M.A. 1961*

5.

Rateable Value per Head of Weighted Population 1956-57* £	Number of Counties	County Boroughs
0 and under 2·5	—	—
2·5 and under 5·0	1	—
5·0 and under 7·5	18	6
7·5 and under 10·0	26	35
10·0 and under 12·5	11	23
12·5 and under 15·0	3	10
15·0 and under 17·5	2	5
17·5 and under 20·0	—	1
20·0 and under 22·5	—	3
22·5 and under 25·0	—	—
	61	83

*Source: New Sources of Local Revenue, R.I.P.A.

Calculate for both the above distributions the Median and Quartile Deviation. Check your results for counties only by graphical methods.

Rating and Valuation Association 1963

6. Calculate for the following distribution the arithmetic mean and the standard deviation:

Age	20–24	25–29	30–39	40–49	50–59	60–69	Over 69
Number	14	140	438	336	157	49	5

I.M.T.A. 1961

7. The following data relates to age at admission to a certain hospital of surgical cases during the past year. Calculate the mean age and standard deviation for both males and females.

Age	Males	Females
20–24	43	17
25–29	87	43
30–34	40	21
35–39	32	16
40–44	39	24
45–49	25	48
50–54	51	52
55–59	86	40
60 and over	91	12
	494	273

I.M.T.A. 1959

8. Age Distribution of Consultants in Hospital Service of England and Wales
1952 and 1958

Age Group	No. of Consultants at Mid 1952	No. of Consultants at Mid 1958
Over 65	79	49
65–62	292	340
61–58	387	708
57–54	553	849
53–50	900	909
49–46	854	1,079
45–42	989	1,345
41–38	1,055	1,023
37–34	796	457
Under 34	157	36
Total	6,062	6,795

Source: Ministry of Health and Dept. of Health for Scotland, Medical Staffing Structure in the Hospital Service H.M.S.O. 1961

(a) Calculate medians and quartile deviations for these age distributions and consider what light these measures, and the data themselves throw on changes in the staffing structure of the hospital service.

(b) Represent the distributions by cumulative diagrams and estimate the medians from them. *B.Sc. (Sociology) 1963*

9. From the following data calculate the mean and the standard deviation of the ages at death.

Deaths from Cirrhosis of the Liver among Males

Age:	25–9	30–4	35–9	40–4	45–9	50–4	55–9
No.:	21	42	109	253	588	1,159	1,805

I.M.T.A. 1960

10. Using the data in Question 9 calculate the median age at death and the quartile deviation for the distribution. Compare these results with estimates of these statistics derived by graphical methods. *I.M.T.A. 1960*

11. Percentage Distribution of General Agricultural Workers and Cowmen in England and Wales according to Total Weekly Earnings July–Sept. 1961

Total Weekly Earnings	General Workers	Cowmen
Under 170s.	9·2	0·6
170s. to 189s. 11d.	21·0	1·7
190s. to 209s. 11d.	18·1	3·8
210s. to 229s. 11d.	18·3	16·0
230s. to 249s. 11d.	13·8	20·5
250s. to 269s. 11d.	7·7	25·9
270s. to 289s. 11d.	5·7	12·3
290s. and over	6·2	19·2

(*Economic Trends, May 1962*)

Compare the earnings of the two groups of agricultural workers in terms of the appropriate averages and measures of dispersion. *I.M.T.A. November 1962*

12. Tuberculosis Cases (Non-respiratory) notified in Scotland 1950
(Females, by age)

Under 1 ..	15	25 and under 35	132
1 and under 5	113	35 and under 45	65
5 and under 10	122	45 and under 65	46
10 and under 15	91	65 and over	15
15 and under 25	229	All Ages ..	828

Source: Report of Department of Health for Scotland 1951. Cmd 8496

Calculate the median and quartile deriation for the above distribution and comment upon the meaning of your results. Verify your estimates by graphical methods and explain any differences between the two sets of results.

B.Sc. (Sociology) 1959

13. The weights of a group of men are as follows:

Weight in Stones				Number of Men
Under 9 ..	..	..	..	14
9 and under 9½ ..	..	..	63
9½ „ „ 10 ..	..	..	84
10 „ „ 10½ ..	..	..	131
10½ „ „ 11 ..	..	..	164
11 „ „ 11½ ..	..	..	175
11½ „ „ 12 ..	..	..	158
12 „ „ 12½ ..	..	..	25
12½ „ „ 14 ..	..	..	18

Calculate the measurements of dispersion which adequately describe these figures. *I.M.T.A. 1960*

CHAPTER VIII

STATISTICAL AND ARITHMETIC ACCURACY

For the layman statistics may often possess a significance far beyond their real importance. For example, a speech containing a large number of statistics is often regarded by audiences as much more impressive than one which concentrates on ideas and principles. Much the same is true of the printed figure, and as many a statistician knows, what was once a hopeful guess in the committee room can all too often later appear to haunt him in a published report. In all probability the latter will be quoted with all the authority of the office from which the guess emanated! There are good statistics and bad statistics; it may be doubted if there are many perfect data which are of any practical value. It is the statistician's function to discriminate between good and bad data; to decide when an informed estimate is justified and when it is not; to extract the maximum reliable information from limited and possibly biased data.

Poor statistics may be attributed to a number of causes. There are the mistakes which arise in the course of collecting the data, and there are those which occur when those data are being converted into manageable form for publication. Still later, mistakes arise because the conclusions drawn from the published data are wrong. The real trouble with errors which arise during the course of collecting the data is that they are the hardest to detect. It is virtually impossible to check whether an interviewer should have ticked *Yes* instead of *No* as the answer to a given question. Like the rest of mankind, interviewers make mistakes; they don't always ask the right questions and they sometimes write down the wrong answer.[1] When the questionnaires and schedules are returned to the Head Office for tabulation a new source of error appears. The answers may be incorrectly transferred from the schedules to the punched cards or tabulations; but good supervision can reduce this risk considerably. Sometimes, however, the answer given on the schedule has to be classified. This is at best an arbitrary procedure and mistakes in classification arise. Once the tables have been prepared detailing the results of the enquiry, their contents are analysed. At this stage too, a great deal more can be read into some statistics than the people who provided them ever dreamed of!

[1] Some of these problems are discussed in Chapter XV.

A weakness frequently encountered in reports which quote published statistics of trade, unemployment and other economic or social subjects, is the failure to consult with sufficient care the source from which a total or figure has been taken. Unemployment figures for Great Britain may be incorrectly related to population figures for England and Wales; pre-1958 employment figures in certain industries may be freely compared with current data yet there is no evidence that the author realises that owing to a re-classification of Ministry of Labour statistics, the two totals may cover different fields. This is especially true of index numbers which are quoted to measure changes in quantities and values over periods of time. For example, changes in the present cost of living can be measured from month to month by the Index of Retail Prices, but this index cannot be directly compared with either the Interim index for the period 1947-52-56 or the pre-1947 Cost of Living index.[1] Earlier, in Chapter II, great emphasis was laid upon the need for verifying definitions and sources of data taken from published documents; no apology is made for returning to this theme because inaccurate extraction of published data is an extremely prevalent disease.

Most people tend to think of values and quantities expressed in numerical terms as being exact figures; much the same as the figures which appear in the trading account of a company. It therefore comes as a considerable surprise to many to learn that few published statistics, particularly economic and sociological data, are exact. Many published figures are only approximations to the real value, while others are estimates of aggregates which are far too large to be measured with precision. For example, the *Monthly Digest of Statistics* contains many series of economic statistics which are expressed to the nearest million pounds, or hundreds of thousands of yards, or thousands of tons. It would be very satisfactory to know that every figure in that Digest was correct to the nearest unit, but in many cases it would be quite impossible to achieve complete accuracy, for some units are always missed out in a count involving hundreds and thousands of units. To achieve such accuracy would also take a great deal of time so that when finally these statistics were published they would be so much out of date as to be useless. In the case of many economic series, early publication is usually more important than precision to the last unit. If action is required on the evidence of these data, the sooner it is taken, the better. In many series, it is not so much the aggregates themselves which are of interest as the pattern of change which emerges over a period of months. A good example is provided by the monthly figures of overseas trade which may fluctuate sharply from month to month. Such fluctuations may be a completely

---

[1] This index is discussed in Chapter XVI.

unreliable guide to the nation's trade, but over a period of months a trend may emerge and it is this that needs to be watched.[1]

Approximate Data

The simplest way of indicating that figures are not given precisely to the last unit is to express them to the nearest 100 or 1,000; or in some cases to the nearest 100,000 or million. Take, for example, the annual mid-year estimates of the population of England and Wales. The figure for 1962, *i.e.* 46,669,000, is based upon the figure for the last census plus the net increase of births over deaths together with net migration. This pre-supposes that the census figure was accurate, and the fact that we are informed that 46,071,604 persons were enumerated in England and Wales on the night of 23rd April, 1961, suggests a quite remarkable degree of accuracy. It is, however, certain that some persons were missed out in the census count, hence the term 'enumerated'. The birth and death registration system in England and Wales is very reliable but the figures for migration are not so good. Hence, the estimate of the population of England and Wales at any date after 23rd April, 1961 until the next census is inevitably subject to a margin of error. For this reason, the mid-year estimates are given to the nearest 1,000.

The widespread desire for precision is reflected in many reports on economic trends which quote figures in great detail, rather than emphasising the trends and movements reflected in the figures. For example, while it may be true that exports of a particular product rose last year to £20,879,169 from £13,998,372 in the previous year, it is much clearer to state that the value of exports rose from about £14 million to approximately £21 million; alternatively, that this year's figure of almost £21 million was half as large again as that for last year. It is important to distinguish the purposes to which such published statistics are to be put; if they are merely inserted to indicate the course of events or approximate magnitude of the variables, then rounded figures which are immediately comprehensible are infinitely preferable to the exact figures. On the other hand, if a detailed analysis is to be carried out then the results may have to be given to the nearest unit.

The practice, described above, of expressing large figures more simply, *i.e.* by dropping the last few digits, is described as *rounding*. This is done with the mid-year estimate of the population, the total being expressed to the nearest thousand, and such rounding implies that the last digit in the rounded figure is correct. The following are other methods available. Assuming the original figure to lie within the limits of 82,500 to 83,500, we may write:

[1] This series is discussed in Chapter XVIII.

(1) 83,000 $\pm$ 500.

(2) 83,000 correct to ·6 per cent.

(3) 83,000 correct to nearest 1,000.

Often, in order to simplify statistical tables, the practice of rounding large figures and totals is resorted to. Where the constituent figures in a table together with their aggregate have been so treated, a discrepancy between the rounded total and the true sum of the rounded constituent figures frequently arises. Under no circumstances should the total be adjusted to what appears to be the right answer. A note to the table to the effect that the figures have been rounded, *e.g.* to the nearest 1,000, is all that is necessary. The same remark applies to percentage equivalents of the constituent parts of a total; if they do not add to exactly 100 per cent, leave them. This has been done in the 1962 column of Table 9 (page 48). The error arises here because each regional percentage has been calculated to two places of decimals and then rounded to the nearest first place. Similarly, in Table 8 the column headed 'Membership' does not add up exactly to the total shown, due to the fact that the numbers in each class have been rounded to the nearest thousand.

Biased and Unbiased Errors

The rounding of individual values comprising an aggregate can give rise to what are known as *unbiased* or *biased* errors. Table 25 below illustrates this. The *biased* error arises because all the individual figures are reduced to the lower 1,000, as in column 3; or as in column 4, where they have been raised to the higher 1,000.

The *unbiased* error is so described since by rounding each item to the nearest 1,000 some of the approximations are greater and some smaller than the original figures. Given a large number of such approximations, the final total may therefore correspond very closely to the true or original total, since the approximations tend to offset each other. This is true of Column 2, which totals 75,000, the same as the true total expressed to the nearest 1,000. It is possible even if rather unlikely, that the 'unbiased' total may be very different from the true total if most of the approximations are in one direction only, *e.g.* a group of figures each rounded to the nearest '000' where most of them lie just above '500'. The larger the number of values, however, the less likely is the total to differ from their true aggregate since the unbiased rounding of each figure will tend to balance out.

With *biased* approximations, however, the errors are cumulative and their aggregate increases with the number of items in the series. 'Biased' errors may arise in a variety of ways. If a number of women are asked to state their ages, and an average is computed, the latter is quite likely to be lower than the 'true' average, since many of the informants will tend to understate their ages by a year or so. Like-wise, an administrative officer in computing his probable staff require-ments in a group of offices may, to be on the safe side, over-estimate his needs for each office. Similarly, data based on readings from an inaccurate slide-rule, thermometer, or similar measuring instrument will be consistently biased in the same direction, *i.e.* either above or below the true figures.

Absolute and Relative Errors

Errors may be measured in two ways: *Absolutely* and *Relatively*. The *absolute* error is the arithmetic difference between the approxi-mated figure and the original quantity. Thus, in Table 25, the absolute error in the total arising from the 'unbiased' rounding in Column 2 is —182.

TABLE 25

EXAMPLES OF ROUNDING AND BIASED ERRORS

Actual Figures (1)	Unbiased (000) (2)	Biased (Lower 000) (3)	Biased (Higher 000) (4)
17,118	17	17	18
613	1	0	1
1,253	1	1	2
8,362	8	8	9
15,443	15	15	16
7,645	8	7	8
11,759	12	11	12
10,509	11	10	11
2,480	2	2	3
TOTAL 75,182	75	71	80

The *relative* error is generally derived by expressing the absolute error as a fraction of the estimated total, *i.e.* 75,000. Thus $\frac{-182}{75,000} =$ — ·0024. The same calculations have been performed for both the biased errors, where the estimated figures are adjusted downward to the nearest 1,000, *i.e.* in Column 3 the relative error is —·0590, when the figures are adjusted upwards, the error is +·0602 (Col. 4). The advantage of relatives is that widely differing quantities can be com-pared in similar terms. Thus, an error of 100,000 in £10 million is the same as 5 in £500, *i.e.* 1 per cent.

Actual Absolute Error	—182	—4,182	+4,818
Actual Relative Error	$\left(\frac{-182}{75,000}\right)$	$\left(\frac{-4,182}{71,000}\right)$	$\left(\frac{+4,818}{80,000}\right)$
	$= -0.24\%$	$= -5.90\%$	$= +6.02\%$
AVERAGE, 8,353·5	8,333·3	7,888·8	8,888·8

Absolute Error in Average

$$\left(\frac{-182}{9}\right) = -20 \cdot \overset{.}{2} \quad \left(\frac{-4,182}{9}\right) = -464 \cdot 6 \quad \left(\frac{4,818}{9}\right) = +535 \cdot 3$$

Relative Error in Average

$$\left(\frac{-182}{9} \div \frac{75,000}{9}\right) \qquad \left(\frac{-4,182}{9} \div \frac{71,000}{9}\right) \qquad \left(\frac{4,818}{9} \div \frac{80,000}{9}\right)$$
$$= -0 \cdot 24\% \qquad\qquad\qquad = -5 \cdot 90\% \qquad\qquad\qquad = +6 \cdot 02\%$$

The *absolute* error in the aggregate of a *biased* series will tend to increase with the number of items. The *relative* error in the total figure will, generally speaking, tend to diminish as the number of items increases.

The remainder of the calculations following Table 25 are simple, involving the calculation of the absolute and the relative errors in the *average*. The results in this case provide an indication of the accuracy, not of the aggregates themselves, but of the *averages* computed from those aggregates. The reader will note by reference to Table 25 that the relative error in both the average and the aggregate is the same. In any series where the individual figures have all been rounded to the same unit, (*e.g.* rounding to nearest '000' means an error $\pm$ 500), then the *average* of the total is probably more reliable than any of the individual figures.

Calculations with Estimates

Wherever any arithmetical calculation involving approximated figures is carried out, and the degree of error in those figures is known, it is possible to estimate the error arising in the final result. Starting with simple *addition:*

Add 56,000, 7,000 and 20,000 correct to 5 per cent., ·5 per cent., and ·05 per cent. respectively:

56,000	5% of which is	2,800
7,000	·5% of which is	35
20,000	·05% of which is	10
83,000		2,845

The aggregate is 83,000 $\pm$ 2,845, *i.e.* correct to 3·43 per cent. Thus, the error in the aggregate is the sum of the absolute errors in the component items.

For *subtraction* the difference will be at a minimum if the estimate of the larger figure is assumed to be below the true figure by the amount of error, and the smaller estimated figure to be subtracted is above its true amount by the amount of error shown. A maximum result is obtained if the reverse of the above case applies. Assuming that 45,000 is to be subtracted from 72,000, and the former figure lies between 44,000 and 46,000, and the latter sum between 71,500 and 72,500, then:

Maximum difference	Minimum difference
72,500	71,500
44,000	46,000
28,500	25,500

The answer is $\dfrac{(28,500 + 25,500)}{2} = 27,000 \pm 1,500$, i.e., correct to 5·55 per cent. With subtraction, as with addition, the error in the answer is equal to the sum of the errors in the individual amounts, i.e. $\pm 1,000$ and $\pm 500 = \pm 1,500$.

When *multiplying* two rounded values together, it is usual to show both the result and the maximum possible error which could have occurred in it. This is derived as follows. The product of two values, 12,500 and 400, is 5 million. Assume that the larger value has been rounded to the nearest 500 and the other value to the nearest 50. Thus the true values might have read 12,749 and 424 which when multiplied together give a product of approximately 5·43 million. The maximum error in the original product of 5 million is thus 0·43 million which is equal to an error of $8\frac{1}{2}$ per cent. Therefore, the answer to the multiplication given above would be written 5 million $\pm 8\frac{1}{2}$ per cent.

With *division* the same principle is applied. If the total 300,000 is divided by 500, the quotient is 600. If the dividend had been rounded to the nearest 10,000 and the divisor to the nearest 10 units, then the maximum error in the above quotient of 600 is derived when we divide the smallest possible divisor into the largest possible dividend, i.e. 496 into 304,999, which gives 615. In other words, there is a maximum error of ± 15 in a figure of 600, which is equal to $\pm 2\frac{1}{2}$ per cent.

It should not be necessary to memorise the various formulae which are sometimes evolved for these operations, usually remembered at the expense of the principles on which they are based. All that needs to be kept in mind is that the maximum error is always possible. Generally speaking, no amount of juggling with figures can increase the accuracy of the result if the original data are liable to error. The result of any calculation involving approximation can be no more accurate than the least accurate of the figures used in the calculation. Thus, in stating the final result, it may be advisable to give it to at least one significant figure less than found in the least accurate factor employed in the calculations. The same point should be borne in mind when calculating means and medians, etc., from grouped frequency distribution where the class-interval is large.

Use of Ratios

Earlier in this chapter the usefulness of relatives instead of absolute quantities for comparative purposes was mentioned. The main feature to be remembered is that a ratio or percentage expresses the variation in the data, irrespective of its actual or absolute size. Thus an expansion in a firm's turnover from £500,000 to £750,000 is the same in relation to the first value as is a rise from £10,000 to £15,000, *i.e.* 50 per cent. rise on the base year, *i.e.* the year in which the £500,000 and £10,000 were earned.

The main danger to avoid in expressing variations in terms of ratios or percentages is the use of two different bases in the comparison, or more generally, of failing to make clear on which base the change has been calculated. Thus, if in Year 1 profits were £25,000, and the chairman made the following statement: 'In Year 2 profits rose 10 per cent., the following year 25 per cent., and last year 33 per cent', the shareholders might be forgiven if they arrived at two very different results:

	Year 1	Year 2	Year 3		Year 4
(i)	£25,000	£27,500	£34,375	and	£45,833
(ii)	£25,000	£27,500	£31,250		£33,333

The first line is calculated on the assumption that each percentage rise is based on the figure of the immediately preceding year; in the second line the percentages are all worked on Year 1 as base year. Whichever method is intended, it should be made clear which is to be the base year.

Such comparative ratios or percentages are a frequent source of confusion. If the two sets of quantities are widely different in absolute size, as in the examples of the sales given above, a mere percentage comparison may be quite misleading in so far as it may tell only half the story, or more seriously, it may suggest that the comparison made is justified when it most certainly is not. Thus, if a school teacher discussing the latest examination results with the head of a large coaching institution, states that 50 per cent. of his candidates obtained distinction in all subjects, and the coaching institution only 5 per cent.; can any conclusions be drawn? The answer is 'no'. It may be that 500 pupils of the coaching institution sat and only six at the teacher's school. In any case it is highly improbable that a direct comparison can be made between the two teaching methods, until more is known of the calibre of the pupils, the amount of study done by the students, and the numbers of staff. The first rule of statistics, 'compare like with like', has hardly been observed.

The averaging of percentages themselves requires care, where the percentages are each computed on different bases, *i.e.* different quantities. The average is *not* derived by aggregating the percentages and dividing them. Instead of this, each percentage must first be multiplied by its base to bring out its relative significance to the other percentages and to the total. The sum of the resultant products is then divided by the sum of the base values as in (Col. 2) below, not merely the number of items.

Suppose the ratio of equity capital to the total capital for each of six public companies is as follows:

Company	A	B	C	D	E	F
%	100	50	50	40	50	25

The average ratio for all six companies is *not* the average of these ratios, *i.e.* 315 ÷ 6, as set out in column (4) below, because this method makes no allowance for the different amounts of the companies' capital as shown in column (2) below. The correct method requires us to work from the actual figures of share capital and equity capital given in columns (2) and (3) below. Then each percentage is 'weighted', *i.e.* multiplied by the figure of total capital, their product added together as in column (5) and that total divided by the total weights. The correct answer is 43·3%, as compared with the answer of $\frac{315}{6} = 52·5\%$ derived by using the incorrect method.

Percentage of Equity Capital in Six Public Companies

(1) Company	(2) Total Capital 000's £	(3) Equity Capital 000's £	(4) Ratio of Col. 3 to Col. 2 %	(5) Col. 2 × Col. 4 000's £
A	50	50	100	50
B	100	50	50	50
C	200	100	50	100
D	250	100	40	100
E	500	250	50	250
F	400	100	25	100
	1,500	650	315	15)650 = 43·3%

The reason for the difference is simple. In column (4) there is an implied assumption that all the companies are of equal size (in terms of their issued capital) and consequently equal importance has been attached to the percentages for the small and the large firms. In column (5), however, correct importance has been given to each firm; *i.e.* the percentage of equity to total capital has been *weighted* in the ratio of their total capitals one to the other.[1]

The same rules apply to the 'averaging' of averages, *e.g.*:

(1)	(2)	(3)	(4)
	No. of	Average Weekly	Products in
Plant	operatives	Output per operative	00's
A	180	540	972
B	140	530	742
C	50	490	245
D	90	500	450
E	110	510	561
F	160	525	840
	730	3,095	730)3,810

522 to
nearest unit

The average output of all workers in the six plants is not the average of the six plant averages, *i.e.* 3,095 ÷ 6 = 516. This is wrong. The correct method is given in the last column; the products of multiplying columns 2 and 3 together and dividing their sum by the total of operatives, *i.e.* column 2. This has the same effect as adding up the output of every one of the 730 operatives in the six factories and dividing the aggregate output so derived by the number of operatives. The true average per operative is thus found to be 522 per week.

Conclusions

While it is true to assert that much statistical work involves arithmetic and mathematics, it would be quite untrue to suggest that the main source of errors in statistics and their use is due to inaccurate calculations. This happens, of course. Lord Randolph Churchill was not the first man to be confused by those 'damned' dots; nor was he the last to get the answer to the wrong place of decimals. Arithmetic apart, the first lesson to be learned in statistics is that many figures are little better than good estimates. The art of the statistician lies in two related fields. First, he learns how to collect his data in such ways as to minimise the risk of errors; second, he learns

---

[1] The subject of 'weighting' is discussed more fully in the chapter on Index Numbers.

to judge the quality and reliability of other people's data, whether they be published in official reports or have been collected by someone else, and thereby minimise the effects of any possible errors.

When using estimated figures, *i.e.* figures subject to error, for further calculation make allowance for the absolute and relative errors. Above all, avoid what is known to statisticians as 'spurious' accuracy. For example, if the arithmetic Mean has to be derived from a distribution of ages given to the nearest year, do not give the answer to several places of decimals. Such an answer would imply a degree of accuracy in the results of your calculations which are quite unjustified by the data. The same holds true when calculating percentages. For example, 'of the 46·7 million estimated home population in England and Wales at 30th June 1962, 14,644,000 were under 21 years of age, *i.e.* 31·14 per cent.'; this answer is silly. It implies that the original estimates of total population and those under 21 were correct to the last unit, instead of having been approximated to the nearest 1,000. It could well be, if there had been heavy migration that year, that both figures would be subject to an even larger margin of error. At most, given these figures, one could say that 31 per cent of the estimated population are believed to be under 21 years of age. Probably it would be safer to say that 'some 30 per cent' are under 21.

A great deal of nonsense is talked these days about economic growth. Year by year comparisons are made of the rate at which Britain's economy is expanding and the National Economic Development Council has set a target of 4 per cent per annum. How sure can the statisticians be that performance matches the target? The key estimate is the gross national product (G.N.P.) which is about £24,000 million, subject to an error which is not less than 3 per cent and could be considerably more than 5 per cent.[1] In other words, the error in the estimate is probably as large as the required annual increase in the gross national product. How much confidence can we have in public declarations by politicians that the G.N.P. this year has risen by 4 per cent, or by 3½ per cent, or any other figure? These data are at best tentative estimates to be used with caution. In the longer run they provide some indication of the trend of the economy. Economic fluctuations from year to year, however, as measured by current statistics, need to be taken with more than the proverbial pinch of salt. The same is true of many other series. Hence, it is not enough for the statistician to know how to manipulate his data; he must also know what the data can stand.[2]

---

[1] These and other official statistical series are discussed in Chapter XVIII.

[2] So important is a thorough understanding of the nature of statistical data in all statistical work, that there is often dispute as to whether a firm should engage a statistician to analyse their data, or teach a member of its own staff, who is familiar with the product and production processes, sufficient statistical theory to enable him to analyse the available data. There are good arguments for both policies.

Questions

ACCURACY AND APPROXIMATION

1.

	Number of cattle slaughtered mln.	Quantity of beef produced thous. tons
1954	0·22	60·9
5	0·21	56·9
6	0·24	65·0
7	0·25	66·6
1958	0·25	66·4

Having regard to the degree of rounding up used in the above figures, comment on the change, if any, over the past five years, in the weight (cwt.) of beef produced per head of cattle.   *I.M.T.A. 1956*

2. A motorist whose car has a broken distance recorder and who, therefore, measures his distances by map-reading, wishes to assess his petrol consumption on a Continental trip. His petrol gauge indicates his fuel consumption in litres with an error of ±10%. He also uses approximate conversion factors of 1 gallon = 4·5 litres and 1 mile = 1·6 kilometres, instead of the more accurate 1 gallon = 4·55 litres and 1 mile = 1·61 kilometres.

 (i) Estimate his probable petrol consumption in miles per gallon, giving limits of error, if he claims to be getting 31 miles per gallon after a trip of exactly 615 miles.

 (ii) If 31 miles per gallon is, in fact, an exact estimate and he states that he has used 95 litres of petrol, within what limits does his true mileage lie?

Institute of Statisticians 1963

3. The table below shows the weekly wage bill and the numbers employed in three subsidiary companies. The figures have been rounded to the nearest £100 for the wage bill and the nearest 10 for the numbers employed.

Company	Weekly Wage bill £	Numbers employed
A	5,000	530
B	5,200	590
C	8,000	800

Calculate the average wage paid by each company and by the group as a whole
What is the maximum possible error due to rounding in the figures for:

(a) the wage bill;

(b) the numbers employed;

(c) the average wage,

for each of the three companies as well as for the group as a whole? Show in detail how you reach your estimates.   *I.C.W.A. 1963*

4. Two measurements, A and B, are each subject to an error of ± 10 per cent. Determine the range of possible error of·

  (i) The sum $(A + B)$,
 (ii) The product (AB),
(iii) The difference $(A - B)$,
 (iv) The quotient (A/B),
       when $A = 375$ and $B = 25$.

In each case express your result both as an absolute and as a relative error.

Institute of Statisticians 1962

5. In a report making use of the following table (but not reproducing it in the text), it was stated only that the mean weight of the particles was 8·49 milligrams with standard deviation 3·891 milligrams.

Discuss the accuracy of these figures and usefulness of the summary statement.

Weight of small chemical particles (in milligrams, to nearest unit)				Number of particles
1	..	..	..	1
2	..	..	..	2
3	..	..	..	4
4	..	..	..	9
5	..	..	..	15
6	..	..	..	20
7	..	..	..	20
8	..	..	..	19
9	..	..	..	18
10	..	..	..	19
11	..	..	..	21
12	..	..	..	20
13	..	..	..	15
14	..	..	..	9
15	..	..	..	4
16	..	..	..	2
17	..	..	..	1
25	..	..	..	1
				200

Institute of Statisticians 1956

6. The data below refer to births and infant deaths in a single year for two areas in which vital registration is defective. The possible error is believed to be within the limits shown. Express the neo-natal death rate for area B as a percentage of that for area A, and indicate the margin of possible error in this result. Do the same for all infant deaths.

	Area A	Area B
No. of births ..	$4{,}168 \pm 5\%$	$6{,}285 \pm 7\%$
Neo-natal deaths..	$392 \pm 6\%$	$614 \pm 4\%$
Post-natal deaths..	$168 + 2\%$	$267 \pm 3\%$

Inter. D.M.A. 1962

CHAPTER IX

REGRESSION AND CORRELATION

So far we have been discussing the description and analysis of one variable. For example, the weekly turnover of twenty retail shops can be set out in tabular form as follows and for this distribution it is possible to calculate the arithmetic mean and the standard deviation.

Retail Shop		1	2	3	4	5	6	7	8	9	10
Weekly Turnover	£	150	200	210	230	260	280	300	320	350	370
Retail Shop		11	12	13	14	15	16	17	18	19	20
Weekly Turnover	£	380	400	410	430	460	470	480	500	520	540

A similar tabulation can be constructed to show the gross profit of each of the same twenty shops and the same statistics calculated. There would, however, probably be some relationship between turnover and the amount of profit. We might ask, for example, whether the profit increases constantly with the turnover? In order to answer this question the data can be arranged, as a preliminary to further analysis, in the form of a table such as that above together with the following data relating to profits:

Retail Shop		1	2	3	4	5	6	7	8	9	10
Profit	£	30	35	40	45	50	50	60	65	70	70
Retail Shop		11	12	13	14	15	16	17	18	19	20
Profit	£	80	75	85	90	100	80	90	100	110	115

There are, however, rather too many figures to judge to what extent turnover and profit are directly connected, *i.e.* that a given increase in sales is accompanied by a specific increase in profits. A simple device for examining the data so as to bring out any relationship between the two variables is the so-called *scatter* diagram. This is an ordinary graph on which turnover is measured along the base and the profit against the vertical axis. For each shop there are two values which together locate a point on the graph. The twenty pairs of values are plotted on the graph depicted in Figure 17A. It is immediately apparent that the points plotted form a clear pattern diagonally across the graph from the bottom left-hand corner to the top right. Such a pattern indicates that the amount of profit tends to rise with the sales volume of each shop.

Lines of Regression

It is possible to define the approximate relationship between profit and turnover in mathematical terms by means of an equation. Any straight line conforming to the path of the points plotted in the graph can be defined by an equation of the form $y = a + bx$. Given this equation and the values of a and b, which are termed constants, by substituting in the equation any value of x we can derive a value for y. Unfortunately, in this particular case, the points plotted do not lie exactly along such line. They are scattered about that line, some above and some below.

Now rearrange the data in the form of a frequency distribution so that for any given turnover the average profit is given.

Turnover x	Observed profit y	Average profit	Turnover x	Observed profit y	Average profit
150—	30	30	400—	75, 85, 90	$83\frac{1}{3}$
200—	35, 40, 45	40	450—	100, 80, 90	90
250—	50, 50	50	500—	100, 110	105
300—	60, 65	$62\frac{1}{2}$	500—	100, 110, 115	$108\frac{1}{3}$
350—	70, 70, 80	$73\frac{1}{3}$			

For each class of turnover an 'average' profit is derived. If these average profits are now plotted with the mid-points of the corresponding classes of turnover, it will be seen from Figure 17B that the approximation of the plotted points to a straight line is much better. The line that has been drawn on the basis of these points, *i.e.* through the means of each group of y values corresponding to each value of x, is drawn in such a way that the squares of the vertical distances between each of the points and the line under or over the point are at a minimum. A method of calculating the equation of such a line is explained later on p. 173. Such a line is called a *regression* line because it is derived from the equation which defines the regression of y upon x. In terms of our data, it measures the extent to which y, *i.e.* profit, is related to the value of x, *i.e.* the turnover.

The rather curious term *regression* needs some explanation since it has no obvious relevance to the data. Regression analysis was introduced by Sir Francis Galton towards the end of the 19th century and

Figure 17
REGRESSION LINES

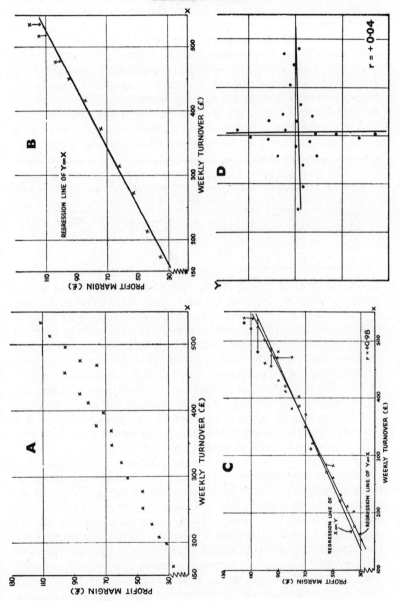

the original data he used related to the heights of fathers and sons. He found that on average tall fathers had tall sons, but the sons tended to 'regress' to the average male height. The term has remained in use for all types of data ever since although the methods now used to define such relationships were developed later by Professor Karl Pearson.

By re-classifying the original data as is done in the table below, another regression line can be derived for which the values for x and y are plotted on the scatter diagram (Graph C) as small circles. In this case, however, it is so drawn that the squares of the *horizontal* differences between the points and the regression line are minimised. This particular line is known as the regression line of x upon y. It yields similar information to the other regression line of y upon x.

Profits x	Observed Turnover y	Average Turnover	Profits x	Observed Turnover y	Average Turnover
30—	150, 200	175	80—	380, 410, 470	420
40—	210, 230	220	90—	430, 480	455
50—	260, 280	270	100—	460, 500	480
60—	300, 320	310	110—	520	520
70—	350, 370, 400	373	110—	520, 540	530

It will be seen from Figure 17c that these two regression lines do not coincide, although they intersect each other. But suppose, when the first regression line was drawn, all the points lay along its path. In other words, the *vertical* distances between the line and the observed values of the dependent variable for each value of the independent variable were zero. If the other regression line were then to be drawn so that the *horizontal* differences were at a minimum, where would it lie? Clearly, it would coincide with the first line; in other words, the regression lines are identical. But this situation will only arise if the relationship between x and y is perfect, *i.e.* any given movement in x is invariably accompanied by the same proportionate movement in y. It then becomes possible to predict from the regression equation the value of y which corresponds with any value that x may take.

Such a perfect relationship, or degree of association as it is sometimes termed, is the exception rather than the rule. The points do not usually lie along a single line and therefore for any set of plotted points it becomes possible to draw two regression lines. At one extreme these lines coincide, that is when the relationship between the two variables is perfect. In Graph C of Figure 17 the two regression lines are very close and therefore the degree of association or their inter-dependence is high. At the other extreme in Graph D, where there is virtually no association whatsoever between the two variables, the regression lines are at right angles to each other.

The two regression lines intersect at the means of the two distributions of x and y values. If the mean turnover is calculated it is found to be £363, while the mean profit is £72. Reference to the diagram will show that these values mark the point of intersection. This fact of intersection at the respective means of the two variables is very important. The degree of association between the variables is measured by the extent to which the variations of each of the observed x and y values correspond, the variations of each x and y value being measured from the respective means of the two variables. If the relationship between the variation in a given value of x and the corresponding variation for the related value of y is fairly regular, then it follows that the two variables are co-related, *i.e.* movements in one variable are associated with given changes in the other The degree of association or relationship is measured by the *coefficient of correlation* represented by the letter r.

The Coefficient of Correlation

In the above illustration the relationship between x and y was such that the pairs of observed values increased together or declined together. In such a case the correlation is described as positive or *direct*. When an increase in one variable is accompanied by a fall in the other, for example, an increase in family incomes is accompanied by a fall in the consumption of cheaper foods such as bread and potatoes, then the correlation is known as *inverse* or negative. The distinction between the two types of relationship is always indicated by the sign, plus or minus, placed before the value of the coefficient of correlation. This coefficient is zero when there is no association between x and y, *i.e.* when the regression lines would be at right angles to each other (Graph D). When the regression lines coincide, however, *i.e.* perfect association, the coefficient has a value of unity. For example, in Graph C the relationship is very close and in consequence r is equal to $+0.98$, whereas in Graph D it is only $+0.04$.

The formula for determining the value of the coefficient is such that the latter will always lie between zero and unity, either positive or negative. It can never be greater than 1. Hence, if the student finds at the end of an exercise that his coefficient exceeds unity, then he will know he has made a mistake in the calculations.

As already explained, the coefficient of correlation is a measure of association. Sometimes, but not invariably, the relationship is causal, *e.g.* the age of a child and its weight; the number of hours of sunshine at a seaside resort and the receipts from the hiring of deck-chairs. The coefficient differs from the equation defining either of the lines of regression in so far as the latter define a unique relationship from which, given the change in (say) x, it is possible to compute the most probable change that will follow in *y*. The coefficient of correlation merely indicates the closeness or intensity of the association without defining it. For example, the correlation coefficient of +0·98 given by the data relating to the 20 retail shops suggests that there is a close relationship between the turnover and profit. If the coefficient had been only +0·3 we should probably have deduced that the relationship was not sufficiently strong – at least from the data available – to justify any use of the relationship for analysis or prediction. Generally speaking, it is customary to compute the correlation coefficient and to ignore the lines of regression and the regression equations unless the relationship between the variables is such that it may reasonably be summarised in the form of a mathematical equation. The next few sections of this chapter are devoted to explaining methods of deriving the value of *r*. In common with all statistics derived from sample data, the larger the sample the more reliable the statistic. In order, however, to simplify the exposition and keep the calculations to a minimum the illustrations are based on typical examination questions rather than realistic data.

Calculating the Coefficient

When the relationship between two variables is linear and the data have not been grouped, and only in such cases, the coefficient of correlation is calculated by means of the following formula:

$$r = \frac{\Sigma\, xy}{n\sigma x\, \sigma y}$$

This formula or method is sometimes referred to as the product-moment, or the Pearsonian coefficient; the first following the method of calculation and the second the name of its discoverer, Karl Pearson.

The symbols *x* and *y* in the numerator (*i.e.* the upper part of the fraction) represent not single values of *x* and *y*, but the deviations of all *x* and *y* values from their respective means, in the same way as we used the deviations from the mean when calculating standard deviations by the short method (see p. 126). To remind us of the meaning of the above formula for *r*, we can write the numerator as $\Sigma\, (x - \bar{x})$ $(y - \bar{y})$ where *x* is any single value of that variable and $\bar{x}$ is the mean of all the *x* values.[1] The same applies to *y*. Note that these 'deviations' will also be used for calculating the two standard deviations which the formula requires.

In the following illustration the data relate to the turnover and the profit margin of the sample of shops discussed earlier. It is required

---

[1] The symbol Σ means the 'sum of' all these products.

to calculate the coefficient of correlation between the size of the weekly turnover and profit margin. The data are given in columns 1, 2 and 3 and the calculations may best be followed if set out in stages. The true means of both distributions are easily found. In the case of the weekly turnover it is £363, and in the case of profit, £72. In columns 4 and 5 the deviations from the means of the values of each variable are set out. In columns 6 and 7 the deviations given in columns 4 and 5 have been squared and summed. The calculation so far is no more than would be required for deriving the standard of deviations of any two distributions. Since the deviations of the individual values in both

CALCULATION OF r BY PRODUCT-MOMENT METHOD

1 Re-tailer No.	2 Weekly Turn-over (x) £	3 Profit Margin (y) £	4 $(x-\bar{x})$	5 $(y-\bar{y})$	6 $(x-\bar{x})^2$	7 $(y-\bar{y})^2$	8 $(x-\bar{x})(y-\bar{y})$
1	150	30	− 213	− 42	45,369	1,764	+ 8,946
2	200	35	− 163	− 37	26,569	1,369	+ 6,031
3	210	40	− 153	− 32	23,409	1,024	+ 4,896
4	230	45	− 133	− 27	17,689	729	+ 3,591
5	260	50	− 103	− 22	10,609	484	+ 2,266
6	280	50	− 83	− 22	6,889	484	+ 1,826
7	300	60	− 63	− 12	3,969	144	+ 756
8	320	65	− 43	− 7	1,849	49	+ 301
9	350	70	− 13	− 2	169	4	+ 26
10	370	70	+ 7	− 2	49	4	− 14
11	380	80	+ 17	+ 8	289	64	+ 136
12	400	75	+ 37	+ 3	1,369	9	+ 111
13	410	85	+ 47	+ 13	2,209	169	+ 611
14	430	90	+ 67	+ 18	4,489	324	+ 1,206
15	460	100	+ 97	+ 28	9,409	784	+ 2,716
16	470	80	+ 107	+ 8	11,449	64	+ 856
17	480	90	+ 117	+ 18	13,689	324	+ 2,106
18	500	100	+ 137	+ 28	18,769	784	+ 3,836
19	520	110	+ 157	+ 38	24,649	1,444	+ 5,966
20	540	115	+ 177	+ 43	31,329	1,849	+ 7,611
	7,260	1,440	0	0	254,220	11,870	+ 53,780

$$\bar{x} = \frac{7,260}{20} = 363. \quad \bar{y} = \frac{1,440}{20} = 72.$$

$$\sigma x = \sqrt{\frac{254,220}{20}} = 112\cdot7. \quad \sigma y = \sqrt{\frac{11,870}{20}} = 24\cdot4.$$

$$r = \frac{\Sigma(x-\bar{x})(y-\bar{y})}{N\sigma x \, \sigma y} = \frac{53,780}{20 \times 112\cdot7 \times 24\cdot4}$$

$$= \frac{53,780}{54,997\cdot6} = 0\cdot98.$$

series have been measured from their means their sum is zero (see columns 4 and 5). Column 8 provides the sum of the cross products. It will be seen that the deviations given in columns 4 and 5 are multiplied together and the products aggregated. The mean of their sum, *i.e.* the total products divided by N, is often referred to as the *co-variance*. The next stage is to substitute these values in the appropriate formula. The numerator is given by the sum of the cross-products in column 8 and is divided by the number of pairs of items, *i.e.* 20, and the product of the two standard deviations. The two standard deviations are easily derived from the data given in columns 6 and 7. It will be noted that the working in this example is from the true means, hence no correction is necessary. Substituting the appropriate values in the formula for the correlation coefficient, the value of the coefficient is derived by simple arithmetic.

The numerator in the formula is based on the need to provide a meaning for 'high' and 'low' values of the coefficient. All the values

		Ia			Ib		
x	$(x-\bar{x})$	y	$(y-\bar{y})$	$(x-\bar{x})(y-\bar{y})$	y	$(y-\bar{y})$	$(x-\bar{x})(y-\bar{y})$
3	− 6	7	−12	+ 72	28	+ 9	− 54
5	− 4	14	− 5	+ 20	24	+ 5	− 20
7	− 2	20	+ 1	− 2	21	+ 2	− 4
9	0	21	+ 2	0	20	+ 1	0
13	+ 4	24	+ 5	+ 20	14	− 5	− 20
17	+ 8	28	+ 9	+ 72	7	−12	− 96
6)54		6)114		+ 182	6)114		− 194
$\bar{x}=9$		$\bar{y}=19$			$\bar{y}=19$		

II.

x	$(x-\bar{x})$	y	$(y-\bar{y})$	$(x-\bar{x})(y-\bar{y})$
3	− 6	20	+ 1	− 6
5	− 4	14	− 5	+ 20
7	− 2	28	+ 9	− 18
9	0	21	+ 2	0
13	+ 4	7	−12	− 48
17	+ 8	24	+ 5	+ 40
6)54		6)114		− 12
$\bar{x}=9$		$\bar{y}=19$		

Note that the values of x are unchanged for all three sets of values of y. In Ia, the high values of y are related to high values of x; in Ib the reverse. In both cases the value Σxy is large in arithmetic or absolute terms. In case II there is no pattern about the values of y in relation to x and Σxy is a negligible quantity, therefore r would also be very small. The standard deviations of the three distributions of y given above will, of course, be the same in each case.

of each variable must be related to a comparable value; in this case the means of the two distributions are the norm for each pair of values. Provided the pairs of x and y values follow some pattern in respect of their variation from their respective means – which will be the case if the two values are related – then positive deviations from the mean for the values of x associated with similar positive deviations of the other variable y means that r will also be positive. If, however, positive deviations of x are related to negative deviations of y, then r will be negative. It will be seen from the example above that if high and low values of x and y are indiscriminately associated (*i.e.* the paired variations from the respective means of both x and y values are not consistent related signs) then the sum of the products of the deviations will be small. Therefore r too, will tend to be small.

When there is any relationship between the variation in x and y values the sum of the cross-products will be large. And, the larger the absolute value of the deviations, the larger will be the sum of these cross-products. The case of the standard deviation may be recalled; this is expressed in the units of the distribution and if the values are large the σ is large, *e.g.* the σ of the weights of adult males in absolute terms is greater than the σ of the weights of adult females; but if related to their respective means, *e.g.* as with the coefficient of variation described on page 129, the dispersion of weight among females is greater. For similar reasons, in the product-moment formula the sum of the cross-products is related to the products of the standard deviations of the two distributions. In effect, the cross-products are 'standardised' so that the value of r is no longer influenced by the absolute value of the factors in the cross-products. In consequence the value of r is a coefficient that has no dimensions.

Students often complain that they cannot remember the formula for calculating the coefficient of correlation. This is understandable, but if they remember the basis of the statistic it should be possible to work it out. It has been explained that the value of r is dependent on the degree to which deviations of x and y move in sympathy. As was shown on page 164 if they do not correspond, then the sum of the cross-products is small; if they do, then the sum of the cross-products is high. On the other hand, the sum of the cross-products may be high because the actual values, and therefore the deviations, are large in absolute terms, *e.g.* they may be expressed in pounds instead of ounces. Secondly, their sum will tend to be greater, the larger the number of pairs of variables. Obviously a statistic which is affected both by the number of observations and the absolute size of the variates is unsatisfactory. This weakness is overcome by relating the sum of the cross-products of the x and y deviations to (*a*) the number of paired observations *i.e.* N and (*b*) a measure of the actual deviations in both variables, *i.e.* the standard deviations of x and y. Since both the co-variance and the standard deviations are affected by the absolute size of the variates by dividing one into the other this distorting factor is eliminated. If the student reader examines the formula for the coefficient in the light of the foregoing, he may not need to rely on his memory so much.

Calculating r using assumed means

In the second example given below the data are derived from a social survey in London. It is proposed to calculate the coefficient of correlation between poverty and overcrowding. For the purposes of the survey poverty was defined as living below a prescribed minimum standard, and overcrowding was defined as living more than two persons per room. These definitions are unimportant for the purposes of the calculation. They are, of course, essential for purposes of interpreting the data. The figures for poverty and over-crowding, *i.e.* the variables x and y, are in the nature of percentages, although they are actually expressed per 200 households. For example, in borough A while 17 out of every 200 households were living in poverty, 36 were overcrowded.

The data are set out in the same way as in the earlier example, but the calculation is complicated by the fact that the standard deviations and the cross-products are worked not from the true means of the two variables but from assumed means. In other words, a correction must be introduced. The student will remember the correction required for purposes of calculating the standard deviation. It is shown

Calculation of co-efficient of correlation between Poverty (defined as living below given minimum standard) and Overcrowding (defined as more than two persons per room) in 12 London boroughs.

Borough	No. per 200 households in Poverty(x)	Over-crowded(y)	$(x\text{-}\bar{x})$	$(y\text{-}\bar{y})$	$(x\text{-}\bar{x})^2$	$(y\text{-}\bar{y})^2$	$(x\text{-}\bar{x})(y\text{-}\bar{y})$
A	17	36	+ 7	+14	49	196	+ 98
B	13	46	+ 3	+24	9	576	+ 72
C	15	35	+ 5	+13	25	169	+ 65
D	16	24	+ 6	+ 2	36	4	+ 12
E	6	12	− 4	−10	16	100	+ 40
F	11	18	+ 1	− 4	1	16	− 4
G	14	27	+ 4	+ 5	16	25	+ 20
H	9	22	− 1	0	1	0	0
I	7	2	− 3	−20	9	400	+ 60
J	2	8	− 8	−14	64	196	+112
K	10	17	0	− 5	0	25	0
L	5	10	− 5	−12	25	144	+ 60
12	125	257	+ 5	− 7	251	1851	+535

Actual means are 10·42 and 21·42 but to avoid calculations with such awkward figures, work from assumed means. Select 10 as the mean of x and 22 as the mean of y. The formula in this case is as follows:

$$r = \frac{\dfrac{\Sigma(x-\bar{x})(y-\bar{y})}{n} - \left(\dfrac{\Sigma(x-\bar{x})}{n} \times \dfrac{(y-\bar{y})}{n}\right)}{\sigma x \ \sigma y}$$

$$\sigma x = \sqrt{\frac{251}{12} - \left(\frac{5}{12}\right)^2} = 4.5 \quad \sigma y = \sqrt{\frac{1851}{12} - \left(\frac{-7}{12}\right)^2} = 12\cdot4$$

$$r = \frac{\dfrac{535}{12} - \left(\dfrac{5}{12}\right)\left(\dfrac{-7}{12}\right)}{4\cdot5 \times 12\cdot4}$$

$$= \frac{44\cdot83}{55\cdot8} = +\ 0\cdot80$$

in the example above. The correction for the sum of the cross-products measured from their true means is quite simple. The cross-products actually given in the final column of this table are based on the deviations measured from assumed means in each distribution. The difference between the true mean and assumed mean for both distributions is given in the working, columns 4 and 5. If the sum of the cross-products in the last column is divided by N, *i.e.* the number of pairs, we obtain the average product. From this we deduct the product of the two errors in the two averages or means divided by n^2. The result is then equal to the sum of the cross-products of the deviations just as if they had been measured from their true means. The subsequent arithmetic is as in the earlier example.

The student-reader will not have failed to note the considerable amount of arithmetic required in the first example on page 163. Quite apart from the fact that it contained twenty pairs of observed values, there was disproportionately more calculation than in the second example. As an exercise the student can apply the principles of the second illustration to the first set of data. Note, for example, that all the values of x are rounded to the nearest £10, similarly the y values are rounded to the nearest £5. Instead of taking the true means of x and y, use a multiple of 10 and 5 for the means of x and y respectively. Take, for example, £350 and £75 as the assumed means. All the deviations of x and y can then be given in multiples of 10 and 5. If, as in the illustration of calculating the standard deviation from an assumed mean in terms of class intervals given on page 127, you use deviations in terms of the class intervals, the calculations will be much easier. Note that the cross-products will then be in units of the product of the two class intervals, *i.e.* £10 × 5. If the actual value of the standard deviation for either variable were required, it would be necessary to convert the figure obtained, by multiplying it by the appropriate class interval. This correction is not required for calculating the coefficient of correlation since the sum of the cross-products making up the numerator in the formula, and the product of the two standard deviations, are both expressed in class-intervals. The answer derived from this calculation will be the same as for the more detailed method. By performing it, the student reader will ensure that he has followed the basis of the calculations.

Calculating r from grouped data

In the third example illustrating the calculation of the correlation coefficient the data are set out in what is known as a bivariate table. It will be noted that the pattern of the figures over the grids is somewhat similar in appearance to the scatter diagram discussed earlier. The data in this example relate to the weekly expenditure on accommodation and food of 33 individuals. All figures are given in shillings. Instead of the pairs of individual values of x and y in the two earlier examples, this illustration comprises two grouped frequency distributions, one of which is read horizontally, *i.e.* expenditure on food, and the other vertically, *i.e.* expenditure on accommodation. The layout of the calculation should be studied with especial care. The four vertical columns to the side of the table and the four horizontal columns below it are the same except that those to the right show the calculations for the y values and those below for x. The column headed f is nothing more than the frequencies derived by cross-adding the frequencies within each cell. Thus reading from the right-hand columns, there are 9 cases where the expenditure on accommodation is 50s. – per week. The second column in each case relates to the deviation from the assumed mean which is 45s. for x and 55s. for y. As before, these assumed means are represented by the symbols $\bar{x}$ and $\bar{y}$, and the third and fourth columns in each case are the sum of the frequencies and deviations and the deviations squared required to compute the difference between the assumed and true mean and the standard deviation. Note at the head of each column the reminder that the deviations are in class-interval units, hence we write $f(y - \bar{y})$ $\div c.i.$

The calculation of the cross-products is more complex, however. As in the two earlier examples, the related pairs of deviations must be multiplied together but, whereas in the earlier examples there was only one of each pair, in the present example there are more than one in several cases. For example, in the bottom left-hand corner of the grid it will be seen that there are four cases in which the expenditure on accommodation is 90s. and over, while the expenditure on food is

Expenditure on Food (x)
(shillings)

Expenditure on Accommodation y (shillings)	10—	20—	30—	40—	50—	60—	f	$y-\bar{y} \div$ c.i.	$f(y-\bar{y}) \div$ c.i.	$f(y-\bar{y})^2 \div$ c.i.
20—					1_{-3}		1	−3	−3	9
30—		1_2					1	−2	−2	4
40—		1_2		2_0	4_{-1}		7	−1	−7	7
50—		5_0		4_0			9	0	0	0
60—			1_{-1}	3_0	2_1	1_2	7	1	7	7
70—				2_0			2	2	4	8
80—								3		
90—	4_{-12}	2_{-8}					6	4	24	96
	4	8	2	11	7	1	33		23	131
$x-\bar{x} \div$ c.i.	−3	−2	−1		1	2				
$f(x-\bar{x}) \div$ c.i.	−12	−16	−2		7	2	−21			
$f(x-\bar{x})^2 \div$ c.i.	36	32	2		7	4	81			

$$r = \frac{\dfrac{\sum f(x-\bar{x})(y-\bar{y})}{n} - \dfrac{\sum f(x-\bar{x}) \times \sum f(y-\bar{y})}{n^2}}{\sigma x \; \sigma y}$$

Sum of the cross-products, *i.e.* $\sum f(x-\bar{x})(y-\bar{y})$: $-3 + 2 + 2 - 4 - 1 + 2 + 2 - 48 - 16 = -64$.

$$\frac{\sum f(y-\bar{y})}{n} = \frac{23}{33} \qquad \frac{\sum f(x-\bar{x})}{n} = \frac{-21}{33}$$
$$= 0.697 \qquad\qquad = -0.636$$

$$\sigma x = \sqrt{\frac{131}{33} - \left(\frac{23}{33}\right)^2} \qquad \sigma y = \sqrt{\frac{81}{33}\left(-\frac{-21}{33}\right)^2}$$

$$= \sqrt{3.969 - 0.485} \qquad = \sqrt{2.454 - 0.405}$$

$$= 1.87 \qquad\qquad = 1.43$$

by substitution we get $\dfrac{-64}{33} - (0.697)(-0.636) = -1.94 + 0.44 = -1.5$.

$$r = \frac{-1.5}{1.43 \times 1.87} = \frac{-1.5}{2.67} = -0.56$$

between 10s. and 20s. The deviations corresponding to this item are +4 and —3. These will be found in the second columns of the calculations beside and below the grid respectively. The product of —3 and 4, *i.e.* the deviations equal to —12, is inserted in the corner of the cell containing the four cases. In the adjacent cell, which shows that there are two cases where the expenditure on accommodation ranges from 90s. while expenditure on food is 20s. and over, the appropriate deviations are —2 and +4, so that —8, *i.e.* the product of these deviations, is inserted in the corner of the cell. This is done for each cell which contains a frequency. It will be noted, however, that all the cells opposite the classes containing the assumed means, 40– in the case of food and 50– for accommodation, have for obvious reasons a product equal to zero.

The next stage is then to multiply the cross-products of the deviations inserted in these cells by the frequency within that cell, due regard being paid to signs. This is done below the table, each product being set out individually, yielding in this case a net sum of —64. Very often the products of the cell frequency and the deviation product are inserted in the cell in the opposite corner to that containing the product of the deviations. This practice can be confusing for the student and entails a double lot of writing since the products have to be summed separately. The student should work these stages through by himself with the example, checking that he fully understands what has been done. The remainder of the calculation is then similar to that in our second example. A correction is required for the fact that the cross-products and standard deviations have both been measured from assumed means. As in the second example, the sum of the cross-products, *i.e.* —64, is averaged over the 33 pairs, and the product of the differences between the assumed and true means of x and y respectively deducted from the average cross-product. That value is then divided by the products of the two standard deviations.

The Significance of r

Correlation analysis has been applied to data from most scientific fields. It has been used to determine the relationship between crop yields and variations in the application of fertilisers; the level of fat-stock prices and its relation to the cost of feeding-stuffs. Engineers and chemists employ correlation to determine the extent to which properties of their products are affected by variations in the production processes. Its use has been extended to psychological tests designed to measure aptitude for particular types of work, *e.g.* accident proneness and temperament. It is largely for this reason, namely the widespread and frequent references to results derived from correlation analysis, that the topic is discussed in this elementary text.

In practice, the analysis is complicated by other than the purely statistical problems of technique.

In common with most statistical techniques, correlation analysis is usually employed on samples. Thus r, like other statistics derived from samples, must be examined to see how far the results may be generalised for the population from which the sample was drawn. Significance tests have also been evolved for the correlation co-efficient.[1] These lie beyond the scope of an elementary text, not because they are difficult to compute, but simply because, like correlation analysis itself, the technique has to be employed with great care and the interpretation of the data, as well as the results, demands a skill and knowledge of the field of enquiry only possessed by the expert. The value of correlation analysis is underlined by the variety of fields in which it finds application, but at all times it is essential to consider the data and ask 'what is the nature of the relationship measured by the coefficient?' Do not assume that the relationship is causal unless there is other evidence to support the assumption. Watch out for 'spurious' correlations; *e.g.* linking two related variables one of which is part of the other. For example, the number of households is obviously dependent on the size of the population and one would get a high value for r. But just what would it mean?

In the illustrative examples given above, the samples were extremely small, although in our final example, which contained 33 paired observations, it may be conceded in theory at least that this is a large sample. The most difficult problem is to interpret the value of the correlation coefficient. Thus the fact that in our final example the correlation coefficient was -0.5 might, since r falls considerably short of unity, lead the reader to assume that the correlation between the two types of outlay is negligible. Unfortunately it is not possible to interpret the coefficient correlation in this way. It is not possible, for example, to say that a value of r equal to 0.9 is very high and more significant than one equal to 0.8 since much will depend upon the size of the sample used. Nor should too much be read into the co-efficient. In the first example, which showed the relationship between the turnover and profit margin, it is apparent and reasonable to assume that these two variables are interrelated, and that the margin is presumably dependent upon turnover, *i.e.* the relationship is causal. The coefficient of correlation tells us nothing about the nature of the relationship; it merely indicates its existence. It is for the statistician to interpret it and deduce its nature and significance. It is in this respect that regression is so useful, since it defines in exact terms the relationship between the two variables. In the second example, the correlation coefficient of $+0.8$ suggests a significant

---

[1] Some of the simpler tests of significance are discussed in Chapter XIII.

relationship between poverty and overcrowding which is probably causal too, *i.e.* people are overcrowded because they are poor. At all times, however, one must beware of drawing dogmatic conclusions from limited data.

One final use of correlation analysis may be mentioned. Generally speaking, the square of the value of r may be regarded as the percentage of the variation in y directly attributable to changes in x. Thus, as far as the first illustration is concerned, approximately 81 per cent of the variation in y is explained by variations in x. This figure is known as the 'explained variance', while the balance of 19 per cent is termed the 'unexplained variance'. This means that as far as the available data are concerned, no precise explanation of the cause of 19 per cent of the variation is given. It may be attributable to any or many of several causes.

Rank Correlation

The methods of correlation that have so far been demonstrated have all been concerned with the measurement of the relationship between series of *numerical values*. It is possible, however, to measure the degree of correlation between two sets of observations or between paired values when only the *relative* order of magnitude is available for each series. For example, suppose a group of students sat two papers in an examination and instead of the actual marks awarded on each paper they were told only their ranking in order of merit. To establish whether the performances on the two papers were correlated or not, the method of *rank* correlation could be used.

The coefficient of rank correlation is given by Spearman's formula:

$$r_r = 1 - \left(\frac{6\Sigma d^2}{n(n^2 - 1)} \right)$$

where d is the numerical difference between corresponding pairs of ranks and n the number of pairs.

Student	Rank in French	Rank in Latin	d	d^2
A	1	3	2	4
B	2	2	0	0
C	3	1	2	4
D	4	6	2	4
E	5	5	0	0
F	6	8	2	4
G	7	4	3	9
H	8	10	2	4
I	9	7	2	4
J	10	9	1	1
				34

In the preceding example, ten students are ranked in order of merit on two examination papers in French and Latin.

Substituting the values derived from the above table in Spearman's formula

$$r_r = 1 - \left(\frac{6 \times 34}{10(10^2 - 1)}\right) = 1 - \left(\frac{204}{990}\right) = + 0.79$$

which suggests quite a strong relationship between performance in the two papers. Here again, however, the sample is so very small that it would be unwise to infer from this evidence that students who are good at French are also good at Latin. This may be true and common-sense supports the thesis, but the statistical evidence here is too slight to confirm it.

As well as in the type of problem just illustrated the coefficient of rank correlation may be calculated for series of *qualitative* instead of *quantitative* data, *e.g.* colour of hair and intensity of emotion measured on a non-numerical scale, or any attribute which cannot be measured numerically. Similarly we could calculate the coefficient of rank correlation for any group of paired observations even if they were numerical values, *e.g.* marks instead of placings in an examination, and the normal coefficient of correlation could be calculated by the product moment formula shown earlier. Significance tests for rank correlation do exist, but they are not relevant in this elementary text.

As with all the techniques described so far, correlation analysis has no value for its own sake. It is useful solely because, if properly used, it permits theories and hypotheses to be verified or rejected on the basis of empirical evidence. At all times it must be remembered that such specialised tools may easily be misapplied and give misleading results.

NOTE TO CHAPTER IX

Calculation of Regression Lines

The line of regression of y upon x is given by the equation $y = a + bx$ if the relationship between y and x is linear, *i.e.* when plotted it gives a straight line. We need first to determine the value of the two constants a and b. It is known that when x has its average value, the best estimate of the corresponding value of y is its own mean. Thus, we have two values for the above equation, $\bar{x} = 363$ and $\bar{y} = 72$.

The value for b, which gives the slope of the regression line, is derived from the equation $b = \dfrac{\Sigma\, xy}{\Sigma x^2}$ where x and y are the deviations of individual values of x and y from their respective means.[1] The numerator requires the deviations for each pair of x and y values to be multiplied together and their products added. This is done in col. 6 of the table opposite. The denominator in the above equation is the sum of all the squared deviations of x values from their mean as given in column 5. This is done in the same way as if we were proposing to calculate the standard deviation of the series of x values.

$$\text{Thus if } b = \frac{\Sigma\, xy}{\Sigma x^2} \quad \text{then} \quad b = \frac{53{,}780}{254{,}220} = 0{\cdot}2115$$

We now have three of the four values in the equation $y = a + bx$, i.e. $x = 363$, $y = 72$ and $b = 0{\cdot}2115$. The fourth value a is easily derived:

$72 = a + 0{\cdot}2115(363)$

$72 = a + 76{\cdot}8 \therefore a = -4{\cdot}8$

Therefore, the complete regression equation of y on x as drawn in Figure 17B reads: $y = -4{\cdot}8 + 0{\cdot}2115\,(x)$. The y' symbol indicates that this value of y is a calculated or predicted value, using this equation.

We can test the equation by substituting some of the known values of x, e.g. 200, and compare the predicted value of y, which is $37{\cdot}5$ when $x = 200$, compared with an observed value of 35. Thus the correspondence is quite close, as we would expect when the value of r is so high.

Just as we have calculated the regression line of y upon x, we can calculate the line of x upon y. In this case, the value of b is given by the equation $b = \dfrac{\Sigma\, xy}{\Sigma\, y^2}$. The calculation of Σy^2 is given in column 7 below and the student can repeat the calculations given above to obtain the regression equation of x upon y. The answer is $x = 36{\cdot}1 + 4{\cdot}53y$.

In the following example it was possible to work out the value of b from the true means of x and y. This is not always the case and just as when calculating the mean or the standard deviation we may have to work from assumed means, so it is possible to derive the regression equations by working from assumed means. In such a case, the formula for b for the line of y on x reads

$$\frac{\Sigma\, xy - \dfrac{\Sigma\,(x)\,(y)}{n}}{\Sigma\, x^2 - \dfrac{(\Sigma\, x^2)}{n}}$$

The corrections $\dfrac{\Sigma\,(x)\,(y)}{n}$ and $\dfrac{(\Sigma\, x^2)}{n}$ are really the same as we should

---

[1] Or in the alternative notation at the top of the columns on page 169 $\dfrac{\Sigma(x - \bar{x})\,(y - \bar{y})}{\Sigma(\,x - \bar{x})^2}$.

x	y	$(x - \bar{x})$	$(y - \bar{y})$	$(x - \bar{x})^2$	$(x - \bar{x}) \times (y - \bar{y})$	$(y - \bar{y})^2$
(1)	(2)	(3)	(4)	(5)	(6)	(7)
150	30	—213	— 42	45,369	8,946	1,764
200	35	—163	— 37	26,569	6,031	1,369
210	40	—153	— 32	23,409	4,896	1,024
230	45	—133	— 27	17,689	3,591	729
260	50	—103	— 22	10,609	2,266	484
280	50	— 83	— 22	6,889	1,826	484
300	60	— 63	— 12	3,969	756	144
320	65	— 43	— 7	1,849	301	49
350	70	— 13	— 2	169	26	4
370	70	+ 7	— 2	49	— 14	4
380	80	+ 17	+ 8	289	136	64
400	75	+ 37	+ 3	1,369	111	9
410	85	+ 47	+ 13	2,209	611	169
430	90	+ 67	+ 18	4,489	1,206	324
460	100	+ 97	+ 28	9,409	2,716	784
470	80	+107	+ 8	11,449	856	64
480	90	+117	+ 18	13,689	2,106	324
500	100	+137	+ 28	18,769	3,836	784
520	110	+157	+ 38	24,649	5,966	1,444
540	115	+177	+ 43	31,329	7,611	1,849
$\bar{x} = 363$	$\bar{y} = 72$	0	0	254,220	53,780	11,870

use when calculating the standard deviations of x and y. Space does not allow us to show all the details, but if the data given in the table were re-calculated taking as the assumed mean of x 350 and of y 75, then the total line at the bottom of the table would read as follows:

$\Sigma(x - \bar{x})$	$\Sigma(y - \bar{y})$	$\Sigma(x - \bar{x})^2$	$\Sigma(x - \bar{x})(y - \bar{y})$	$\Sigma(y - \bar{y})^2$
+ 260	— 60	257,600	53,000	12,050

Substituting in the above equation for b, we get:

$$b = \frac{53,000 - \dfrac{(260)\ (-60)}{20}}{257,600 - \dfrac{(260)^2}{20}} = 0 \cdot 2115$$

Since we are working from assumed means a further correction is required to derive the regression equation:

$$y - \bar{y} = b(x - \bar{x})$$
$$y - 72 = 0 \cdot 2115(x - 363)$$
$$y = 72 + 0 \cdot 2115x - 76 \cdot 8$$
$$y = -4 \cdot 8 + 0 \cdot 2115x$$

Questions

REGRESSION AND CORRELATION

1.

1959			Monthly Index of Inland Goods Transport	Monthly Index of Industrial Production
			monthly average 1958 = 100	
January	..	..	95	101
February	..	..	104	104
March	..	..	98	103
April	..	..	106	108
May	..	..	101	105
June	..	..	105	109
July	..	..	100	96
August	..	..	95	92
September	..	..	106	110
October	..	..	111	115
November	..	..	112	117
December	..	..	103	109

(*Economic Trends*, Feb. 1960 (page iv).)

Calculate the correlation coefficient for these two series of indices and briefly comment on the result.   *I.M.T.A. 1960*

2. Total weekly income and weekly expenditure on fuel and light of 11 old men living alone:

shillings

Income	..		..	49	56	48	67	42	51	46	52	50	57	65
Expenditure on fuel and light				15	11	11	9	8	5	9	9	3	4	11

Source: National Old People's Welfare Committee: Over Seventy (1954)

Compute the coefficient of correlation between the total weekly income and weekly expenditure on fuel and light. Discuss what conclusions you could draw from your result.   *B.Sc.* (*Sociology*) *1956*

3. Explain carefully what is meant by correlation. The following table gives the mean soil temperature (F°) at 3 in. and the number of days for the germination of winter wheat.

Soil temperature F°	..	38	39	40	41	42	43	44	45
No of days for germination	41	37	33	27	26	20	19	19	

Construct a scatter diagram and calculate and draw a line of best fit. State with reasons any conclusion you could reach.   *Institute of Statisticians 1960*

4. The following observations were made on 10 exceptionally heavy children.

Weight in lbs.	..	121	122	124	123	124	126	127	129	131	133
Height in ins.	..	59·0	54·5	61·5	54·5	60·0	64·0	61·0	54·5	60·0	61·0

Compute the coefficient of correlation between height and weight.

B.Sc. (*Sociology*) *1954*

5. What do you understand by 'regression'. Explain and illustrate graphically, how it may be illustrated by graphical methods. Discuss, with examples, the usefulness of the concept in the study of social phenomena. *Inter D.M.A. 1961*

6. Explain carefully the meaning of 'regression' and the difference between the two coefficients of regression that may be used to describe the same data. Describe what each is intended to measure, and point out the relations between them. Illustrate your answer from the following figures, where x = age at which full time education ceased, and y = annual income (£'s) at age 40:

x	15	15	16	18	20	21	22
y	500	520	720	950	1400	1800	1750

Inter D.M.A. 1962

7. The following are the gestation times and birth weights of 10 infants.

Gestation time (days)	Birth Weight (lbs)
240	6·5
250	5·5
255	7·0
260	9·0
270	9·0
275	8·0
285	7·0
285	10·0
290	9·5
310	10·5

Calculate the correlation coefficient between gestation time and birth weight, and comment on your result. *Final D.M.A. 1961*

8. Using the data in Question 7, compute the regression line of weight on gestation time. The normal gestation time (believed to be most favourable to the survival of the baby) is 280 days. What, according to your equation, is the mean birth weight corresponding to this gestation time? What purpose would be served by computing the other regression line, of gestation time on weight. *B.Sc. (Sociology) 1959*

9. The figures below show the infant mortality rates and social class indices (based on occupation) for ten British towns:

Infant mortality rate	Social class index
18	107
22	93
21	98
23	95
26	103
31	77
28	99
31	86
19	97
21	107

Source: General Register Office

Calculate the coefficient of correlation between these two sets of figures. Explain carefully how your result is to be interpreted. *Final D.M.A. 1963*

10. What are the main differences between the product moment correlation coefficient and a rank correlation coefficient for two characteristics each measured on a number of individuals? Calculate a coefficient of rank correlation between cinema admissions and TV licences per 1,000 of population for ten towns from the following data:

Town	1	2	3	4	5	6	7	8	9	10
Cinema admissions (1,000's) ..	12·8	10·9	13·7	12·4	8·8	13·3	10·1	11·6	10·6	9·4
TV licences per 1,000 population	18	97	39	76	83	42	106	21	84	55

Institute of Statisticians 1959

11. Indicate, and illustrate with a numerical example (though without carrying through the computations), the techniques of correlation and regression analysis. Explain how the results of such analyses can be used, and with what qualifications.

B.A. (Soc.) 1963

12. (*a*) Preferences of two groups of students for different examination subjects.

	Order of Preference among	
	Male students	Female students
Theories and methods of sociology ..	4	7
Statistical methods in social investigation..	6	8
Social institutions	7	4
Social philosophy	2	6
Social psychology	5	3
Social structure	1	2
Social history	8	5
Criminology	3	1
Principles of economics	9	10
Applied economics	10	9

By means of Spearman's rank correlation coefficient, examine these data for evidence of any similarity in the preferences of the male and female students. Comment on this method and discuss your findings.

(*b*) Explain briefly the computation and use of the ordinary (product moment) correlation coefficient. What are the main precautions to be taken in the interpretation of correlation coefficients generally? *B.A. (Soc.) 1960*

13. Below are given the entries in a table of values for a function *y* of *x*. Unfortunately, an ink blot obscures the entry for *y* corresponding to a value 1·10 for *x*. By a graphical method determine this missing value for *y*.

x	*y*
1·00	0·000
1·02	0·202
1·04	0·408
1·06	0·618
1·08	0·833
1·10	?
1·12	1·275
1·14	1·503
1·16	1·735

Institute of Statisticians 1958

CHAPTER X

THE CONSTRUCTION OF INDEX NUMBERS

Every reader of this book will know that in each year since the war, prices of the goods we buy have risen in greater or lesser degree. The price of a loaf of bread has risen from 4d. to 1s. 2d.; a hundredweight of coal from 2s. 6d. to 15s.; a ready-made suit from £3 to £12. These price changes give an indication of the extent to which prices have risen, but generally speaking it is more convenient to indicate the fall in the real value of money by a single measure. For example, it might be said that since 1939 the cost of consumer goods as a group has risen by 300 per cent., *i.e.* prices are now four times what they were then. This is a useful device because the degree of change in the prices of various goods differs; for example, bread as quoted above had risen by 250 per cent., coal and the ready-made suit by 500 and 300 per cent. respectively. It is much more convenient if all these changes can be expressed by a single figure. The figure used for comparing changes as between different points of time in what is sometimes referred to as 'the price level' is described as an *index* number of prices.

In view of what has been said it follows that the term 'price level' is a misnomer. In any period of time individual prices move differently, but the *average* movement can be calculated and it is this average which is measured by the index number. In fact, an index number is really no more than an average, with both its advantages and shortcomings. The most frequent use of index numbers is for measuring the change over time in *prices of* selected goods, commodities, assets such as securities and so on. For example, there is an Index of Retail Prices, an Index of Wholesale Prices, and one for prices of stocks and shares quoted on the Stock Exchange. Price indices are specially prepared to measure changes in particular groups of prices. There is no index suitable for measuring the 'general level of prices', but there is an index for measuring the cost of goods and services which the average family in the United Kingdom buys as part of its mode of life. This is often termed a 'cost of living' index. Index numbers are specially designed for a particular purpose. For example, changes in the volume of industrial output are measured each month by the index of industrial production; changes in the prices of exports and imports from and into the United Kingdom by indices of import and export prices. All these various index numbers are described in later chapters.

Measuring Price Changes

All prices do not change to the same extent over a given period of time. Some prices rise (or fall) more than others, *i.e.* they move relatively to one another. The difficulty arises when these relative changes have to be 'averaged'. It is possible to get different answers to what appears to be the same question according to the method used for measuring price changes. Suppose we have four commodities which in 1954 and 1964 cost per pound weight as follows: A 5s.–7s 6d.; B 10s.–12s. 6d.; C 15s.–£1; D £1–£3. In the earlier year the four items could have been purchased for £2 10s. and by 1964 they cost £5. The proportionate increase in the total cost of these goods is given by expressing the actual change in cost over the total cost in the earlier year, *i.e.* £2 10s. divided by £2 10s. which is equal to unity. It is customary to express such changes in percentage terms, so that if the 1954 expenditure is termed the 'base', or 100, then the corresponding value for 1964 is 200, or a 100 per cent. increase. Such an index, which is no more than the percentage change in the *aggregate* expenditure on a collection of goods at different points of time, is sometimes referred to as a *simple* aggregative type of index. As will be seen below, it does not really justify the title of an index.

Now instead of taking the actual price of each commodity, calculate the percentage increase in its cost. A is then 50, B 25, C $33\frac{1}{3}$, and D 200 per cent higher. If these percentages are added together they total $308\frac{1}{3}$, which apportioned over the four items represents an increase of approximately 77 per cent. in the prices. This, it will be noted, differs from the figure of 100 per cent. derived by using simple aggregate expenditures. Generally speaking, the simple aggregative type of index derived by relating two aggregate expenditures is not used in practice because the index may be distorted by any single *large* absolute change which swamps all the other movements. This, it will be recalled, is much the same as saying that the arithmetic mean is unduly affected by extreme values and may sometimes be unsatisfactory as a measure of a given distribution.

It is nevertheless quite possible to use the actual prices of the goods purchased to calculate a price index, without converting them into percentages. To make such an *aggregative* index, as it is termed, the various prices have first to be multiplied by the physical quantities purchased of each commodity. The products of the prices and quantities are summed and the two totals expressed as ratios of one another.

TABLE 28

CALCULATION OF AGGREGATIVE TYPE INDEX

Commodity	Prices		Quantity	Product of Quantity × Price	
	1954	1964	lbs.	1954	1964
	s.	s. d.		s.	s.
A	5	7 6	4	20	30
B	10	12 6	6	60	75
C	15	20 0	3	45	60
D	20	60 0	1	20	60
	50s.	100s. 0d.	14	145s.	225s.

The calculations can be followed in the above table. In this aggregative index, the average increase in the prices of the commodities is derived by dividing the product of the 1954 quantities and prices into the corresponding product for 1964, *i.e.* $\frac{225}{145}$. This gives an increase of 55 per cent. Alternatively the product 145 may be termed base 100, then the product 225 is equal to an index of 155 or 55 points higher.

Each of the three 1964 figures calculated so far shows a different increase over 1954. Since the arithmetic is correct, we can only ask which is the 'correct' figure. The simple so-called aggregative index giving an increase of 100 per. cent is quite unsatisfactory. If the prices for the various constituent items vary widely in absolute terms the 'index' will always be distorted by the extreme values. In practice, this method is discarded. The second index required the calculation of the percentage change for each price; by this means the distorting effect of the absolute differences in the prices between the two dates is eliminated. The arithmetic average of the percentage increases might be used for a very simple rough and ready index, but it assumes that each price change is just as important as any other and this, generally speaking, is not true. For example, in a cost of living index a 10 per cent. increase in the price of bread would be much more serious for the family budget than a similar increase in the price of biscuits, even if in absolute money terms the latter increase was greater. But if the percentage changes are weighted in accordance with the relative expenditure on each commodity to the total expenditure, then we get the same results as if we had multiplied the prices at the two dates by the actual quantities purchased· In other words, the *aggregative* index in Table 28 which gave an increase of 55 points, *i.e.* 1954 = 100 and 1964 = 155, may be directly compared with an index based on the products of the percentage change in each price multiplied by its appropriate share in the total outlay. In this case we are using 'value' instead of 'quantity' as the weight and the 'value' is the amount spent on each commodity in 1954. Table 29 shows the total outlay on each commodity in 1954 and these figures appear in the column headed 'value' weights of Table 29 below.

TABLE 29

METHODS OF CALCULATING INDEX OF WEIGHTED RELATIVES

Com-modity	Prices		'Value' weights	Method A		Method B			
				Per-centage change in prices	Products of % change & weights	Price Relatives		Relatives × Weights	
	1954	1964				1954	1964	1954	1964
	s.	s. d.							
A	5	7 6	20	+ 50	1000	100	150	2000	3000
B	10	12 6	60	+ 25	1500	100	125	6000	7500
C	15	20 0	45	+ 33⅓	1500	100	133⅓	4500	6000
D	20	60 0	20	+ 200	4000	100	300	2000	6000
	50	100 0	145		8000	400	708⅓	14500	22500

Method A. Weighted percentage change in prices between 1954 and 1964 = $\frac{8000}{145}$ = 55 (to nearest unit).

If 1954 prices = 100, 1964 prices = 155.

Method B. Weighted average of 1954 price relatives = 14500

,,   ,, ,, 1964   ,,   ,,   = 22500.

But if 1954 prices = 100, then 1964 prices = $\frac{225}{145} \times 100 = 55$ (to nearest unit).

It will be seen that in columns headed 'A', by multiplying the percentage change in each price by the 'value' weight, their product of 8,000 divided by the sum of the value weights 145, gives an increase as between the two sets of prices in the period 1954-64 of 55 per cent. The more conventional method is to convert the prices at each date into 'relatives'. Thus if the base year price, i.e. 5s. in 1954 is termed 100, then the corresponding figure for 7s 6d. in 1964 is 150 or 50 per cent. higher. This figure is referred to as a *price relative*. In the columns headed 'B', each price relative is multiplied by its corresponding weight, and the total products for 1954 expressed as a ratio of 1964. As is expected, the same result is achieved.

Since we have seen that the same result can be derived by either of two methods:

(a) an aggregative index using actual prices and quantities purchased as shown in Table 28; or

(b) an index of weighted relatives using price relatives and 'value' weights as in Table 29;

the question poses itself, 'which is the better method?' Clearly, the answer is that they are equally efficient in so far as they produce the same result. From the purely practical point of view the aggregative index seems simpler since it avoids the need for converting prices into relatives. On the other hand, the data available for the construction of the index may be in the form of relatives. This is not so often the case with price indices as with indices such as that measuring industrial production. In this case changes in the output of individual industries may be based on physical output, numbers employed, or on total sales. To combine the changes for a number of industries using different bases of measurement therefore necessitates the use of relatives.

Index Number Notation

The arithmetic processes described so far are usually expressed in simple symbols. Since they are extensively and commonly employed to indicate the type of index used, it is important that the student reader should understand them. If p represents a price, the base year of 1954 is written $_0$ and the year 1964 as $_1$, then p_0 represents a base year price and p_1 the price of the same article in the other year. The symbol Σ, it will be recalled, is known as large sigma and indicates summation. Thus instead of writing $\dfrac{p_1^1 + p_1^2 + p_1^3 + p_1^4 + \ldots \ldots p_1^n}{p_0^1 + p_0^2 + p_0^3 + p_0^4 + \ldots \ldots p_0^n}$ where the numbers $1, 2, 3, 4 \ldots n$ represent different goods, we can simply abbreviate to $\dfrac{\Sigma p_1}{\Sigma p_0}$. This is the formula for the so-called simple aggregative 'index'. But this type is only useful if the prices are weighted by quantities. In our example the quantities were those purchased in the base year, so that if q_0 represents the base year weight, then the index I is merely the average of a series of such weighted prices, i.e., $I = \dfrac{(p_1 q_0) + (p_1 q_0) + (p_1 q_0) \ldots}{(p_0 q_0) + (p_0 q_0) + (p_0 q_0) \ldots}$ or $\dfrac{\Sigma p_1 q_0}{\Sigma p_0 q_0}$.

The same symbols are used to indicate an index based upon the price relatives. The price relative for any article or goods is derived by relating the base year price to the other year, i.e., $\dfrac{p_1}{p_0}$. Thus when there are a large number of relatives, we get $\dfrac{p_1^1}{p_0^1} + \dfrac{p_1^2}{p_0^2} + \dfrac{p_1^3}{p_0^3} + \dfrac{p_1^4}{p_0^4} \ldots \ldots \dfrac{p_1^n}{p_0^n}$ which can be abbreviated to $\dfrac{1}{N}\Sigma \dfrac{p_1}{p_0}$

i.e. the sum of the relatives divided by their number. When the price relatives are weighted by reference to quantities, then we write each relative $q_0 \frac{p_1}{p_0}$. Their aggregate is represented by $\frac{\Sigma q_0 \frac{p_1}{p_0}}{\Sigma q_0}$. Using 'value' weights given (see Table 29) by the product of the base price (p_0) and the quantity purchased in the base period (q_0) a base year weighted price relative is written $\frac{p_1}{p_0}(p_0 q_0)$. Since the index comprises a number of such weighted relatives, then $\frac{\Sigma\left(p_0 q_0 \dfrac{p_1}{p_0}\right)}{\Sigma p_0 q_0} = \frac{\Sigma p_1 q_0}{\Sigma p_0 q_0}$

Arithmetic or Geometric Mean?

It was stated earlier that an index is merely a form of 'average' and in the foregoing simple illustrations the arithmetic mean has been used. It is equally possible to use the geometric mean. This statistic is the nth root of the product of n values. For example, if three numbers are multiplied together, their geometric mean is the cube root of their product; for ten numbers it is the 10th root of their product.[1] We can apply this formula to the data in the above illustration. For purposes of this calculation logarithms are employed, but merely because they greatly simplify the calculation. Note too that the difference between the logarithms for any two prices of a single commodity gives the anti-logarithm of the relative change. In other words the use of logarithms saves the trouble of working out the price relatives for every commodity in the index.

Com-mod-ity	Prices		Logarithms		Value weights	Logarithms × Weights	
	1954	1964	1954	1964		1954	1964
(1)	(2)	(3)	(4)	(5)	(6)	(7)	(8)
A	5	7·5	0·6990	0·8751	2	1·3980	1·7502
B	10	12·5	1·0000	1·0969	6	6·0000	6·5814
C	15	20	1·1761	1·3010	4·5	5·2925	5·8545
D	20	60	1·3010	1·7782	2	2·6020	3·5564
					14·5	15·2925	17·7425
							15·2925
						14·5)	2·450
							0·1690

anti-log. 0·1690 = 1·48

Since the difference between the logarithms of each price at the two dates gives the relative increase. the appropriate weights are the *value*

[1] This is explained more fully in a note at the end of this chapter.

based weights, not the *quantity* weights. To simplify the arithmetic the value weights in column 6 of the illustration have simply been divided by ten. The reader will appreciate that it is not the absolute size of the weights which is important, but their ratio one to the other; 6:2 is the same as 60:20. As may be seen in the example the logarithm of each price is multiplied by the corresponding weight. The difference in the sums of the logarithms of these products for each year is then divided by the sum of the *weights*, not the number of prices and this anti-logarithm gives the relationship between 1964 and the base year 1954. In this case it is 1·48 so that if we call 1954 the base year = 100, then for 1964 the index is 148.

It will be noted that this index is smaller than the index of price relatives based upon the arithmetic mean; this is a characteristic of the geometric mean since it is less affected by the larger values than the arithmetic mean. In theory it would be possible to use any 'average', *e.g.* either the median or mode of a distribution of price changes, as an index. In practice, however, the choice lies between the geometric and arithmetic means of the weighted prices or their relatives. Neither has any marked advantage over the other. Generally speaking, an index based on price relatives often uses the geometric mean while aggregative indices often employ the arithmetic mean, but this is by no means an invariable procedure.

An exposition of the relative merits of these two types of mean lies beyond the scope of this book, for quite apart from the fact that it introduces the highly complex subject of index number theory, it is also subsidiary to the more fundamental problems of index number construction which are now discussed. For the student with some algebra, interested in this aspect of economic statistics, there are several suitable texts for study.[1]

The basic problems of index number construction are three in number. The first concerns the nature of the index, *i.e.* the functions it is to perform and the choice of suitable prices or values to be included in it. The second problem is to determine the best weighting system, while the third involves the selection of the base period. None of these particular points is simple to solve in practice and the final index is usually the product of compromise between theoretical standards and the standard attainable with the given data. To simplify the text, references will be to price indices and the base period will usually be assumed to be a year. All the points made are equally relevant to indices measuring other than price changes and where the base period is shorter or longer than a year.

[1] See appropriate chapters in *Applied General Statistics* by Croxton and Cowden, or *Applied Statistics for Economists* by Karmel, P. H., or *Economic Arithmetic* by Marris, R.

The Nature of the Index

It is helpful to remember when dealing with index numbers that they are specialised tools and as such are most efficient and useful when properly used. A screwdriver is a poor substitute for a chisel, although it may be used as such. All index numbers are designed to measure particular groups of related changes. For example, the Index of Retail Prices in the United Kingdom measures the monthly change in the cost of a collection of goods and services bought by the 'average' household. Note that an index does not cover all such changes, merely a selection. Thus, if one household does not buy fruit at all, when the index goes up because fruit prices rise, that family's real income cannot be determined by reference to that index. The prices of the articles included in the index are those charged at certain types of shop, not every shop. In some cases the allowance for house rent in the index is far below what some families in 'middle class' circumstances pay; just as is the proportion of expenditure on alcohol and tobacco usually well above that applicable to families with young children. In other words, the index is an 'average' of certain household expenditure which determines the standard of life of the average household in the United Kingdom; it may not apply to any single household. Not all households are included in the index; those where the head of the household earns over £30 a week and homes where the old age pension and other state welfare benefits account for more than three-quarters of the total income are excluded. Despite all these limitations which are discussed in Chapter XVI the public continues to refer to the index as a 'cost of living' index which is indiscriminately applied to all wage and salary earners, whatever their domestic circumstances.

The Board of Trade's Index of Wholesale Prices illustrates the change in attitude to index numbers in recent years. The old pre-war index was designed to measure changes in the 'general level of prices'; it is now recognised that the 'general level' is an over-simplification – it is itself an average of a large number of changes in differing directions and of varying degree. The Board of Trade therefore now prepares a whole series of indices, each especially constructed to measure small groups of related prices, *e.g.* the cost of building materials (see Chapter XVIII).

Once the decision has been taken as to the purpose of the index the choice of 'representative prices' to be included in the index must be made. The more prices that are included, the longer it takes to calculate the index each month. Furthermore, not all prices are readily available and items may therefore be omitted on this account. Nor is it true to argue that an index with a large number of items is automatically a better index than one based on a few prices. For example, the Board of Trade's old wholesale price index consisted of 258 quotations for 200 commodities; the sensitive index of commodity prices prepared by *The Economist* uses only 21 prices.[1] Each served its declared purpose quite well. In the index of Retail Prices only those articles which are regularly and extensively purchased are included. Even so this involves a very large number of price quotations taken from all regions of the United Kingdom, from all types of town and district, as well as from all types of shop![2] How much better for its purpose the index is in consequence of this great amount of detailed work, than if it were to be based on a handful of articles and services from a few shops might well be debated at length! Unfortunately, the government cannot use what might appear to be a 'makeshift' index for this purpose; but it does not follow that the latter would not be so much less satisfactory in the longer period than the present one!

The Choice of Weights

As we have already seen, an index based on a simple average of price relatives or aggregate prices is virtually useless. The individual prices or their relatives must be weighted. What determines the weights? They can be of two kinds. The first are the so-called *quantity* weights used in the aggregative type of index (Table 30) and the others are *value* weights derived by reference to the actual outlay on the particular item which are applied when using the price relatives (Table 31). The actual size of the weights is unimportant; what is important is the *relative* weights. For example, in the current index of Retail Prices the food prices are given a weight of 314 against a weight of 66 for fuel and light. This indicates that food absorbs roughly five times as much of the average household's weekly expenditure as does fuel and light. Therefore a change in the price of the former is five times as significant as the same *price* change in the latter. To show the effect on the household's income by means of an index, the food item must be weighted five times as heavily as that for fuel and light, since the expenditure on the former is five times as great and takes proportionately more of his weekly income. Whether the weights are 5:1, 310:60 or any other numbers is irrelevant provided they give the relationship between the items. For example, up to 1947 the old cost of living index weights totalled 100; in the new index they add up to 1,000. That is *not* significant. But the fact that in the old index food accounted for 60 per cent. of the weighting and in the new current index only 31 per cent., *is* important.

The next aspect of the weighting problem is to decide between what is termed *base* and *current* year weighting. For example, in the

[1] See *Economist* of January 7, 1961 for description of the current index, 1958 = 100.
[2] See Chapter XVI for details.

index of industrial production the weights have been determined by reference to Census of Production data relating to 1958. These weights are applied to output relatives which indicate the changes in monthly output for a wide range of industrial products. It would be possible, but hardly practicable, to weight the relatives by reference to current census of production data derived each year from either a census or sample survey. If current year weighting is used, *i.e.* 1964, then all the indices for the earlier years must be revised so that they remain comparable one with another and with the latest year. The nature of the weighting is indicated in the formula. For a price index using *base* year weighting we can write $\dfrac{\Sigma p_1 q_0}{\Sigma p_0 q_0}$; the same formula for a *current* year weighted index is written $\dfrac{\Sigma p_1 q_1}{\Sigma p_0 q_1}$. The reader will observe that the difference is indicated by the subscript q_1, instead of q_0.

The two different weighting systems will give different results, although the actual differences in most cases will not be great. The formula for the base year weighted index is often termed a *Laspeyre* index, and the current year weighted index a *Paasche* index. These two names have been used for about half a century and are now used to convey to the reader the nature of the index being used. Once again, however, the reader needs to be reminded that the difference between these two types of index is more apparent than real, at least for most practical purposes. In practice, the base year weighting has the great advantage that the weights are constant throughout the life of the index. Furthermore, the work in calculating the index is much less than with current weights. The major disadvantage of base year weighting is that it may become out of date, in which case the efficiency of the index declines. It is for this reason that Laspeyre type indices are revised at fairly regular intervals, although this means that comparability of the index over the longer period is virtually impossible.

The difficulty over one current year weighting arises because the relative importance of the items comprising the index is continuously changing. If there were no change in the amounts of different commodities bought from year to year, then of course the one original set of weights would serve as both current and base year! Naturally, when prices of the constituent articles in the index change, so does demand and, strictly speaking, the weight needs to be changed. Because people tend to spend less on goods when their prices are rising, the use of the Paasche or current weighting produces an index which tends to understate the rise in prices, just as the Laspeyre index overstates it since the base year weights reflect an outmoded purchasing pattern.

But as already stated, the difference between the two formulæ is of interest primarily to the theorist and much has been written on the problem of designing the 'ideal' index. For example, the suggestion has been made that the geometric mean of the Laspeyre and Paasche indices should be used as an index.

In practice, however, the base year weighted Laspeyre type index remains the most popular for reasons of its practicability. The Paasche type index can only be constructed when up to date data for the weights are available. This is exceptional; only one Paasche type index was prepared officially in the United Kingdom – the *average value* index of imports and exports – but even this is no longer published. This was practicable because the monthly overseas trade returns are available with a delay of only a month or two so the weighting could be continually revised.

Selecting the base year

At first sight this should be the easiest part of constructing an index number, not least since it would appear to be logical to take as the base year the first year for which the index is constructed. With a new index this is valid, but with every Laspeyre type index the time comes when the base year must be revised. If the revisions are fairly regular, as for example is the case with the price and volume indices of imports and exports of the United Kingdom, then to some extent the base tends to choose itself. But if the intention is to try and keep a single series of indices for a longer period than a few years, the base period may be difficult to select. This is because the ideal base would be the so-called 'normal' year, but with economic data who is to decide what is normal? At best depression or boom years – in respect of the particular phenomenon measured by the index – can be avoided. Much of the wrangle over the relative success of the two political parties' post-war economic policies could be explained by basing their data on different base years. If a depression year is taken as the base, most later years show an improvement; if a boom year is taken the rate of expansion in the period following appears very slow.

The tendency nowadays is to keep the Laspeyre type of index up to date by regular revision. For example; the index of industrial production was first produced in 1948 with 1946 as a base. In 1952, with the results of the 1948 full census of production available, the index was revised and in 1958 rebased on to 1954 and in 1962 was revised again to 1958 in the light of the census of production data for that year. The index of retail prices has also been revised three times since the end of the war, as have the import and export trade indices. The reason for the periodic revisions of many indices of economic data is probably the desire to use a reliable index for relatively short period analysis of economic trends. After all, it is only the historian who wants to look back over the longer period; the economist is usually concerned with the immediate short run future. For much historical economic analysis it is usually practicable to link, if only approximately, the old and new index. This can be done quite effectively by calculating for the last two or three years of the old series of index numbers, an index on the new basis. This has been done in the case of the 1962 revision of the index of industrial production, annual indices having been worked back for a few years to overlap with the old series.

The only index of prices which covers any long period of years is the Sauerbeck index of wholesale prices which is still compiled by *The Statist* in its original form. This dates back over a hundred years, although its base period is 1867-77. This is a rather simple type of index. It covers only basic commodities which do not change in character over the years. But to calculate, as could be done from the records, an index of (say) exports from the United Kingdom for the past century would be pointless. Many goods exported today were unknown in the 19th century, just as many exports of the last century have ceased to be important today. In other words, what purpose would be served by preparing such a long-period index?

One method of overcoming weaknesses in an index from an outdated or frequently changing base is to use a *chain-base* index. In this case, the index for each period is based on the index of the comparable period immediately preceding it. A particular advantage of this type of index is that it is easy to introduce new items. With a fixed base index of the Laspeyre type, to alter the composition of the index would necessitate the re-calculation of the index for all previous years. Against the chain-base type of index is the point that it is really only suitable for the short period. If changes in the component items are frequent, the index may in the later years reflect quite different price movements than the figures in the earlier period.

Conclusions

It will be apparent that an index number, whether it be of prices, of physical quantities or any other measure, is an arbitrary and imperfect measure. At best it will perform the task for which it is designed if every care has been taken to include the relevant constituent items and to weight them correctly. Important though weighting may be, it is still more important to ensure that relevant values are included in the index and that the quotation, *i.e.* price, from month to month, is comparable, than to devote undue attention to calculating precise weights. Provided the latter are approximately correct, the index should reflect the pattern and trend of change in the data.

Further illustrations

Most index numbers involve many values and much calculation. The highly simplified illustrations on pages 181 and 182 illustrate the basic principles involved, but the actual construction of an index involves a great deal of mechanical work which can hardly be reproduced here. However, the two illustrations which follow may enable the reader to see how two important index numbers are calculated after the basic data have been processed. The first is the famous Sauerbeck index prepared by *The Statist* and the other is the Index of Retail Prices.

The Statist Index of Wholesale Prices comprises 45 commodities divided into two main groups, Food and Materials, each of which in turn consists of three sub-groups of prices. The actual number of prices used to construct each index is given in the column headed 'Number of Commodities in Index'. There is no weighting by reference to values or quantities but the 'General Average' index is weighted in so far as there are more quotations for certain commodities than for others. The index for any single commodity is very

TABLE 30

CONSTRUCTION OF 'THE STATIST' INDEX NUMBER FOR 1962*

1866–77 = 100

Commodities	Number of Commodities in Index	Total Numbers		Index for 1962
		1867-77	1962	
General average	45	4,500	16,531	367
Food	19	1,900	5,724	301
Vegetable food	8	800	2,487	311
Animal food	7	700	2,391	343
Sugar, coffee and tea ..	4	400	846	213
Materials	26	2,600	10,807	416
Minerals	7	700	3,790	541
Textiles	8	800	3,228	404
Sundry materials ..	11	1,100	3,789	344

* *Source: J.R.S.S.*, 1963, Part IV, p. 566.

simple to obtain. The average price for the month is expressed as a price relative of the average price of the commodity in the eleven year base period. Thus, in the index for vegetable food, the price relative for each of the eight commodities is calculated and their average gives the index for the group. This is repeated for each group in the index. The 'general average' is the simple arithmetic mean of all the 45 price relatives. The monthly indices are based upon the end-of-month quotations, but the annual indices are derived from the average of the 52 weekly quotations, so that the annual index for any year does not necessarily coincide with the average of the 12 months' indices for that year.

The figures under the main column headed 'Total Numbers' are quite clear. For 1867-77 it is apparent that the figure shown is merely the product of the number of commodities each expressed as base 100, *e.g.* the 'general average' index comprises 45 commodities each of which in the base period was expressed as 100. The relatives for the appropriate commodities in each group having been worked out as described above, they are summed and inserted in the second column under that same heading. The index for 1962 is derived by dividing the number of commodities into the 'total numbers' for those commodities, *e.g.* Food = 5,724 ÷ 19 = 301. This method yields the general average index as well; it is not merely the average of the various group indices.

A full account of the method of the construction of this index, together with annual indices for each group back to the beginning of the century, as well as average prices and monthly indices for recent years, is published annually in the *Journal of the Royal Statistical Society.*

TABLE 31

CALCULATION OF INDEX OF RETAIL PRICES FOR 18TH FEBRUARY, 1964*

Commodity Groups				Weights	Group Index	Weights × Indices
Food	..	..	..	314	105·4	330 956
Alcoholic Drinks	..			63	103·5	65 205
Tobacco	..	..	..	74	100·0	74 000
Housing	..	..	..	107	111·1	118 877
Fuel and Light	..	..		66	110·2	72 732
Durable Household Goods	..		..	62	101·3	62 806
Clothing and Footwear	..		..	95	104·2	98 990
Transport and Vehicles	..		..	100	100·7	100 700
Miscellaneous	..	..	..	63	103·2	65 016
Services	..	..	..	56	105·2	58 912
All items	..	..	..	1,000	104·8	1048 194

* *Source: Ministry of Labour Gazette*, March, 1964.

The Index of Retail Prices is fundamentally similar in construction to *The Statist* index described briefly above, although it covers a very much wider range of goods and services. The index is compiled monthly, the prices used being those prevailing on the Tuesday nearest the middle of the month. For each article and service included in the index the current price is expressed as a relative of the price at the base date, 16th January, 1962. As will be seen from Table 31 above, there are ten group indices the weighted average of which is the 'all items' index which is customarily used as a measure of changes in the cost of living. Each group index is separately calculated by deriving the percentage change in the price of each article included in that group, *e.g.* food, and then weighting the change by reference to the share of the total outlay on that group absorbed by the article. The weighted relatives are then averaged (arithmetic mean) to yield the group index. Each group index in its turn is then weighted in accordance with its relative importance in the entire household budget. The current weights are shown in the second column of Table 31. The products of the weights and the group indices are aggregated and then averaged to give the 'all items' index.

The index is published monthly in the *Ministry of Labour Gazette* as well as in the *Monthly Digest of Statistics* in the form of first to the third columns of Table 31, *i.e.* the ten group indices and 'all items' index all expressed to one place of decimals. A detailed account of the construction of this index is published by H.M. Stationery Office.[1] and subsequently revised.[2]

NOTE TO CHAPTER X

Calculating the Geometric Mean

The arithmetic mean is calculated by *aggregating* the values in a distribution and dividing them by their number or frequencies. The geometric mean is derived by *multiplying* together all the values and then extracting the relevant root of the product of those values. Thus, for the following series, 4, 6, 9, the geometric mean is: $\sqrt[3]{4 \times 6 \times 9}$ $= \sqrt[3]{216} = 6$. The root to be calculated depends on the number of values in the series; thus, with three values, it is the cube root. The principle can be summarised as follows: 'the G.M. is the nth root of the product of n items'.

It will be apparent that if n is a large number, even a dozen values in the series, the problem of computing the twelfth root of the product of the twelve values by simple arithmetic is likely to be a tedious operation, if not impossible. Thus for calculting the G.M., logarithms are used.

By obtaining the logarithm of each value, and aggregating the logarithms, the same purpose is being served as by multiplying the original numbers together. Having aggregated the logarithms, their sum is divided by the number of items, *i.e. n*. This in turn has the

[1] Index of Retail Prices. Method of Construction and Calculation. H.M.S.O. 1959.
[2] Report on Revision of the Index of Retail Prices, Cmnd. 1657, H.M.S.O. 1962.

same effect as calculating the nth root of the product of the original values. The quotient of the sum of the logarithms divided by n is then looked up in the tables of anti-logarithms, which yield the equivalent in ordinary values, the G.M. of the original series. In other words, find the arithmetic mean of the logarithms and convert the answer back into natural numbers.

The following examples illustrate the procedure. Example 1 illustrates the calculation of the G.M. of a simple series of unweighted values; the second shows the calculation for an ungrouped weighted series, as might arise in the calculation of a simple index number.

(1) Calculate the G.M. of the following series: 20, 58, 87, 130, 170, 250;

i.e., $\sqrt[6]{20 \times 58 \times 87 \times 130 \times 170 \times 250}$.

Values	Logs
20	1·3010
58	1·7634
87	1·9395
130	2·1139
170	2·2304
250	2·3979
	11·7461

G.M. = Anti-log of $\dfrac{11·7461}{6}$ = anti-log of 1·95768.

which converted into original units = 90·7, or 91 to nearest unit.

(2) Calculate the G.M. of the following weighted frequency distribution:

Relatives	110	125	92	100	160	84
Weight ..	4	1	3	10	5	8

$= \sqrt[31]{110^4 \times 125^1 \times 92^3 \times 100^{10} \times 160^5 \times 84^8}$

Relatives	Frequency or Weight	Logs of Indices	Weight × Logs
110	4	2·0414	8·1656
125	1	2·0969	2·0969
92	3	1·9638	5·8914
100	10	2·0000	20·0000
160	5	2·2041	11·0205
84	8	1·9243	15·3944
	31		62·5688

G.M. = Anti-log of $\dfrac{62 \cdot 5688}{31}$ = anti-log of $2 \cdot 0183$.

G.M. = $104 \cdot 3$, or 104 to nearest unit.

As with the Arithmetic Mean, so with the G.M., the formula is often given in algebraic notation. Thus with the G.M. described as the nth root of the product of n values the usual form is:

$$g = \sqrt[n]{x_1 \times x_2 \times x_3 \dots x_N}$$

If weights are to be introduced as in the second example above, then

$$g = \sqrt[w]{x_1^{w1} \times x_2^{w2} \times x_2^{w3} \dots x_n^{wn}}$$

where $w =$ the total of weights used and $w^{1 \dots n}$ are individual weights.

Since the computation involves logarithms, the first form may be written:

$$\log g = \frac{\log x_1 + \log x_2 + \log x_3 \dots + \log_n}{n}$$

$$= \frac{\Sigma \log x}{n} \text{ or } \frac{1}{n} \Sigma (\log x)$$

The weighted series is then written:

$$\log g = \frac{w_1 \log x_1 + w_2 \log x_2 + w_3 \log x_3 + \dots w_n \log x_n}{\Sigma w}$$

$$= \frac{\Sigma w \log x}{\Sigma w}$$

where Σw represents the total weights.

The reader may care to transpose figures from the examples in the preceding pages to test the above formula.

Questions

THE CONSTRUCTION OF INDEX NUMBERS

1. Explain the following terms:

(i) Price relative;
(ii) Base period;
(iii) Quantity weighting.

Give examples of two official index numbers with which you are familiar, briefly explaining uses to which they may be put.

2. What is the purpose of 'weighting' in the construction of index numbers? Describe the basis and use of the weights in the British index of retail prices.

Final D.M.A. 1963

3. The data given in the table below are the component items of the Interim Index of Retail Prices for 15th January, 1952.

Calculate an index for:

(a) Items (1) to (4) combined; and

(b) All items combined.

Item	Price Index	Weight
(1) Food	149·7	348
(2) Rent and Rates ..	104·2	88
(3) Clothing	147·1	97
(4) Fuel and Light ..	140·1	65
(5) Household durable goods	136·6	71
(6) Miscellaneous goods ..	137·3	35
(7) Services	123·9	79
(8) Drink and Tobacco ..	108·5	217

Institute of Statisticians 1957

4. What are the main problems to be solved in the construction of an index number? Illustrate your answer by reference to either the Index of Industrial Production or the official series of index numbers of wholesale prices.

Intermediate D.M.A. 1961

5. The following figures relate to all the exports of a certain country over a period of 10 years:

	Value of exports in $ based on		
	(1)	(2)	(3)
Year	Actual prices and quantities exported each year	Actual quantities exported but prices in year 1	Actual prices but quantities exported in year 1
1	100	100	100
2	107	98	110
3	112	97	119
4	115	94	132
5	116	91	140
6	115	90	150
7	112	89	162
8	107	86	177
9	100	83	171
10	91	82	191

(a) Column 2 provides an index for quantities exported each year, using year 1 as base. Calculate an alternative index for quantities, using year 1 as base..

(b) Compare your own index with that given in the table above, explaining why any differences may have arisen. State, with reasons, which type of index number you prefer here. *Institute of Statisticians 1963*

6. The following table shows consumers' expenditure on certain categories of goods in 1954 and 1960 in £ millions at current market prices, and also the expenditure in 1960 revalued at 1954 prices. Using 1954 as base, calculate a price index and a quantity index for 1960 consumers' expenditure.

	Expenditure at market prices (£ millions)		1960 expenditure revalued at 1954 prices (£ millions)
	1954	1960	
Food	3,738	4,860	4,248
Drink and tobacco ..	1,673	2,141	1,990
Housing, fuel and light ..	1,536	2,252	1,669
Household and durable goods	1,239	1,906	1,756
Clothing	1,205	1,632	1,537
Other items	2,559	3,540	2,933
	11,950	16,331	14,133

Intermediate D.M.A. 1962

7.   *Prices of Home-grown Grain*

(Shillings per qr.)

	1949	1950	1951	1952	1953
Wheat ..	100	110	123	129	138
Barley ..	92	100	124	116	102
Oats ..	58	60	73	74	68

Compute index numbers of prices of home-grown grain taking 1949 as the base year, using an arithmetic average and weighting the grains as follows:

Wheat ..	..	6
Barley ..	..	3
Oats ..	..	1

Institute of Statisticians 1955

8. From the following figures of price and expenditure calculate a 'cost-of-living' index showing the change in the average level of prices between 1956 and 1961. State what type of index you have used. What other index numbers are there and what are their relative merits?

Item	Unit	Average expenditure per head per week		Average price per unit	
		1956	1961	1956	1961
		s. d.	s. d.	s. d.	s. d.
Bread	Large loaf	2 6	1 11	10	1 1
Meat	lb.	4 5	4 3	2 4	3 10
Fish	lb.	4	1 5	2 0	2 10
Butter	lb.	10	1 9	4 0	3 2
Milk	Pint	2 7	3 4	8	9
Sugar	lb.	1 3	1 3	8	8
Potatoes	lb.	1 6	1 8	4	5
Tea	lb.	9	1 1	5 0	6 0

I.C.W.A. 1963

9. Give a brief outline of the principles underlying the construction of the Ministry of Labour Index of Retail Prices. Discuss the main recommendations contained in the *Report on Revision of the Index of Retail Prices* (Cmnd. 1657, March 1962). *Intermediate D.M.A. 1963*

10.

			Annual Average Prices				*Weights*
			1956		*1960*		
			s.	d.	s.	d.	
Wheat	..	.. cwt	30	3	26	11	7
Barley	..	.. cwt	27	5	28	0	6
Oats	..	.. cwt	24	7	27	3	1
Potatoes ..		.. ton	362	6	261	6	8
Sugar Beet		.. ton	128	6	119	0	3
Hops	..	.. cwt	649	6	571	0	1

(Ann. Abstract of Statistics, 1961)

Calculate the percentage change in price for the above group of crops on the basis of the weighted geometric mean. *I.M.T.A. November 1962*

CHAPTER XI

TIME SERIES

Numerical data, which have been recorded at intervals of time, form what is generally described as a time series. Thus the annual sales of a shop, the quarterly output of coal or the monthly total of passengers carried by a bus company; all these are time series. Undoubtedly the most popular form of presenting such data is in the form of a graph as was shown in Chapter V. Graphs, however, have only a limited value in statistics. While they enable data to be presented in simple and easily intelligible form, they do not add anything to our knowledge of the data. They are of little value for analysis, although a graph does sometimes help to bring out the inter-relationship between two or more time series.

For the economist and business man a study of past events is an aid to making judgments concerning the future. Statistical techniques have been evolved which enable time series to be analysed in such a way that the influences which have determined the form of that series may be ascertained. If the regularity of occurrence of any feature over a sufficiently long period could be clearly established then, within limits, prediction of probable future variations would become possible. Thus a decline in capital investment is sometimes regarded as heralding the initial stages of a recession. If this assumption can be statistically tested and verified in the light of past experience, then the authorities responsible for economic policy possess a useful piece of knowledge to aid them in their contra-cyclical policy.

In practice, the economist and statistician are the first to admit that analysis of trade conditions is extremely complicated since the economic life of a nation is subject to so many complex forces and influences, any one of which it is impossible to isolate. It cannot be too strongly emphasised, then, that the elementary techniques described in the following text appear deceptively simple. This field of study remains the undisputed preserve of the experts.

Types of fluctuation

Most economic time series may be regarded as composed of four constituent elements which are set out below. In passing it should be noted that not all series combine all four elements, *e.g.* not all trades are seasonal although many are. Note too that not all time series concern economic data. A college may maintain records of examination marks gained by students over their period of study; such a time series would hardly show the same type of fluctuation as that expected from a series based on the quarterly output of motor cars or monthly consumption of electricity. In other words, one applies the methods of time series analysis only to such data as justify their employment, *i.e.* where some useful lessons may be gained for the future. Most economic series covering long periods may be analysed into the following constituent parts:[1]

(1) secular trend – or simply the 'trend';

(2) cyclical changes;

(3) seasonal variations;

(4) irregular or spasmodic fluctuations.

The *secular trend* is the course which the data have followed over a a considerable period. In other words, despite temporary deviations from the course, *i.e.* both large and small fluctuations, there is a clearly-marked tendency in a given direction. For example, Table 32 (p. 203) gives the quarterly receipts of income tax collected in the period 1956-63. Although quarterly fluctuations in the series are quite prominent, as can be seen in Figure 18, the *trend* of tax receipts is continuously upward with the passage of time. If a trend can be determined, then the rate of change or progress can be ascertained, and tentative estimates concerning the future made accordingly. The period covered in this example is rather short to warrant the term 'trend' which, ideally, should be restricted to a definite continuous movement which has been observed over several decades. There is a tendency nowadays, however, to employ the term to indicate the main course followed by the series at a particular point of time.

Cyclical fluctuations are far more complex, since their causation differs from period to period. In practice, they are the most difficult of all to anticipate for purposes of effective economic offsetting action. The term 'trade-cycle' would imply a systematic regularity in its appearance, but economists have established a variety of 'cycles', with durations of 3, 5, 7, 9, or 11 years. It is only in recent years that economic opinion has crystallised on the *basic* features of the trade cycle – but despite almost monumental research, particularly in the United States, all that has been proven is that there is no such phenomenon as *the* trade cycle, every one is different. In practice, with most economic series one may at best hope to establish some approximation to a cyclical pattern, *e.g.* in the periodicity of the fluctuations, while the amplitude of the fluctuations will probably show even greater variation.

---

[1] Modern statistical theory rejects this highly simplified classification of economic fluctuations. But there is no simpler way of explaining the basic structure of such movements, provided it is remembered that they are all closely interwoven, and their complete separation is in practice virtually impossible. Modern analysis emphasises the seasonal movements about relatively short period trends.

Seasonal variations, however, are somewhat simpler to deal with. It is common knowledge that many industries are more active at certain periods of the year than at others, *e.g.* the dress and fashion trade anticipates the Spring demand, the toy manufacturers the Christmas season, the motor car industry the Easter and summer holidays, while the building and constructional trades are slack in the winter months. If a definite periodicity for any occurrence can be established and, more especially, the extent of the average fluctuation at that time can be determined, the change in conditions may be anticipated to some degree and provision made to offset any disturbance it might otherwise cause. For example, the Treasury and the Bank of England undertake special offsetting measures to counteract the effect on the clearing banks of such large transfers of tax money during the first quarter of the year as are reflected in the data in Table 32 on page 203.

Irregular or random fluctuations: The economic life of society would be very much simpler if reliable forecasts concerning the future course of business activity were possible. Unfortunately, even with the most advanced existing techniques the forecasting of economic trends remains little more than intelligent estimating, since extraneous and unexpected factors continue to appear and upset the best-laid calculations. In the interpretation of any time series, apart from establishing the trend and the extent of the regular seasonal deviations from it, the economist or statistician tries to isolate not only the random but also the unusual fluctuations. For example, import statistics are made up from Customs records; in the event of a dock strike lasting many weeks, the goods would be landed late and the normal seasonal movement in import statistics would then be replaced by a single swollen figure for the month in which the strike ended. If the seasonal movement were to be computed for a long period, including that year, the distorted figure would have to be adjusted.

Locating the Trend

Table 32 shows the amount of income tax paid to the Inland Revenue each quarter during the years 1956-63 inclusive. The quarterly totals given in Column 3 are plotted on a linear scale graph in Fig. 18, which conveys almost immediately two clear impressions.

(1) The direction of the plotted curve is upwards from left to right, *i.e.* throughout the period apart from seasonal variations, the amount of tax paid has increased continuously.

(2) There appears to be a regular rise and fall in the amount of tax paid within the space of each year, and throughout the whole

period these movements appear almost identical from one year to another.

At the outset it can be stated that the first observation concerns the *trend*; the second refers to the marked *seasonal* movement characterising this series. As a first approximation, the trend line may be drawn 'free-hand', sketching through the middle of the fluctuations so that the seasonal fluctuations are ignored but indicating thereby the course of the data in the period. Unfortunately this method requires not merely a moderate degree of artistic skill, but a considerable knowledge of the data. The point is that while, in theory, the trend line should pass through the middle of the fluctuations as shown in Fig. 18, the latter are not usually as regular as in this particular series. Superimposed upon the seasonal movement, which by itself may well be fairly regular in amplitude, there are the random fluctuations which, in varying degrees according to their causation, reduce or increase the amplitude of the seasonal fluctuations. Thus for a series in which the fluctuations are not as consistent in their periodicity or amplitude as in the example given, the trend line must inevitably be affected by the various fluctuations; it certainly cannot be drawn completely independently of the larger fluctuations in the series. The practice of sketching in a smooth freehand line means that the influence of the fluctuations which may not be entirely seasonal or random may be ignored.

A method of deriving the trend line less subject to the personal evaluation of the data than is the freehand method, is therefore needed. There are two methods which are extensively used, and both are based on the notion that the line should pass through the middle of the fluctuations. In other words, a smooth trend line is achieved by virtue of the fact that the oscillations above the line of trend may be offset by those below the trend. The two methods are, first, the method of moving averages, and second, fitting the trend line by algebraic methods. The former is especially well suited to monthly or quarterly series, although it may be used for annual series covering a long period. The second method is more useful for analysing shorter series. Both techniques, however, are no more than approximate methods, but they will normally produce a result approximating closer to the facts than any intuitive method, except perhaps where the latter method is used by an experienced practitioner who is familiar with the data and the factors influencing them.

The Method of Moving Averages

This method has two features. It is very simple, even if rather tedious to use, and it is especially useful in providing a breakdown, albeit an arbitrary one, of the time series into the four component parts defined earlier.

The method is to ascertain a number of plotting points through which the trend line should pass. These points are obtained by the method set out in Table 32, Cols. 4-6, *i.e.* by selecting a number of consecutive values and averaging them so that the variations in the individual values are reduced. The number of values utilised for determining the average depends on the periodicity of the fluctuations. The periodicity of these movements is usually measured by the time between the recurrent 'peaks' shown on the graph of the original values. In this example no real problem of selecting the correct period can be said to exist, the data clearly requiring the average of four consecutive values. In contrast, however, for many other series the determination of this figure may be quite difficult. Thus, data relating to business activity, *e.g.* a series giving the number of company liquidations over several decades, would cover a number of business

TABLE 32

AMOUNT OF INCOME TAX PAID IN EACH QUARTER U K 1956-1963 (£ million)

(1) Year	(2) Quarter	(3) Quarterly Totals	(4) Sum of (3) in Fours	(5) Sum of (4) in Pairs	(6) Centered Trend (5) ÷ 8	(7) Total Fluc-tuations from Trend (3) — (6)	(8) Seasonal Vari-ations	(9) Residual Fluc-tuations (7)—(8)
1956	March	1,213	—	—	—	—	—	—
	June	260	2,010	—	—	—	—	—
	September	286	2,132	4,142	518	— 232	— 243	+ 11
	December	251	2,153	4,285	536	— 285	— 298	+ 13
1957	March	1,335	2,189	4,342	543	+ 792	+ 806	— 14
	June	281	2,185	4,374	547	— 264	— 263	— 3
	September	316	2,221	4,406	551	— 235	— 243	+ 8
	December	253	2,239	4,460	557	— 304	— 298	— 6
1958	March	1,371	2,258	4,497	562	+ 809	+ 806	+ 3
	June	299	2,279	4,537	567	— 268	— 263	— 5
	September	335	2,297	4,576	572	— 237	— 243	+ 6
	December	274	2,325	4,622	578	— 304	— 298	— 6
1959	March	1,389	2,304	4,629	579	+ 810	+ 806	+ 4
	June	327	2,327	4,631	579	— 252	— 263	+ 11
	September	314	2,216	4,543	568	— 254	— 243	— 11
	December	297	2,213	4,429	554	— 257	— 298	+ 41
1960	March	1,278	2,252	4,465	558	+ 720	+ 806	— 86
	June	324	2,285	4,537	567	— 243	— 263	+ 20
	September	353	2,429	4,714	589	— 236	— 243	+ 7
	December	330	2,496	4,925	616	— 286	— 298	+ 12
1961	March	1,422	2,545	5,041	630	+ 792	+ 806	— 14
	June	391	2,565	5,110	639	— 248	— 263	+ 15
	September	402	2,721	5,286	661	— 259	— 243	— 16
	December	350	2,737	5,458	682	— 332	— 298	— 34
1962	March	1,578	2,794	5,531	691	+ 887	+ 806	+ 81
	June	407	2,846	5,640	705	— 298	— 263	— 35
	September	459	2,821	5,667	708	— 249	— 243	— 6
	December	402	2,859	5,680	710	— 308	— 298	— 10
1963	March	1,553	2,835	5,694	712	+ 841	+ 806	+ 35
	June	445	2,824	5,659	707	— 262	— 263	— 1
	September	435	—	—	—	—	—	—
	December	391	—	—	—	—	—	—

Source: Cols 1-3 only Monthly Digests of Statistics.

'cycles'. Probably it would be necessary to experiment with 5, 7, 8, or 9 years moving averages before a suitable trend line could be clearly established, *i.e.* when the fluctuations of the individual values in the series from the trend are reduced to a minimum.

Figure 18

INCOME TAX RECEIPTS QUARTERLY 1956-63 WITH TREND

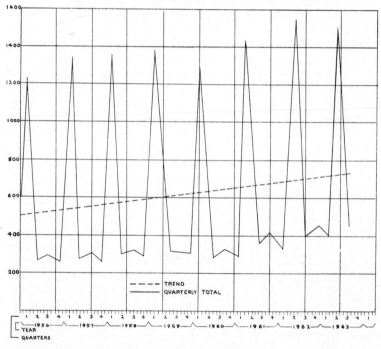

If the four values, *i.e.* those for the four quarters of 1956, are averaged, the average would be located between the values of the second and third quarters. It is then necessary to bring this average value into line with an actual value in the series. This is done by 'centering' the moving average, *i.e.* by finding the average of the second, third, fourth and fifth items, and adding this to the average of the first four values. Then if the aggregate of the two averages is halved, the moving average will be 'centered', *i.e.* the average of the two averages will lie against the third figure of the series, instead of between the second and third. If the period averaged covers an odd number of months or years (almost invariably the latter), the problem of centering does not arise. The average then lies against the middle item, *e.g.* a nine-yearly average would be placed against the fifth value. It should be noted that 'centering' is only really necessary if the calculation of the extent of the seasonal or cyclical fluctuations is required; it is not absolutely necessary for the sole purpose of drawing the 'trend' line, as plotting points can be derived from the 'mid-values' of successive averages.

A simpler way of working, used in Table 32, is to aggregate the first four values, then the second to fifth values inclusive, yielding successive totals of 2,010 and 2,132 (Col. 4). By adding them, as in Col. 5, and dividing by 8, the moving average is centered (Col. 6). By continuing this process throughout the series, the succession of averages in Col. 6 is obtained.

The series of moving averages plotted on the graph (Fig. 18) are the trend line values which yield the dotted line passing through the graph of the original data. This dotted line is described as the trend or line of trend. The subtraction of the trend values from the actual values of the series in Col. 3, gives the total fluctuations about the line of trend shown in Col. 7 of Table 32.

Derivation of Seasonal Movement

The purpose of analysing time series is not always the determination of the trend by itself. Interest may be centered on the seasonal movement displayed by the series and, in such a case, the determination of the trend is merely a stage in the process of measuring and analysing the seasonal variation. If a regular basic or underlying seasonal movement can be clearly established, forecasting of future movements becomes rather less a matter of guesswork and more a matter of intelligent forecasting. Thus, if grain shipments are known to fall 30 per cent from the August level by the following December, provision for this change and all that it implies for ports, etc., can be made by those concerned. If in the event the fall is sharper still, there is still the consolation that some part of it was correctly forecast. Lastly, before proceeding to the actual statistical techniques, it should be emphasised that the conditions in all the periods of time for which data are available were probably different. The decision as to what constitutes *normal* and the extent of the 'abnormal' fluctuation may be quite arbitrary.

Since, as with the data given, the seasonal movements are a compound of the true seasonal variation and a random or irregular element it is necessary to separate the 'normal' seasonal movement from the 'residual' variation, *i.e.* the abnormal or unexpected movements. The first stage in this process is completed by averaging the actual movements or deviations from the trend (given in Table 32, Col. 7) for each of the four quarters. The method of deriving the seasonal movement is illustrated in Table 33. The deviations from the trend are given in the previous table (Col. 7).

TABLE 33

CALCULATION OF MEAN SEASONAL MOVEMENT
(Data from Table 32, Col. 7)

	Quarters			
	March	June	September	December
1956	—	—	— 232	— 285
1957	+ 792	— 264	— 235	— 304
1958	+ 809	— 268	— 237	— 304
1959	+ 810	— 252	— 254	— 257
1960	+ 720	— 243	— 236	— 286
1961	+ 792	— 248	— 259	— 332
1962	+ 887	— 298	— 249	— 308
1963	+ 841	— 261		
Total	+ 5,651	—1,834	—1,702	—2,076
Less Adjustment*	+ 10	+ 10	+ 10	+ 10
	+ 5,641	—1,844	—1,712	—2,086
Seasonal Movement.. ..	+ 806	— 263	— 243	— 298

* Note that the difference in sums of the negative and positive totals is + 39, *i.e.*, 5,651 — 5,612; therefore to balance the quarterly figures it is apportioned over each quarter, *i.e.* + 10.

Before the average seasonal fluctuation can be calculated and graphed, a further adjustment is usually necessary. It may be remembered that the sum of the individual deviations of a series of values from the mean of that series was equal to zero. Since the values on which the line of trend is based are also averages, the sum of the fluctuations from that line should also equal zero. If they are totalled, as in Table 33 (+ 5641 — 1834 — 1702 — 2076 = 5651 — 5612) there is a difference of + 39 between the sums of the positive and negative movements. Such a difference frequently arises since the effects of large but irregular seasonal fluctuations cannot be entirely eliminated by the moving average method. Since the source of the difference is unknown, it is eliminated by arbitrarily averaging it over the four quarters and taking it from each quarter's total. The adjusted totals of each quarter's fluctuations from the trend line are then divided by the number of years covered. The quotients represent the *average* seasonal variation in each of the four quarters during the period covered by the data and are inserted in Col. 8 of Table 32. The residual movements are derived by subtracting the seasonal variation (Col. 8, Table 32) from the total fluctuation from the trend *i.e.* Col. 7, Table 32.

The moving average method of deriving the trend of values and the seasonal fluctuations over a period is the main elementary method available for this work. It can be applied to series covering several decades provided the data can still be treated as a single series, the moving average in such a case being used to eliminate the so-called trade cycle. When the cycle is regular both in its periodicity and in the amplitude of fluctuations the moving average method will completely eliminate the fluctuation and yield a straight trend line. Such a cycle is the exception rather than the rule, hence, skill and knowledge in the selection of the period used for averaging is necessary where the cycle is not so clearly defined, *e.g.* is the cycle to which the data under scrutiny are subject one of 3, 5, 7, 9, or 11 years; how many cycles are superimposed on one another?

The moving average method, however, has certain disadvantages:
  (i) the trend line cannot be derived for the period covering the whole series. The trend line falls 'short' at both ends, and if the cycle is one of, say nine or eleven years, this may constitute a marked gap where the data cover only two or three cycles;
 (ii) the difficulty already explained of establishing a definite periodicity in the fluctuations. Different views on this will result in differing trend lines. Unless, therefore, the seasonal or cyclical movement is definite and clear-cut, the moving average method of deriving the trend may yield only a rather unsatisfactory compromise line;
(iii) since the 'trend' values are arithmetic averages, any extreme individual variation affects them unduly. If the seasonal variations vary considerably in extent from year to year, the trend may appear as a series of humps rather than a smooth line, *i.e.* it is not possible to entirely eliminate the seasonal variation.

Fitting the Trend by Algebraic Methods

An alternative method of defining the trend is to estimate it by mathematical methods. In Chapter IX on Correlation, it was stated that linear distributions could be summarised by a simple equation of the first degree: $y = a + bx$. This tool is extensively employed for fitting the trend to time series which approximate to a straight line trend. As in linear correlation where the relationship is represented by a straight line, the equation is of the first degree. Many time series when plotted (as well as the relationships in correlation analyses) are curvilinear, the curve then being described by more complex equations. The principle, however, remains the same.

For quite a considerable body of phenomena, even in the economic field, data may reflect a continuous and fairly consistent rate of change or inter-relationship which may occasionally be described by a mathemical equation. Too much should not be read into this. No economic data will ever conform exactly to some scientific law expressed mathematically. Some data, however, do conform approximately to this concept and for certain analytical work, *e.g.* statistical testing of trade cycle theories, it is useful to be able to define the trend in such a form instead of using the foregoing moving average method which merely describes instead of analysing the data. Especial care is required, and this frequently constitutes the stumbling block to the use of this technique, to ensure that the basic conditions are unchanged throughout the period to which the data relate. Thus in the example depicting the growth of air passenger traffic the period 1950-1962 is selected rather than say 1937–1950 since the war years can hardly be linked with the post-war period to judge the degree of expansion and the probable future trend.

To illustrate the technique a simple example is given based on the data in Table 34 which illustrate the growth in the air passenger traffic of the United Kingdom between 1950 and 1962.

TABLE 34

PASSENGER MILES FLOWN IN SCHEDULED FLIGHTS FROM U.K. AIRPORTS 1950–1962
(Million passenger miles)

Year	Passenger-miles flown	Year	Passenger-miles flown
1950	794·0	1957	2,421·9
1951	1,065·0	1958	2,571·1
1952	1,242·5	1959	3,090·8
1953	1,434·2	1960	3,959·3
1954	1,515·4	1961	4,531·0
1955	1,801·4	1962	4,876·8
1956	2,102·3		

In order to 'fit' a curve to the data given in the above Table the values of the terms a and b in the equation $y = a + bx$ must be computed. The method employed is detailed in the calculations below and is known as fitting a curve by the 'method of least squares'. The best fit is obtained when the sum of the squares of the deviations from the trend line is at a minimum; hence the name of the method. There is only one line which meets this condition.

The method employed in the calculation opposite can be set out in successive stages.

CALCULATION FOR FITTING TREND LINE TO TIME SERIES

(1) Year	(2) Passenger-miles flown (to nearest million) y	(3) Years in sequence x	(4) Col. 2 × Col. 3 xy	(5) x^2	(6) y^t
1950 ..	794	1	794	1	424·7
1951 ..	1,065	2	2,130	4	756·6
1952 ..	1,243	3	3,729	9	1,088·4
1953 ..	1,434	4	5,736	16	1,420·3
1954 ..	1,515	5	7,575	25	1,752·1
1955 ..	1,801	6	10,806	36	2,083·9
1956 ..	2,102	7	14,714	49	2,415·8
1957 ..	2,422	8	19,376	64	2,747·6
1958 ..	2,571	9	23,139	81	3,074·5
1959 ..	3,091	10	30,910	100	3,411·3
1960 ..	3,959	11	43,549	121	3,743·1
1961 ..	4,531	12	54,372	144	4,075·0
1962 ..	4,877	13	63,401	169	4,406·8
Totals ..	31,405	91	280,231	819	—

(1) The original data, the years within the period under review, and the annual totals are set down in columns 1 and 2.

(2) Col. (3) headed 'x' represents the 'plot' along the abscissa (horizontal axis) and is written as a consecutive series 1 to 13 inclusive. 'x' may be referred to as the independent variable.

(3) 'y' in the equation is the 'dependent variable', *i.e.* the annual miles flown. In Col. (4) headed 'xy' the products of Cols. 2 and 3 are given, x in this case being not the year but its position in the series 1 to 13.

(4) In Col. 5 the values in Col. 3, *i.e.* the years in sequence 1–13 are squared.

(5) The values in Cols. 2, 3, 4 and 5 are next totalled, and the aggregates are then substituted for the symbols in the following equations which are employed to derive the trend line:
(i) $\Sigma y = na + b\Sigma x$
(ii) $\Sigma xy = a\Sigma x + b\Sigma x^2$
Thus the equations become:
(i) $31,405 = 13a + 91b$
(ii) $280,231 = 91a + 819b$

(6) The solution of these simultaneous equations proceeds as follows: either a or b must be eliminated to derive the value of the other

8

symbol. By multiplying the first equation by 7 the following results are obtained:

(i) $219,835 = 91a + 637b$

(ii) $280,231 = 91a + 819b$

The difference betweent the two equations:

$60,396 = 182b$

$$\therefore b = \frac{60,396}{182} = 331\cdot84.$$

(7) Substituting $331\cdot84$ for b in the original equation:

$31,405 = 13a + 91b$

$31,405 = 13a + 91\,(331\cdot84)$

$31,405 = 13a + 30,197\cdot44$

$31,405 - 30,197\cdot44 = 13a$

$+ 1,207\cdot56 = 13a$

$92\cdot89 = a$

(8) The basic equation which yields the straight line curve is calculated by substituting for a and b the values $+ 92\cdot89$ and $331\cdot84$ respectively. Thus for 1950, the value of y^t is derived as follows

$$y^t = 92\cdot89 + 331\cdot84 = 424\cdot73$$

and for 1959: $y^t = 92\cdot89 + 331\cdot84\,(10) = 3411\cdot29$

The estimated values (y^t) for each successive year are set down in Col. 6 and these may be compared with the original values given in Col. 2. Fig. 19 shows the original values plotted as crosses on the graph; the straight line passing through the positions given by the values in Col 6.

For future reference, however, it will be necessary only to write: 'Trend of Passenger-miles flown from United Kingdom Airports in million miles 1950–62 $y = 15\cdot97 + 331\cdot84x$. Year of Origin 1949', a statement which not only describes the character of the data, so that the approximate mileage flown in any intermediate year can be calculated, but also defines precisely the trend.

It will be realised that only two values for y^t are required for the drawing of a straight line. The values for y^t for the whole range of x values are calculated only so that the annual fluctuations from the trend may be derived, as well as enabling the actual values for each year to be compared with the 'trend'.

The student may wonder why in the note concluding Chapter IX explaining the calculation of the equation of a regression line, an apparently different method was used from that above to obtain the line of trend. Actually the same result can be obtained by either method; but for calculating an estimated line of trend it is simpler to use the above short-cut method.

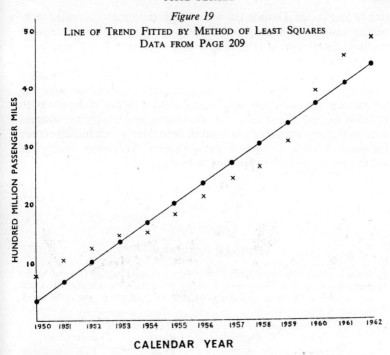

Figure 19

LINE OF TREND FITTED BY METHOD OF LEAST SQUARES
DATA FROM PAGE 209

CALENDAR YEAR

Conclusions

Time series analysis often requires more knowledge of the data and relevant information about their background than it does of statistical techniques. Whereas the data in some other fields may be controlled so as to increase their representativeness, economic data are so changeable in their nature that it is usually impossible to sort out the separate effects of the various influences. Attempts to isolate cyclical, seasonal and irregular, or random movements, are made primarily in the hope that some underlying pattern of change over time may be revealed. It is noteworthy that analysis of cycles up to 1914 does reveal what appear to be regular waves of alternating boom and slump. Some of the older trade cycle theories, such as the sun-spot and harvest variations, which are themselves natural phenomena, recognised that economic activity seems to follow a rhythmic pattern. But in present times, when economic forces are often repressed and government action is so widespread in its effect, the fluctuations of the economy become increasingly unpredictable.

The cyclical pattern of activity revealed over the major part of the last century, on which some of the above statistical ideas are based, gave an impetus to the use of statistical analysis for forecasting trends and cycles. Despite the use of both simple and complex techniques, ranging from the simple averaging of seasonal movements to multiple correlation of related series, *e.g.* coal output, rail transport, and steel production, it remains a regrettable fact that 'scientific forecasting' has not added any laurels to statistical science. The ideal situation would be to have 'indicators' which could be used for forecasting the direction and amplitude of future changes. Both economists and statisticians, by examining and analysing business data over many decades have tried to determine which indicators are the most reliable and consistent. Whether the results justify the volume of work entailed remains to be seen.

Questions

TIME SERIES

1. It has been said that the incidence of measles exhibits a regular two-year cycle. Examine this suggestion, using the data below. Investigate the nature of the trend (if any) and estimate the average magnitude of the fluctuations about the trend. Comment on your results.

Incidence of measles per 1.000 of population

Year	England	Wales	London
1940	10·18	10·26	1·77
1941	10·53	10·54	4·86
1942	7·75	3·39	8·62
1943	9·86	10·28	9·17
1944	4·29	2·23	2·98
1945	11·70	11·23	9·03
1946	4·05	1·89	7·35
1947	9·36	10·24	5·39
1948	9·55	5·95	9·17
1949	9·16	5·60	8·54
1950	8·45	7·33	6·57

Intermediate D.M.A. 1961

2. Quarterly sales of a certain fertilizer over the period 1954–1958, in thousands of tons, were as follows:

	Year				
	1954	1955	1956	1957	1958
1st Quarter	48	50	68	93	84
2nd ,,	52	46	34	56	61
3rd ,,	16	22	26	16	29
4th ,,	35	40	35	45	48

By means of moving averages, compute the trend and estimate the seasonal effects, and hence forecast sales for each quarter of 1959.

Institute of Statisticians 1959

3. Explain what is meant by 'moving average' and state its uses. Calculate the twelve-month moving averages of the following monthly outputs (in tons) from a factory.

Month	Monthly Output (in tons)		
	1953	1954	1955
January	116	119	122
February	118	125	132
March	122	122	128
April	116	118	121
May	131	125	121
June	119	125	119
July	99	87	
August	92	100	
September	119	121	
October	128	122	
November	127	130	
December	110	116	

Institute of Statisticians 1956

4.

	Home sales of television sets (thousands)		
	1955	1956	1957
January	85	74	128
February	85	61	107
March	103	50	85
April	151	99	112
May	126	59	100
June	83	68	92
July	82	75	111
August	82	108	156
September	260	177	233
October	303	295	277
November	179	218	232
December	145	150	169

Calculate the trend of the above series after removing the seasonal variation.

I.M.T.A. 1959

5. The following data give the mean seasonal humidities (in grains of moisture/lb. air) in a manufacturing department. Plot the data on a chart and by means of a suitable moving average remove the seasonal variation from the data and plot the residual trend line.

Season			Year		
	1956	1957	1958	1959	1960
1st Quarter	34	36	33	31	32
2nd ,,	44	41	49	51	48
3rd ,,	61	62	67	65	61
4th ,,	43	42	41	41	40

Institute of Statisticians 1962

6. What are the constituent parts of a time series that have to be separately eliminated in order to arrive at the residual variations? Briefly describe the process involved?

I.M.T.A. November 1960

7.

Live Births, England and Wales, 1913-56
Excess of Males over Females, per 1,000 Females Born

Year	Excess of Males	Year	Excess of Males	Year	Excess of Males	Year	Excess of Males
1913	38	1924	47	1935	56	1946	60
14	35	25	45	36	54	47	61
15	40	26	41	37	56	48	61
16	49	27	42	38	51	49	61
17	44	28	44	39	56	50	60
18	48	29	43	40	53	51	60
19	60	30	44	41	53	52	55
20	52	31	49	42	63	53	59
21	51	32	50	43	64	54	59
22	49	33	46	44	65	55	59
23	44	34	55	45	61	56	55

From Annual Abstract of Statistics

Describe the method of trend fitting by moving averages. Illustrate by applying the method to the data above and indicate to what extent you think it is appropriate in this case. Why does the excess of males vary? Comment on what appears to be the nature of the variation since 1913.

B.Sc. (Econ.) 1957

PART II

ELEMENTARY SAMPLING THEORY
AND PRACTICE

CHAPTER XII

THE BASIS OF SAMPLING

Introduction

Throughout the discussion of averages and measures of dispersion it has been assumed that summarisation and description of the various types of frequency distribution was the primary object of such exercises. Such an impression would be confirmed by the description of these statistics in the opening chapter as *descriptive* statistics. Most of the frequency distributions used to illustrate these statistics comprised a complete count of all the members or units in the 'population', used here in its statistical sense. For example, Table I enumerated the output of each individual worker in a given factory; Table 6 classified by income size the 21·6 million income-earners known to the Inland Revenue, while Tables 8 and 9 gave the aggregate membership of trade unions and the population of England and Wales respectively. A large volume of statistical information is collected by means of such periodic enumerations, known as *censuses*. There is, for example, a census of population, of the electors, *i.e.* persons aged 21 and over who are entitled to vote, as well as censuses of production and of distribution. A census entails a complete count of every individual unit of the population (in the statistical sense of the term) and is usually a considerable undertaking. Quite apart from the cost thereof and the large volume of resources such as enumerators, clerical assistance, etc., that is required, the main problem is the time which it takes to process the data. Hence the results appear several years after the collection of the data.

It is not often, however, that the statistician has access to data based upon an up-to-date census. Such enquiries are expensive in both time and money. The population census is held only at ten year intervals and the full scale census of production at three year intervals, although the Inland Revenue can produce detailed statistics such as those in Table 6 at yearly intervals, but with some delay. These enquiries are extremely important from the administrative point of view and it is to serve such needs that they are carried out. Any benefits which the statistician derives from access to such data are merely incidental to the main object. In brief, the statistician can very seldom hope for an enumeration of all the members of a population in which he is interested. Thus when the B.B.C. Audience Research unit wishes to know viewers' reactions to last night's TV programmes it cannot hope to ask all viewers for their opinions. It would be very costly and would take so much time that many viewers would have forgotten all about the programme by the time they were asked for their opinions.

The solution to the problem is for the Research Unit to interview only a small proportion of the viewing public and on the basis of their findings to infer the views of the viewing public as a whole. Clearly, the group of viewers to be interviewed must be representative of all types of viewer, for example, both working class and middle class households must be included. Such a survey of a small proportion of the relevant population, in this case all households with B.B.C. television who actually watched the programme, is known as a *sample* survey. Exactly the same procedure is followed by the public opinion polls which measure the fluctuations in the public's support for the policies of the political parties. Experience shows that before a general election such polls can usually forecast the outcome quite accurately although they may be based upon a sample of less than one in 10,000 voters.

In themselves sample results or *sample statistics* are usually of little real value or interest. They are collected because the statistician is interested in the parent population from which the sample is drawn. For example, a sample mean provides an estimate of the population mean. The theory of sampling makes it possible not only for inferences and conclusions to be drawn from sample data but enables the statistician to make precise probability statements about the reliability of such conclusions. Unless it were possible to generalise sample results in this way, sampling would be of negligible value to the statistician. At best they would provide some indication of the nature of the population from which the sample was drawn. But without modern statistical theory any such inference would be little better than a guess. Statistical sampling theory is based on the mathematical theory of probability and, on one or two issues, there is still some disagreement between the experts. The next few sections of this chapter seek merely to provide a highly simplified outline of the ideas underlying this branch of statistical theory.

The Normal Curve

Everybody has some idea of the meaning of the term 'probability' but there is no agreement among scientists on a precise definition of the term for the purpose of scientific methodology. It is sufficient for our purpose, however, if the concept is interpreted in terms of relative frequency, or more simply, how many times a particular event is likely to occur in a large population. When we say the probability of obtaining 'heads' as a result of tossing a coin is equal to $\frac{1}{2}$, we mean that if the coin were tossed a large number of times the proportion of heads to be expected is one half. Similarly, if a set of ten coins were tossed simultaneously we should expect to get five heads rather than any other number. If we continued to toss the ten coins it would not be a matter for surprise if 3, 4, 6 or 7 heads were recorded quite frequently and from time to time we might even get all the coins falling heads and on other occasions all tails. But in the long run, assuming all the coins to be true, we should expect to record five heads more frequently than any other score.

It can be demonstrated mathematically that where there are only two alternative outcomes to an event, *e.g.*, heads or tails with a spun coin, and the coin is so tossed that chance alone determines which way it falls, then if the experiment is repeated a very large number of times, the distribution of results can be predicted. Take, for example, the experiment of tossing ten coins simultaneously and noting the number of heads. In the early stages one head might follow six heads and then three heads; no particular order would be apparent. But as the number of tosses grew, so the distribution of frequencies of particular numbers of heads would take on a definite pattern. Such a frequency distribution could be represented graphically with the values 0–10 heads marked off along the base and the corresponding proportionate frequencies marked off along the vertical axis. It would resemble a histogram similar to that depicted in Figure 14 on page 82.

The same experiment could be performed tossing sets of 100 coins or even a thousand coins at a time. With such large values to be plotted and if the experiment were continued for a long time the resultant histogram would have an outline resembling tiny serrations rather than the clearly defined steps of the histogram in Figure 14. In fact, for very large numbers of tosses the histogram would cease to resemble that histogram but would approximate to a smooth curve similar to that portrayed in Figure 20. This curve is bell-shaped and symmetrical. These characteristics reflect the fact that the maximum frequencies are recorded by the mean value of the distribution, *i.e.*, five heads, while other values which deviate by only a small amount from the mean also occur quite frequently. But it is very evident that as the deviation between any recorded value and the mean increases, so the frequency of that value declines. There are very few cases indeed where all the coins came down heads and there are equally few where they were all tails.

While it may not be immediately apparent, the reader will on reflection realise that this curve portrays the distribution of a certain sample statistic. Each set of ten (100 or 1,000) coins tossed is merely a sample of an infinitely large population of such tosses. The mean

number of heads recorded in each sample is an estimate of the proportion of heads we are likely to get if every coin in existence could be
tossed, *i.e.* 5 out of 10, implying that the chances of heads or tails with
a spun coin are in fact 50:50.

Figure 20

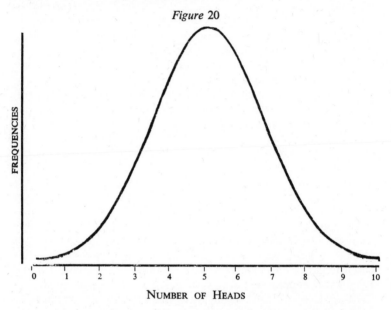

NUMBER OF HEADS

If instead of tossing coins a very large number of samples, each
consisting of 50 adult males of British birth, had been taken and for
each sample the mean height was obtained, then the distribution of
sample means would approximate closely to the curve depicted in
Figure 20. In other words, a particular value would occur more
frequently than any other; this would correspond to the mean of the
distribution of the sample statistics. Other sample means would also
be obtained but, as with the coin tossing experiment, the greater the
divergence of any statistic from the mean value of the distribution, the
less frequently it would occur. It can be shown mathematically that
the mean of such a distribution of a very large number of sample
means is equal to the population mean. Such a distribution is known
as a *sampling distribution*, *i.e.* a distribution of sample statistics such
as sample means, proportions, standard deviations, etc. The curve
based upon this type of distribution is known as the *Normal curve*.
This is only one of several distributions relevant to sampling theory,
but it is the most widely used and is of fundamental importance.
While it is always bell-shaped and symmetrical about its mean, its

actual shape is determined by the standard deviation of the distribution. For example, two Normal distributions can have the same mean, but if for one the standard deviation is much larger than that for the other, then the curve of the former distribution will be flattish

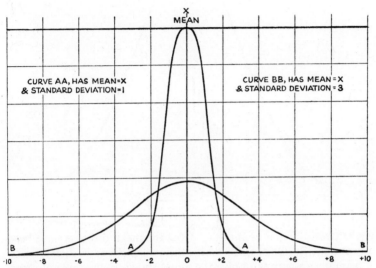

FIG. 21   NORMAL CURVES WITH THE SAME MEAN BUT DIFFERING STANDARD DEVIATIONS

and for the other more peaked as in Figure 21. In other words, the Normal distribution can resemble a tall single-peaked narrow curve or a flattish long-tailed broad based curve. It is, however, always unimodal, *i.e.* a single peak, and symmetrical about that peak.

The Normal curve has one other highly important characteristic. No matter what its shape, *i.e.*, broad or narrow, the area beneath the curve is distributed in a particular way. From Figure 22 it can be seen that if vertical lines termed 'ordinates' are drawn from the base to the curve at intervals of one standard deviation from the mean ordinate, the proportion of the area under the curve bounded by the ordinates drawn at one, two and two and one half standard deviations on either side of the mean is approximately 68, 95 and 99 per cent. respectively.

So important is the Normal distribution in statistical theory that specially prepared tables give the proportion of the area under the curve enclosed between the mean ordinate and any ordinate up to a range of three standard deviations from it. Since the Normal curve is symmetrical, it is sufficient to state the areas in this way, *i.e.* for one side of the curve only, rather than as has been shown in Figure 22 below,

Figure 22
DISTRIBUTION OF AREA UNDER NORMAL CURVE.

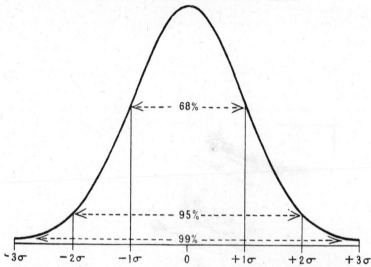

where two ordinates equi-distant from the mean ordinate have been drawn on each side of it. To derive the area enclosed in such a case it is merely necessary to double the figure shown in the table of areas. The reason for setting the table out in this way is that sometimes the statistician is concerned with one side only of the Normal distribution. The following table merely reproduces a few important values taken from the complete table, but it will serve to show what sort of information such a table provides.[1] The distance along the base between

x/σ	1·96	2·33	2·58	3·09
Area	0·4750	0·4900	0·4950	0·4990

the mean ordinate and the other is always measured in units of the standard deviation of the distribution and is written x/σ.

If ordinates are drawn on *both* sides of the mean at intervals corresponding to the above values of x/σ, then the percentages of the total curve enclosed between those ordinates are 95, 98, 99 and 99·8 respectively. It will be seen that these percentage figures are derived by doubling the figures in the lower row of the table above and expressing the result in percentage terms. For most purposes the only values of x/σ that interest the statistician are those between 1·96 and

[1] Many of the more advanced texts reproduce the entire table of areas of the Normal or probability curve in an appendix; e.g. Croxton and Cowden, *Applied General Statistics*, or Yule and Kendall, *An Introduction to the Theory of Statistics*.

3·09 and of those it is the 1·96 and 2·58 values which are customarily used in practical work. The reasons for concentrating on these few values will become apparent to the student during the next few pages.

Now, instead of visualising the Normal curve as an area divided into parts, consider it as a large collection of sample statistics, *e.g.* a very large number of sample means. Clearly, the mean of all the sample means is the best estimate we have of the mean of the population from which the samples have been drawn. The point has been made that the mean of such a sampling distribution corresponds to the population mean. It is apparent that within a range of 1·96 standard deviations about the mean of the Normal distribution 95 per cent. of all sample means are to be found. Similarly, 99 per cent. of the sample means lie within a range of 2·58 standard deviations about the population mean, while within 3·09 standard deviations about the mean we find 99·8 per cent. of the sample means.

In practice, however, there is no time to take such a large number of samples from any particular population, usually we have to be satisfied with a single sample. Fortunately, because the sampling distribution of the statistic is known, the chances of a single sample mean deviating by a given amount from the mean of such a distribution, *i.e.* the population mean, are known. Since 95 per cent. of all sample means will lie within a range of 1·96 standard deviations about the mean of the distribution of sample means, it follows that the chances of the mean based on a single sample differing by more than 1·96 standard deviations from the population mean are only 5 in 100, or 1 in 20.

Standard Error of a Statistic

For any sampling distribution we can calculate the mean and standard deviation. The variation between large numbers of sample means, which produces the dispersion about the mean of the sampling distribution, arises from what are termed *random sampling rerors*. The term 'error' in this context does not mean mistake; it means variations from the true value which are entirely attributable to chance. The 'standard deviation' of such a sampling distribution is known as the *standard error*. This is computed for any sample statistic and is a measure of the precision of that statistic as an estimate of the corresponding population parameter. The standard error of a sample mean is given by the formula $\sigma/\sqrt{n}$. While the formula requires the standard deviation of the population, since this is unknown we substitute the sample standard deviation. Fortunately, where the sample is large, it can be shown mathematically that the inaccuracy introduced by the substitution is unimportant.

From the above formula it is apparent that the standard error of the mean is dependent upon two factors, the variability in the population and the size of the sample. It is self-evident that the variation between sample statistics is dependent largely upon the dispersion of the characteristic within the population. For example, if the weights of all adult males were within a range of 10 lb. about a mean of 150 lb., then clearly no sample mean could be below 140 or above 160 lb. Since, however, the actual weights of units from this population range from about 100 to 200 lb. or more, it follows that sample means of any value between these two limits are possible. Thus the standard error of the sample mean must be dependent in part on the variability within the population which is measured by the standard deviation.

There is also a relationship between the standard error and the size of the sample. Increased precision in the sample estimate of a parameter can be obtained by increasing the size of the sample but, as is apparent from the above formula for the standard error of the mean, it is not a direct relationship. Since the formula uses the square root of the sample size, it follows that in order to halve the standard error of the mean, the sample size must be increased fourfold. It is for this reason that given the standard error of a statistic based on a large sample, an increase in its precision can be achieved only by considerably increasing the size of the sample. Since this is an expensive procedure, the statistician endeavours to select that size of sample which will give the maximum precision in the sample estimate consistent with a given outlay of funds. In other words, the statistician decides first upon the level of precision he requires, *i.e.* the standard error he is prepared to accept, and then he calculates the size of the sample required.

The above points can be simply illustrated. Assume a sample of 100 adult women with a mean height of 63 inches and a standard deviation of 2 inches. The standard error of the mean is obtained by substituting the appropriate values in the formula $\sigma/\sqrt{n} = 2/\sqrt{100}$ equals 1/5th inch. It can be inferred from these data that the assertion that the population mean lies within the range of 63 inches + or —2/5ths inches is likely, in the long run, to be true for 95 per cent. of the samples taken. Suppose that greater precision is required in the estimate of the population mean, *e.g.* the standard error should be one tenth of an inch, which is half the present error. Substituting in the equation we get $1/10\text{th} = 2/\sqrt{x}$ which yields a value of 400 for x. Thus, to reduce the standard error by half, the sample had to be quadrupled.

Just as it is possible to estimate the population mean from sample data so an estimate of a population *proportion* from a sample can be made. For example, a sample of 900 electors reveals that 45 per cent.

intend to vote Conservative in the forthcoming election. What is the probable proportion of the total electorate which will vote Conservative? First, we must derive the standard error of the proportion which, like that of the mean, is based upon the distribution of sample proportions and is given by the formula $\sqrt{\frac{p \times q}{n}}$. The symbols p and q represent the proportions of the sample possessing or not possessing the relevant characteristic, in this case the intention to vote Conservative. With a coin, the odds of obtaining a head were one half and those for tails also one half, so that certainty was represented by $\frac{1}{2} + \frac{1}{2} = 1$. In the same ways, so $p + q$ equals 1 and the value of q is derived from the difference between 1 and the value of p. In this case $1 - p$ equals 0·55. The formula for the standard error of the proportion after substitution of the symbols reads

$$\text{s.e.} = \sqrt{\frac{(0.45) \times (0.55)}{900}} = 0·017$$

The calculation gives a result of approximately 1·7 per cent. As with sample means, so the distribution of sample proportions is such that 95 per cent. of a distribution of sample proportions will lie within a range of two standard errors about the population percentage. Given this knowledge it may be inferred with 95 per cent certainty from the above data that the proportion of potential Conservative voters in the electorate is $45 \pm 2(1·7)$ per cent. or between 41·6 and 48·4 per cent. Using the above formula the student can now calculate for himself the size of sample required to give a result within the limits 44 to 46 per cent.[1]

Confidence limits

In presenting the results derived from sample data it is the invariable rule to show both the statistic and its standard error. Thus the effect is to set limits about the sample statistic and, by convention, the most usual limits are those set by 1·96 times the standard error. This signifies that there is a 95 per cent probability that the sample statistic does not diverge by more than those limits from the corresponding population *parameter*, as it is termed. The statistician can also give his estimate of the parameter at the 99 per cent. level, in which case the interval about the sample statistic is increased from 1·96 to 2·58 standard errors. For present purposes the 95 per cent. limits may be interpreted as signifying that there is a 95/5, or 19 to 1 chance that the population mean lies in that interval.

Strictly speaking such a statement is erroneous since the population mean either lies within the specified limits, or it does not. In the

[1] The answer is 9,900. The student unable to agree this answer should turn to page 265, end of Ch. XIV.

first case the probability is defined as unity; in the other as zero. In other words, we cannot talk of a 95 per cent. probability of it lying within the specified interval. While it would be too much to assert that this is mere hair-splitting, the statistician would insist on stating that the intervals have been so selected that, if a very large number of samples were taken from the population, for 95 per cent. of these samples the limits would include the population mean.

For present purposes, however, it is sufficient if the student remembers that it is the population parameter which has a single fixed value and the sample statistic which may deviate from it. By calculating the standard error of the statistic the amount of deviation between the two values to be expected can be calculated together with the probability of its occurrence. The limits set by the standard error are termed confidence limits and the setting of the 95 per cent. confidence interval (*i.e.* the range between the two limits) implies that the sample estimate will, in the long run, prove to be a good estimate of the parameter 95 per cent. of the time. The 99 per cent. interval signifies that the estimate will, in the long run, be true 99 times in 100, *i.e.* it implies a greater degree of confidence in any statement based upon the sample data.

Summary

Sampling is fundamental to all statistical analysis. Sample statistics are merely estimates of the corresponding population parameters and individual sample statistics will often differ from that value by what are known as random sampling errors. Mathematical theory demonstrates that the distribution of such random sampling errors tends to be Normal and observation supports the theory. Knowledge of this distribution in respect of sampling variation enables the statistician to make rigorous inferences concerning the population on the basis of sample data. A mistake which is commonly made is to assume that it is the population which is normally distributed. This is not the case; very few populations are so distributed. Generally speaking, both the distribution and size of the population are irrelevant in sampling; it is only the statistics derived from successive samples that in the long run distribute themselves normally.

It cannot be too strongly emphasised that conclusions based on the evidence of sample statistics are valid only if the sample has been selected in such a way that every unit in the population has a known and calculable chance of selection. Such samples are *random* samples, and even if the adjective is later omitted in the text the implication of the term 'sample' as used by the statistician is that it has been selected by methods which ensure randomness. Some of the more generally employed methods of choosing a sample are discussed in Chapter XIV.

Questions

THE BASIS OF SAMPLING

1. A coin has fallen 'heads'. What is the probability of it falling 'heads' at the next toss? Give reasons.   *I.M.T.A. 1959*

2. What do you understand by the 'Normal curve'? Describe some of its properties and indicate a type of observation that would give a Normal curve when plotted.

3. If X is distributed in a Normal distribution with mean 100 and standard deviation 3 units, within what limits would you expect 95 per cent of the averages of four values of X to lie?

4. Using the Normal distribution, find the probability of obtaining:

(*a*) at least 85 'tails',
(*b*) 90 or fewer 'heads',
    in a random tossing of two hundred coins.

What conclusion would you draw if 80 heads and 120 tails were obtained?
Institute of Statisticians 1958

5. 'The ultimate object of sampling is to generalise about the total population.' Explain this statement.   *I.M.T.A. 1960*

6. Explain carefully what is meant by the standard error of a percentage. For what purposes is the standard error used?   *B.Sc. (Sociology) 1954*

CHAPTER XIII

TESTS OF SIGNIFICANCE

In the last chapter it was explained that even if nothing is known about a population it is possible on the basis of a random sample from that population to derive reliable information about its nature. This important statistical technique is possible because of our knowledge of what are termed *sampling distributions* of the various statistics. The *standard error* of a sampling distribution, *e.g.* sample means, indicates the deviation from the population mean which can be expected with a given frequency in the means of a large number of random samples. Since it is known that the proportion of a large number of sample statistics lying within two standard errors of the population parameter is 95/100, it follows that the probability of a deviation between the sample statistic and the mean of the distribution, *i.e.*, the population parameter, greater than twice the standard error of the statistic is approximately only one in twenty. In practice, this means that when a solitary sample is drawn, and this is all that is usually practicable, the statistician can be reasonably confident that it is representative of the population and not a collection of the extreme items in that population. This knowledge of sampling distributions is used not only in problems of *estimation*, as they are termed. There is a second group of problems which have to be resolved on the basis of sample data; these are known as the *testing of hypotheses*.

Testing an Hypothesis

Let it be assumed that there is a widespread and generally accepted belief that the electorate is evenly divided in its support for two political parties. A public opinion poll based upon a sample of 900 respondents indicates that Party A has a six point lead over the other, *i.e.* the distribution of support is not 50:50 but 53:47 in every hundred voters. On the evidence of these sample data are we justified in rejecting the generally held view that the parties enjoy equal support? To answer this question we assume that the original statement is correct and by means of a statistical test determine whether or not the assumption is justified. In making the assumption that the electorate is equally divided between the two parties we are, to use the statistician's terms, setting up an *hypothesis*. In this case the hypothesis is that Party A and Party B each enjoy 50 per cent of the electorate's support. The hypothesis is then tested by means of a *test of sig-*

nificance. This is based upon our knowledge of sampling distributions and, in this particular case, the test enables the statistician to determine the probability of drawing a sample of 900 respondents from a population in which the respondents are divided 50:50 and getting a sample result of 53:47. In other words, from such a population what are the chances of drawing a sample of 900 in which supporters of Party A account for 53 per cent of the votes?

The test consists of calculating the standard error of the proportion in a population in which $p = q = \cdot50$ and n is 900. The formula is $\sqrt{\frac{pq}{n}}$ and by substitution we get $\sqrt{\frac{\cdot50 \times \cdot50}{900}}$ which equals approximately 1·7 per cent. Given this standard error it can be inferred that in such a population about 95 out of every 100 samples will, in the long run, produce sample proportions within the limits set by two standard errors about the mean proportion, *i.e.* 50 ± 3·4 per cent. From this it may be inferred that it is highly probable that a sample drawn from such a population would show that Party A enjoyed 53 per cent of the respondents' votes. In short, the sample result is quite consistent with the generally held belief that the parties enjoy equal support. Hence, the significance test supports the hypothesis in the sense that it gives no grounds for its rejection, because the sample gives a result which, in the long run, would correspond with that from some 95 per cent of all samples. If, however, the sample poll had shown Party A as enjoying 55 per cent of the electoral votes, then the conclusions drawn would have been rather different. Using the same standard error of 1·7, it is obvious that the sample result differs from the 50 per cent, which is assumed to be the population parameter, by almost 3 standard errors. From what we know of the sampling distribution of this statistic such a difference has a one in four hundred chance of appearing as the result of random sampling variation. Hence, it seems reasonable to conclude that it is not the result of random sampling errors but is, to use the statistical term, a *significant* difference. In other words, the evidence no longer supports the hypothesis that the sample could have been drawn from our defined population. It is much more reasonable to conclude that the sample came from a population in which the voters were not evenly divided. Hence the hypothesis is *rejected.*

While the basis of such tests is our knowledge of the sampling distribution in question and the actual calculation of the significance test, the decision as to whether or not the hypothesis should be *rejected* or *accepted* is ultimately a matter of judgement. By deciding in advance whether he will test the hypothesis at the 5 or 1 per cent level, the statistician is merely setting up a convenient rule of thumb by which to work. It cannot be sufficiently stressed that the test does not provide an answer which can be accepted with 100 per cent confidence. When the hypothesis in question is accepted at the 5 per cent level, the statistician is running the risk that, in the long run, he will be making the wrong decision about 5 per cent of the time. By rejecting the hypothesis at the same level he runs the risk of rejecting a true hypothesis on 5 out of every 100 occasions. By testing at the 1 per cent level he seeks to reduce the chances of making a false judgement, but some element of risk remains that he will make the wrong decision, *i.e.* he may accept where he ought to have rejected, or *vice versa*.

In brief, the test of significance is no more than a guide to action; the decision as to whether or not to proceed on the evidence of the sample data is made by the statistician.

S.E. of Differences between Means

Assume that the mean consumption of beef in the North of England as given by a sample enquiry among 845 households is 50 lb. per annum with a standard deviation of 13 lb., while a similar survey in the Home Counties covering 1,440 households yields a mean consumption of 48 lb. and a standard deviation of 12 lb. On the evidence of these data can it be inferred that beef consumption is greater in the North than in the Home Counties? In the language of statistics, is there a *statistically significant* difference between Northern and Southern households, or is it more likely that the difference of 2 lb. between the sample means can be attributed to chance, *i.e.* random sampling errors? When a difference such as this cannot reasonably be attributed to sampling errors, it is described as *significant*. Since, as we have seen, we can never be 100% certain about inferences drawn from sample statistics, we have to state at what level we consider the difference to be significant, *e.g.* either at the 95% or 99% level of confidence.

The question posed above can be phrased as follows: 'Is it reasonable to assume that these two samples were drawn from the same population?' The statistician answers this question by testing the hypothesis that the difference between the sample means can reasonably be attributed to sampling variation and hence the difference between them is not *significant*. More simply, it implies that the samples are from the same population. This is known as the *null hypothesis*. The result of the significance test will either strengthen or weaken confidence in the hypothesis; it can never prove *conclusively* that it is false. Thus we shall either *accept* or *reject* the hypothesis at a given level of confidence.

In this illustration we want to know what are the chances of a difference of 2 lb. occurring between two random samples drawn from the same population. To answer this question we first calculate the *standard error of the difference between means,* using the formula:

$$\sqrt{\frac{\sigma_1{}^2}{n_1} + \frac{\sigma_2{}^2}{n_2}}$$

Substituting the known values of the n and σ^2 in the formula we get:[1]

$$\sqrt{\frac{13^2}{845} + \frac{12^2}{1,440}} = \sqrt{0\cdot2 + 0\cdot1} = \sqrt{0\cdot3}$$

which yields an answer of 0·547 lb. This is the standard error of the difference. From our knowledge of the sampling distribution it follows that the chances of a difference between sample means greater than twice the standard error is less than 5/100, while a difference of more than two and a half times as great is likely to arise by chance only once in a hundred samples. The observed difference between the sample means is 2 lb. which is almost four times as great as the standard error. Such a difference could arise by chance only once in about ten thousand times. In other words, the likelihood that the two samples have been drawn from a single population is so small that the null hypothesis may reasonably be rejected. The observed difference is *statistically significant* and it can be assumed, in this case with considerable confidence, that the average northern household eats more beef than its southern counterpart.

S.E. of Difference between Proportions

The same ideas can be illustrated by an example involving a difference between proportions. Assume that a market research agency learns from a survey of 1,000 households that 30 per cent of them regularly purchase a particular branded product. A few months later the same agency carries out a similar survey and learns that 34 per cent of the households sampled purchase this product. Can it be concluded that there has been a four points increase in the product's popularity between the two surveys, *i.e.* a significant increase, or is it reasonable to attribute the difference to sampling variation? The *standard error of the difference between proportions* is given by the following formula:

$$\sqrt{\frac{p_1 q_1}{n_1} + \frac{p_2 q_2}{n_2}}$$

and substituting the observed values for p, q, and n we get

$$\sqrt{\frac{0\cdot3 \times 0\cdot7}{1000} + \frac{0\cdot34 \times 0\cdot66}{1000}} = \cdot021$$

which yields a result of 2·1 per cent. The observed difference of 4 per

---

[1] Note that the sub-scripts 1 and 2 to n and σ^2 in the formula merely serve to distinguish the two samples.

cent between the proportions is not quite twice as large as the standard error. We conclude, therefore, that the probability of random sampling errors accounting for the difference between the two percentages is just over 5/100. In other words, the statistician will be rather less than 95 per cent. confident that the change in the percentages is significant, *i.e.* that it reflects an increase in popularity of the product. In such circumstances the statistician will probably err on the side of caution and accept the hypothesis that the difference is not significant.

It is of the utmost importance to realise that such significance tests can never finally or absolutely prove or disprove a hypothesis, they merely provide evidence which the statistician must evaluate for himself. If the statistician rejects the null hypothesis, he does so because the evidence adduced by the significance test is such as to weaken his confidence in it to such an extent that any other conclusion would be unreasonable. For example, in the first illustration of the difference between the means of meat consumption of northern and southern households, where it emerged that the observed difference was only likely to arise by chance once in about ten thousand samples, the statistician would have no hesitation in rejecting the null hypothesis. Unfortunately, the results achieved by such tests are often similar to that obtained in the second illustration, *i.e.* close to the level of confidence at which it is decided that a result is significant.

It must be emphasised that there is nothing sacrosanct about the 5 per cent level any more than there is about the 1 per cent level. Experience has shown statisticians that the 5 per cent level is in the long run adequate for most of their needs; but if the experiments so require, then the significance of the results can be tested at a higher level. For example, a steelmaker selling a special alloyed steel, upon the toughness of which lives will depend, will want to know that the quality of his product is consistently above a certain level. By special significance tests based on sampling procedures he will ensure that the chances of sending defective goods will not be greater than, say, 1 in 1,000. A manufacturer of a consumer article may have a similar system of quality control and be well satisfied to know that the chances of despatching defective goods is not more than 1 in 20. The former will test at the ·01 per cent level, the latter at the 5 per cent level.

The application of such significance tests can be illustrated by a further simple example. Let us assume that over a long period a pottery has been accustomed to an average rate of defective products from the kilns of 10 per cent. A new firing technique is demonstrated and preliminary tests show that in 100 samples the rate of loss from this method is 7 per cent. The manager of the works believes that this improvement can be directly attributed to the new firing technique and is anxious to adopt the new firing methods. The board of the company is sceptical. Can the statistician help to resolve this problem? The standard of error of a proportion, it will be re-called, is given by the formula $\sqrt{\frac{pq}{n}}$ where n is the sample size, p the proportion of the sample possessing the characteristic and q the balance of the sample. The statistician must test the hypothesis that a variation in the proportion of defectives in samples equal to the observed difference of 3 per cent may quite often arise by chance. The tests will show what in fact are the chances of such a difference being encountered. Substituting the known values in the formula $\sqrt{\frac{0.07 \times 0.93}{100}}$ we get a standard error of 2·7 per cent. The observed difference between the proportions of defectives is only 3 points, which is little more than the standard error. It can be concluded that the chances of a loss rate as low as 7 per cent from the old firing technique are quite high, *i.e.* this loss rate can easily arise by chance. While further tests should be carried out with the new method, the available evidence strongly supports the argument that the new firing technique will not significantly reduce the proportion of defectives.

The same test can be illustrated by another example. A random sample of 2,000 households possessing television sets capable of receiving both B.B.C. and I.T.V. programmes reveals that 1,050 declare a preference for the latter programmes and 950 prefer the B.B.C. May it be concluded from the survey that a majority of all households which can receive both programmes prefer I.T.V.? There are two ways of answering this question. We can assume that the population of households are evenly balanced in their preferences and that the difference revealed by the sample is due to sampling errors. If this is the case, then the proportion preferring I.T.V. is 1,000/2,000 or 50 per cent. We can calculate the chances that in two samples of 2,000 households drawn from a population in which the proportion with a given characteristic is 50% we might get sample proportions as divergent as 950/2,000 and 1,050/2,000, *i.e.* 47½ and 52½%. The standard error of a proportion is given by the formula $\sqrt{\frac{pq}{n}}$ which, when the above values are substituted, reads $\sqrt{\frac{50 \times 50}{2,000}}$ The result is equal to approximately 1·1 per cent. If a large number of random samples of 2,000 households were taken, we may therefore expect 95/100 sample proportions ranging between $50 \pm 2.(1·1)\%$. The actual observed sample proportions of 47½ and 52½% are outside the limits set by the test at the 5 per cent level, but within the limits of $50 \pm 2·58 (1·1)\%$. In other words, the difference between the samples is significant at the 5 per cent level, but it is not significant at the 1 per cent level. If a new advertising policy involving substantial expenditure on T.V. advertising, etc., is contemplated it might be advisable to take further samples before a decision is taken.

Chi-square test

An alternative method of solving this last problem is to use another test known as the *chi-square* (pronounced ki) test. This is written as follows χ^2 the name being derived from the Greek letter '*ki*'. This test is used when an observed distribution of frequencies must be compared with the expected distribution on the basis of some hypothesis. For example, in the above case, if we set up the hypothesis that the programmes are equally popular, the expected distribution is 1,000 households to each programme; the observed distribution is 1,050 and 950. The formula for chi square is written $\chi^2 = \Sigma \frac{(O - E)^2}{E}$. The data and calculations can be presented in the form of a table set out hereunder:

	Observed	Expected	Differences	Differences^2
I.T.V. ..	1,050	1,000	$+50$	2,500
B.B.C. ..	950	1,000	-50	2,500

The formula requires us to derive for each cell the difference between the observed (O) and the expected (E) frequencies and then square them. In this case the differences are 50 and when squared they become 2,500. These figures are then divided by the expected frequency for that particular sub-group. In each case we get 2,500/1,000 which gives 2·5 in each of the two cells and these, as the formula indicates by the summation sign (Σ), are then added together, *i.e.* 2·5 + 2·5 = 5. This is the value of χ^2. We now compare this value with the value of χ^2 which for this particular number of sub-groups is given in specially prepared tables. We learn there that the relevant value of χ^2 is 3·84 at the 5 per cent level and 6·64 at the 1 per cent level. In other words our calculated value of χ^2 is above the value which could be expected to arise by chance 95/100 times, so that our hypothesis that the difference can be attributed to chance is rejected at the 5 per cent level, but not at the 1 per cent, since values of χ^2 up to 6·64 could be expected with such samples about once in one hundred times.

The chi-square test is a highly important and useful statistical test of significance. When we were testing differences between means and proportions we used tests based upon the standard errors of the corresponding sample statistics. These were all based upon the fact that we could assume that their distribution was Normal. Actually, the Normal distribution, important though it is in statistical theory, is only one of several distributions encountered by the statistician. The distribution of χ^2, for example is a special one but, as with the Normal distribution, special tables have been worked out which we can employ to interpret the values of χ^2 even if we cannot demonstrate the mathematical basis of the distribution.

The value of the χ^2 test lies in its application to the comparison of several frequency distributions. We set up the hypothesis that the distributions of frequencies are the same for both or all groups (there may be several), any differences evident between the observed distributions being due to sampling errors. The following example may help to explain both the purpose of the test more clearly, as well as demonstrating the method of calculating the value of χ^2.

T.V. Owners Classified by Income of Head of Household

	Under £750	£750–2,000	Over £2,000	Total
I.T.V. ..	320	160	20	500
B.B.C. ..	280	190	30	500
	600	350	50	1,000

The above data show the distribution of two samples of 500 viewers apiece, one group having declared a marked preference for the B.B.C. programmes and the other for I.T.V. Each sample is classified by reference to the income of the head of the household. For example, those over £2,000 may be regarded as 'upper' class, those with £750 – 2,000 as 'middle' class, while those earning less than £750 per annum we may define as 'working' class. At first sight the larger number of working class households revealing a preference for I.T.V. might be construed as signifying that the lower income groups as a whole prefer I.T.V., and the middle and upper classes B.B.C. programmes. Using this test, we can establish whether or not the differences between what are called the observed and expected frequencies may be attributed to chance, *i.e.* sampling errors, or whether they are *statistically significant*.

We assume for purposes of our test that the two samples have been drawn from the same population and that the difference between the distributions of frequencies is entirely the result of random sampling errors. We set up, in other words, the null hypothesis, *i.e.* that there is no difference in the distribution by income groups of households preferring I.T.V. or B.B.C. and that the observed difference between these two samples can be explained by random sampling errors.

We assume that the two samples are taken from the same population so the best estimate we can get of the distribution of the population among these three classes is obtained by aggregating the samples. The combined samples provide a distribution between the three classes of 600:350:50 and the two samples of 500 apiece are then apportioned on this basis between the three classes. The resultant distribution of each sample between the three incomes classes would be 300, 175 and 25 as set out hereunder:

EXPECTED DISTRIBUTION

	Under £750	£750-2,000	Over £2,000
I.T.V. ..	300	175	25
B.B.C.	300	175	25
	600	350	50

The next part of the calculation gives the differences between the observed and expected frequencies within each of the six cells:

I.T.V. ..	− 20	+ 15	+ 5
B.B.C. ..	+ 20	− 15	− 5

These differences are squared and then divided each by the expected frequency in that particular cell. The reader will see that the upper line of differences is the same as the lower except for sign, but when they are each squared the signs all become positive. In this case there is therefore no need to sum all six values of $\frac{(O-E)^2}{E}$ individually, just the top three and then double their aggregate.

$$\chi^2 = 2\left[\left(\frac{20^2}{300}\right) + \left(\frac{15^2}{175}\right) + \left(\frac{5^2}{25}\right)\right]$$

$$= 2\left[\frac{400}{300} + \frac{225}{175} + \frac{25}{25}\right]$$

$$= 2[1\cdot333 + 1\cdot286 + 1\cdot0]$$

$$\chi^2 = 2(3\cdot619) = 7\cdot24$$

The value of χ^2 comes to 7·24. Reference to the table of values of χ^2, an excerpt from which is reproduced below, shows that when $P = \cdot05$ and $n = 2$ (explained below), *i.e.* at the 5 per cent level, the value of χ^2 is equal to 5·99. This signifies that a value for χ^2 as great as 5·99 could arise by chance in samples of this size as often as 5 in a 100 times. The value of χ^2 when $P = \cdot01$ and $n = 2$ is 9·21. Since the calculated value of χ^2 for the above data is 7·24 it follows that such a value could occur more frequently than one in a hundred times as a result of sampling variation but not as often as one in twenty. On the evidence of this test the null hypothesis is rejected at the 5 per cent level, but not at the 1 per cent level. There is, therefore, on the evidence of sample data and the significance test, sufficient reason for believing that the distribution of households according to income is different for those who prefer I.T.V. from those who prefer B.B.C. programmes.

Note that the greater the value of χ^2, the more likely it becomes that the differences between the observed and expected frequencies cannot be attributed to chance. The student may already have realised that the value of χ^2 depends on two factors. The first is the actual size of the frequencies, or more precisely the differences, but these are of course directly related to the absolute size of the cell frequencies. By relating the cell differences to the corresponding expected frequencies any distorting effect on the value of χ^2 due to the absolute size of the cell frequencies is eliminated. The second factor is the number of cells. The more cells we have, the larger must be the aggregate of $\frac{(O-E)^2}{E}$ since the squared difference is always positive. The table of the values of χ^2 takes account of the number of cells as may be seen from the following small part of a table of values of χ^2:

DISTRIBUTION OF CHI-SQUARE[1]

Degrees of Freedom (n)	Probability level: P			
	0·05	0·02	0·01	0·001
1	3·84	5·41	6·64	10·83
2	5·99	7·82	9·21	13·82
3	7·82	9·84	11·34	16·27
4	9·49	11·67	13·28	18·46

The column on the left contains the values of n, which is dependent on the number of cells. The row across the top shows the various levels of probability to which the values of χ^2 for given values of n relate. For example, the column headed 0·05 gives the values of χ^2 if we are testing our hypothesis at the 5 per cent level; 0·01 values for the 1 per cent level, etc. When $n = 2$, a computed value of χ^2 must exceed 5·99 if it is to be regarded as significant at the 5 per cent level; if it is as much as 13·82, then such a result signifies that the null hypothesis can be confidently rejected, there being only 1 in 1,000 chance of it being true.

The symbol n refers to what the statistician calls the number of 'degrees of freedom'. This is a very important concept but at this

[1] This table is abridged from Table III of Fisher: Statistical Methods for Research Workers, published by Oliver and Boyd Ltd, Edinburgh, by permission of the author and publishers.

level of statistical knowledge it is not necessary to understand its precise meaning. All that is necessary is to be able to check a computed value of χ^2 against the tables. But for this purpose we must know how to obtain the appropriate value of n. It can easily be derived by the simple equation $n = (c-1)(r-1)$ when c represents the number of vertical columns of data and r the horizontal rows. In the above example, $c = 3$ because there were three classes of households and $r = 2$ since there were two main groups of viewers. Thus $(c-1) = 2$ and $(r-1) = 1$; therefore, $n = 2 \times 1 = 2$.

A particularly interesting illustration of the use to which the chi-square test may be put is provided by some recent research into the relationship between lung cancer and smoking. The following table shows the most recent amount of tobacco consumed regularly by women smokers before the onset of their present illness. The patients are divided into two categories, those with lung carcinoma and the so-called *control* group, who are similar in all respects with the former group except that they have diseases other than cancer.

Disease Group	Number of Patients smoking daily			
	1 +Cig.	5 +Cigs.	15 + Cigs.	Total
Lung-carcinoma patients ..	7	19	15	41
Control patients with diseases other than cancer ..	12	10	6	28
	19	29	21	69

Source: Smoking and Carcinoma of the Lung, British Medical Journal, 1950, ii, 739

A study of the data suggests that the number of lung carcinoma patients increases with the daily intake of tobacco, whereas in the control group the majority are in the lowest daily consumption group. To test whether there is any association with the amount of smoking and the incidence of lung cancer we use the χ^2 test. We start by setting up the hypothesis that among the above patients in the two groups there is no association between cancer and smoking, *i.e.* the null hypothesis. If the hypothesis is valid, then the distributions in the two groups of patients should be similar and the divergences apparent in the table should reasonably be attributable to random sampling errors.

The first stage is to calculate the expected distribution in the two groups over the range of tobacco consumption on the basis of the distribution in the totals. Thus the expected distribution as between the two groups of patients will be given by the ratio 41:28 for each class of tobacco consumption. Thus, the 19 cases in the class smoking 1 + cigarettes daily are apportioned as follows: $41/69 \times 19$ and $28/69 \times 19$, giving 11·3 and 7·7 respectively. The same fraction is applied to the next two classes of smokers containing 29 and 21 cases respectively. The remainder of the calculation is set out in the table below.

CALCULATION χ^2 FOR DATA IN TABLE ON PAGE 238

Observed	Expected	O — E	$(O — E)^2$	$\dfrac{(O — E)^2}{E}$
7	11·3	— 4·3	18·49	1·6
19	17·2	+ 1·8	3·24	0·2
15	12·5	+ 2·5	6·25	0·5
12	7·7	+ 4·3	18·49	2·4
10	11·8	— 1·8	3·24	0·3
6	8·5	— 2·5	6·25	0·7
				$\chi^2 = 5·7$

The resultant value of χ^2 is 5·7 and there are 2 degrees of freedom. The value of chi-square at the 5 per cent level where n = 2 is, as may be seen from the table on page 237, given as 5·99. It seems, therefore, that while the test provides evidence of a relationship between smoking and lung cancer, the result of the test does not allow us to reject the null hypothesis. Nevertheless, the result is very close to being statistically significant. The authors of the paper from which these data have been extracted, Professor Sir Austin Bradford Hill and Dr Richard Doll, explain the somewhat indecisive result by drawing attention to the very small samples and the fact that although the test is not conclusive, it nevertheless follows the pattern observed in data for male smokers where statistically significant relationships have been established. Apart from the contemporary interest in this particular social issue, this apparently inconclusive piece of evidence illustrates the type of problem which the statistician so often encounters, *i.e.* the test is not quite conclusive and thus he must either assemble new data or make up his mind as to his future policy in the light of the test and his basic knowledge of the field of enquiry.

Conclusions

Since nearly all statistical analysis is based on sample data, it will be apparent that significance tests form a most important branch of statistical technique. The few examples given in this chapter represent only the more important tests of general application. The primary purpose of this chapter is not so much to explain how the student can imitate the professional. It is first to explain something of the statistician's work in the vitally important field of sample statistics, and second to enable the reader to interpret the published results of research in fields ranging from the social sciences to engineering and chemistry where these methods are extensively employed.

Some Worked Examples

With the exception of some rather specialised statistics used in demography, *i.e.* the study of population, which are discussed in Chapter XVII, the foregoing discussion of significance tests brings to a close the sections of this book dealing with statistical methods. This would seem therefore the appropriate stage to give the reader an opportunity of revising and testing his understanding of some of the statistical methods discussed so far, more especially tests of significance. For the prospective examinee it should be pointed out that examination problems on tests of significance often require the candidate to perform some more basic calculations, *e.g.* the mean and standard deviation, to provide the data for the significance test. Of the four questions appended hereunder, the first and third are typical of this type of problem, while the second and fourth which require only the specific test to be calculated are also typical of their kind. The student is recommended to work through the four questions in turn and then to compare his workings and answers with the illustrations which follow.

1.   Distribution by Age of the Mothers of Triplets born in
England and Wales, 1958

Age of Mother	Under 20	20–24	25–29	30–34	35–39	40–44	All ages
Number of Mothers ..	1	13	29	25	19	3	90

Source: Registrar General's Statistical Review for England and Wales 1958.

Calculate the arithmetic mean and standard deviation of the above distribution. Assume that these mothers are a random sample of all mothers who had babies in 1958 and test whether their mean age is significantly different from that of all mothers, namely 27·84 years. Interpret your result.

2. A marriage survey was undertaken in Great Britain in 1960, covering a random sample of persons aged 16-59. The non-response rate was 18 per cent. The following table is based on the answers to one question, which was addressed only to married respondents.

Attitude to birth control	Respondents married in 1930–39	1950–60
	%	%
Full approval	63·5	69·0
Qualified approval	8·0	5·7
Entire disapproval	16·2	13·7
A neutral attitude	9·3	7·6
No opinion	3·0	4·0
Total	100·0	100·0
Number of respondents	635	739

Source: Griselda Rowntree and Rachel M. Pierce, 'Birth Control in Britain', Population Studies, Nov. 1961.

Test whether these data provide evidence that a higher proportion of persons married in 1950-60 than of those married in 1930-39 express full approval of birth control.

3.   *Annual Net Rents of a Sample of Households in England, 1959*

Annual net rents	Households in:	
	Decontrolled private (unfurnished) tenancies	Council tenancies
	%	%
£9 or less	1	1
£10–19	10	1
£20–29	12	8
£30–39	10	14
£40–49	6	17
£50–59	8	16
£60–69	8	20
£70–79	4	12
£80–99	11	9
£100 or more	30	2
Total	100	100
No. of households interviewed for whom annual net rent was known	118	283

Source: D. V. Donnison, C. Cockburn and T. Corlett: Housing since the Rent Act, 1961.

The above data are based on random samples. Test the hypothesis that there is no difference in the average annual net rent paid by the two groups of households.

9

4.
Professional Status and Social Origin of a Random Sample of 2,369 Teachers, 1955

Social Origin	Headmasters and Deputy Heads	Assistants	Total
Non-Manual	423	993	1,416
Manual	261	692	953
Total	684	1,685	2,369

Source: W. Scott, Fertility and Social Mobility Among Teachers, Population Studies, 1958.

Use the χ^2 test to determine whether these data provide evidence that a teacher's professional status is associated with his social origin.

CALCULATIONS

Example 1.

Age of Mother	f	$d'=5$	fd'^1	fd'^2
Under 20	1	−2	−2	4
20 — 24	13	−1	−13	13
25 — 29	29	—	—	—
30 — 34	25	+1	+25	25
35 — 39	19	+2	+38	76
40 — 44	3	+3	+9	27
	90		+57	145

Note: (i) The class 'Under 20' is assumed to be '16 and under 20', since 16 is the lowest age at which a girl may marry. Admittedly, babies have been born to girls younger than this, but in this distribution almost any reasonable assumption is acceptable since the frequency of that class is so small, *i.e.* 1 unit only.

(ii) The arbitrary origin for calculating the deviations is taken as the mid-point of the class 25 – 29, *i.e.* 27·5 (the variable is taken as continuous since the upper limit may actually be only 1 day off the lower limit of the next class and the deviations expressed in units of 5 years).

The arithmetic mean $= x^1 + \dfrac{\Sigma fd}{\Sigma f} \times$ c.i.

$$= 27\cdot5 + \left(\frac{+57}{90}\right) 5 = 27\cdot5 + 3\cdot16 \text{ or } 31 \text{ to the nearest year}$$

$$\sigma = \sqrt{\frac{\Sigma fd^{\cdot2}}{\Sigma f} - \left(\frac{\Sigma fd}{\Sigma f}\right)^2} \times \text{ c.i.}$$

$$= \sqrt{\frac{145}{90} - \left(\frac{+57}{90}\right)^2} \times 5$$

$$= \sqrt{1\cdot61 - (\cdot63)^2} = \sqrt{1\cdot61 - \cdot40}$$

$$= \sqrt{1\cdot21} = 1\cdot1 \times \text{ c.i.} = 1\cdot1 \times 5 = 5\cdot5 \text{ years}$$

The standard error of the sample mean *i.e.* s.e $_x$ =

$$\frac{\sigma}{\sqrt{n}} = \frac{5\cdot5}{\sqrt{90}} = \frac{5\cdot5}{9\cdot5} = 0\cdot578$$

To test whether the sample mean ($\bar{x}$) 30·66 years is significantly different from the population mean ($\bar{X}$), calculate:

$$\frac{\bar{X} - \bar{x}}{\text{s.e.}_{\bar{x}}} = \frac{27\cdot84 - 30\cdot66}{0\cdot578} = \frac{2\cdot82}{0\cdot578} = 4\cdot87$$

The ratio of the observed difference to the standard error is very much larger than that corresponding to the customary levels of significance, *i.e.* 1·96 or 2·58, equal to the 5% and 1% level respectively. The evidence justifies the rejection of the null hypothesis. The inference is that mothers of triplets are significantly older than the average of all confinements.

Example 2.

Attitude to birth contro		Respondents married in 1930–39	1950–60
		%	%
Full approval ..	..	63·5	69·0
Qualified approval	..	8·0	5·7
Entire disapproval	..	16·2	13·7
A neutral attitude	..	9·3	7·6
No opinion ..	..	3·0	4·0
Total ..	..	100·0	100·0
Number of respondents		635	739

The standard error of the difference between the proportions expressing 'full approval' in the two samples is calculated from the formula:

$$\text{s.e.} = \sqrt{\frac{p_1 q_1}{n_1} + \frac{p_2 q_2}{n_2}}$$

$$= \sqrt{\left(\frac{63\cdot5 \times 36\cdot5}{635}\right) + \left(\frac{69\cdot0 \times 31\cdot0}{739}\right)}$$

$$= \sqrt{3\cdot65 + 2\cdot9} = \sqrt{6\cdot55} = 2\cdot56$$

$$\frac{\text{Actual difference}}{\text{Standard error}} = \frac{69\cdot0 - 63\cdot5}{2\cdot56} = \frac{5\cdot5}{2\cdot56} = 2\cdot1$$

The ratio of the actual difference to the standard error, if it were equal to 1·96, would signify that the difference between the proportions was significant at the 5% level. The actual ratio is 2·1. Hence it is reasonable to conclude that there has been a statistically significant increase in the proportion of respondents who fully approve of birth control. Since no information about the non-respondents is given, no firm conclusions regarding possible bias are possible. In fact, the 18% rate is not unduly high for such a survey.

Example 3.

Annual Net Rents		Households in					
		Decontrolled private unfurnished tenancies			Council tenancies		
	d'	f	fd'	fd'^2	f	fd'	fd'^2
£9 or less ..	-4	1	-4	16	1	-4	16
10 — 19 ..	-3	10	-30	90	1	-3	9
20 — 29 ..	-2	12	-24	48	8	-16	32
30 — 39 ..	-1	10	-10	10	14	-14	14
40 — 49 ..	—	6	—	—	17	—	—
50 — 59 ..	$+1$	8	8	8	16	16	16
60 — 69 ..	$+2$	8	16	32	20	40	80
70 — 79 ..	$+3$	4	12	36	12	36	108
80 — 99 ..	$+4.5$	11	49.5	222.7	9	40.5	182.2
£100 or more ..	$+6.5$	30	195	1267.5	2	13	84.5
		100	$+212.5$	1730.2	100	$+108.5$	541.7

Decontrolled tenancies: $\bar{x} = 44.5 + \dfrac{212.5}{100} \times 10 = 65.75$ shillings

$$\sigma^2/\text{c.i.} = \frac{1730.2}{100} - \left(\frac{212}{100}\right)^2 = 17.30 - 4.52 = 12.78$$

$$\sigma = \sqrt{12.78} \times \text{c.i.} = 3.57 \times 10 = 35.7 \text{ shillings}$$

Council tenancies: $\bar{x} = 44.5 + \dfrac{108.5}{100} \times 10 = 55.35$ shillings

$$\sigma^2/\text{c.i.} = \frac{541.7}{100} - \left(\frac{108}{100}\right)^2 = 5.42 - 1.18 = 4.24$$

$$\sqrt{4.24} \times \text{c.i.} = 2.06 \times 10 = 20.6 \text{ shillings}$$

Difference between sample means $= 65.75 - 55.35 = 10.4$ shillings.

Standard error $= \sqrt{\dfrac{\sigma_1{}^2}{n_1} + \dfrac{\sigma_2{}^2}{n_2}}$

$$= \sqrt{\frac{35.7^2}{118} + \frac{20.6^2}{283}}$$

$$= \sqrt{\frac{1274}{118} + \frac{424}{283}}$$

$$\sqrt{10.8 + 1.5} = \sqrt{12.3} = 3.4$$

$$\frac{\text{Actual difference}}{\text{Standard error}} = \frac{10.4}{3.3} = 3.1$$

The actual difference is more than 2·58 times as large as the standard error and at that ratio the difference would be significant at the 1 % level. We conclude that the difference is significant.

N.B. Note that in this question the concentration of frequencies (30%) in the final class makes the calculation of the mean of the 'uncontrolled tenancies' extremely uncertain. By assuming, as has been done in the calculation of $\bar{x}$ and σ, that the final class interval is 100 — 139, the mean and the s.d. are perhaps somewhat inflated. This problem does not arise with the distribution of council tenancies which is fairly symmetrical about the mean, and the proportion in the final class is so small, *i.e.* 2%. Note the selection of the mid point of the opening class £9 or less has been taken as £4½; *i.e.* the class-interval is assumed to be £0–9. This is obviously very unlikely but is justified for three reasons; (i) only 1 % of the frequencies in each distribution lie in that class; (ii) the element of estimation is already considerable and this will hardly affect the result either way; and (iii) it greatly facilitates the calculation in terms of class interval units. The student who has doubts on these points may care to re-calculate $\bar{x}$ and σ of the 'decontrolled' tenancies taking the opening class as £5—9 with d' equal to 3¾.

Example 4.

Social Origin	Headmasters and Deputy Heads	Assistants	Total
Non-Manual	423	993	1,416
Manual ..	261	692	953
Total	684	1,685	2,369

The hypothesis is that a teachers' professional status is not associated with his social origin. In other words, his chances of promotion are independent of his social background. If this hypothesis is valid, then the distribution of the two academic groups, *i.e.* headmasters, etc., and assistants, as between the two social groups should be similar and the best estimate of their distribution is given by the distribution of the totals.

Thus, if the hypothesis is correct, we would expect the number of 'non-manuals' in the 'headmasters' class to be $\dfrac{1,416}{2,369} \times 684$ and the number of 'manuals' to be $\dfrac{953}{2,369} \times 684$. Likewise, the expected distribution of 'assistants' would be $\dfrac{1,416}{2,369} \times 1,685$ and $\dfrac{953}{2,369} \times 1,685$.

The value of $\chi^2 = \Sigma \dfrac{(O — E)^2}{E}$:

Frequencies		O — E	(O — E)²	$\dfrac{(O — E)^2}{E}$
Observed	Expected			
423	409	+ 14	196	0·49
261	275	— 14	196	0·71
993	1,007	— 14	196	0·19
692	678	+ 14	196	0·29
			$\bar{x}^2 =$	1·68

Number of degrees of freedom: $n = (c - 1)(r - 1) = 1$.

Table of χ^2

n	$P = 0.1$	0.05	0.01
1	2·7	3·8	6·6
2	4·6	6·0	9·2
3	6·3	7·8	11·3
4	7·8	9·5	13·3

The calculated value of $\chi^2 = 1.68$ is considerably smaller than the value which would be obtained if the differences between the observed and expected values occurred by chance 5 in 100 times, *i.e.* 3·8. Only if the calculated value were as high as 3·8 could the differences be regarded as significant (at the 5 % level) and the hypothesis rejected. The small value of χ^2 clearly does not provide sufficient evidence that the differences are due to anything but random sampling errors. Thus, the hypothesis is accepted and we conclude that a teachers' professional advancement is not generally related to his social origins.

Questions

SIGNIFICANCE TESTS

1. One often reads in survey reports phrases such as 'the difference between these proportions is statistically significant'. Explain what is meant by such a statement, and indicate the precautions to be taken in its interpretation. *Final D.M.A. 1961*

2. One group of 60 patients is treated with drug A, while another group of 80 patients is given drug B. After 14 days, 12 members of the A group were free of infection; 20 members of the B group were likewise free. The doctor in charge is convinced that treatment with drug B is superior. Comment on this view, explaining the basis of any statistical test you may employ. *I.M.T.A. 1961*

3. Calculate from the following distribution of 14-year-old schoolboy intelligence test markings the arithmetic mean and standard deviation. On the assumption that this group is a random sample, estimate the standard error of the mean and explain its usefulness.

I.Q. ..	Under 80	80–	90–	100–	110	120–	130–	140–	150–
No. ..	10	54	72	108	136	102	63	12	3

I.M.T.A. 1962

4. The following data relate to measurements of components produced by two different machines. Is there any significant difference between the output of the two machines?

Machine	Number measured	Average size	Standard deviation
A	1,200	31″	5″
B	1,000	30″	4″

I.C.W.A. 1963

5. In a sample survey of a group of regional hospitals, 480 out of 720 male in-patients declared themselves satisfied with their hospital diet. The corresponding figures for women were 420 out of 560. Is there any significant difference in this respect between the sexes? *I.M.T.A. 1959*

6. Out of a sample of 860 married couples in a United States town, 9·3% had no living children while 9·8% had four or more.

Among those with no living children 62·6% of the husbands were classified as 'happily married' while in the families with four or more children the percentage of 'happy' husbands dropped to 32·6.

Is the difference statistically significant?

How would you interpret these figures?

Source: Social and Psychological Factors affecting Fertility – volume VII.

B.A. Makerere University 1961

7. Age distribution of 1,139 Fellows of the Royal Statistical Society, 1957.

Age	No. of Fellows
20-24	14
25-29	140
30-39	438
40-49	336
50-59	157
60-69	49
Over 69	5
Total	1,139

Source: K. Gales, A Survey of Fellows, 'Journal of the Royal Statistical Society',
1958. *Final D.M.A. 1962*

The figures above show that about 12% of the Fellows were aged 25-29. On the assumption that the figures are based on a random sample of all the Fellows, compute the standard error of this percentage, and explain carefully how you would interpret it.

8. Attendance at Infant Welfare Centres by Wives of Professional and Salaried Workers and of Agricultural Workers.

		Wives of Professional and Salaried Workers	Wives of Agricultural Workers
Attending Centres	.. (%)	44·5	35·1
Others ..	.. (%)	55·5	64·9
Total ..	.. (%)	100·0	100·0
Sample number ..	..	569	245

Source: Maternity in Great Britain (1948)

On the assumption that these data are derived from random samples, test the hypothesis that there is no difference between the two occupational groups as regards the proportion of wives attending welfare centres. Explain carefully the interpretation of your result. *B.Sc. (Sociology) 1959*

9.　　　　Fresh fruit consumption by a sample of men students
at Edinburgh University

Oz. of fruit consumed per week	Men living at home	Men living in lodgings
0–10	9	16
11–20	14	18
21–30	9	5
31–40	11	3
41–50	8	1
Over 50	10	4
	61	47

Source: Kitchin, Passmore, Pyke and Warnock, 'British Journal of Social Medicine', 1949.

Test whether there is a significant difference between the average fresh fruit consumption by men living at home and that by men living in lodgings. Explain the meaning of your result. *Final D.M.A. 1963*

10. The following data are derived from daily reports on Wall Street during periods of rising prices of securities in the last months of 1955:

Description of market conditions	Rise in index of share prices*		Total
	Under 2 points No. days	2 points and over No. days	
Steady　..　　..	10	7	17
Firm　.. 　　　..	6	15	21
Total　　..	16	22	38

* *Dow Jones index, 30 industrial shares.*

Test for association between description of market conditions and extent of rise in the index.

Table of χ^2	P = 0·2	0·1	0·05	0·01
Degrees of freedom　1	1·6	2·7	3·8	6·6
2	3·2	4·6	6·0	9·2
3	4·6	6·3	7·8	11·3
4	6·0	7·8	9·5	13·3

B.Sc. (Econ. 1956)

CHAPTER XIV

SAMPLE DESIGN

In statistical work the purpose of sampling is to gain information about the nature and characteristics of the population from which the sample is drawn. In the two preceding chapters we have seen that a sample statistic is no more than an estimate of the corresponding population parameter. The extent to which the sample statistic may differ from the parameter based on a complete count or census is measured by the sampling error. The statistician is concerned with the *precision* of the sample statistic and modern sampling theory enables him to measure the precision or reliability of the statistic. While the relative size of the sample as compared with the population from which it is drawn is quite irrelevant, we know from our sampling theory that by increasing the sample size it is possible to increase the precision of the result.

The majority of sample surveys are, however, quite expensive undertakings. To increase the sample size to the extent required to achieve any substantial gain in precision would entail heavy additional expenditure. Furthermore, it would mean more interviews, more schedules to edit, more data to process. Thus, the interval of time between the survey and the production of its results would be increased. Since one of the objects of sampling is to save time and money, merely increasing the sample size would be a self-defeating policy. For such reasons statisticians in recent times have devoted much effort to devising methods whereby the precision of sample results may be improved. This can be done by using specially constructed samples, but underlying all these *sample designs*, as they are known, is the basic principle of random sampling. However complex the sample design may be, the selection of the sample units must ultimately be made in accordance with some random process. The designing of samples is a highly specialised branch of statistics; the design appropriate to one survey may be quite unsuited to another. The purpose of this chapter is to outline in simple terms the basic notions underlying the choice of sample used in various types of sample enquiry.

The Sampling Frame

Before the sample can be drawn, the units of the population under review must be defined. For example, the Electoral Register contains the names of all those entitled to vote in the United Kingdom; each local authority has a rating roll on which all rateable units in its area, *i.e.* houses, flats, shops, etc., are listed; every school will have registers containing the names of all the children enrolled for attendance. Such lists, card indexes, etc., form what is known as the *sampling frame*. It is imperative that, before drawing a sample, the statistician examines his sampling frame to ascertain to what extent it is adequate for his purpose. Since a random sample has been defined as one in which every unit in the population has a calculable chance of being included, if the sampling frame is deficient in respect of a number of units, then those units cannot possibly have any chance of inclusion and any sample drawn will be biased. That there are mistakes in a sampling frame of any size is inevitable. With lists relating to a human population, *e.g.* the Electoral Roll, some of the people listed thereon are dead. According to Dr Yates, 'all frames are likely to suffer to a greater or less extent from various defects, *i.e.* inaccuracy, incompleteness, duplication and being out of date.'[1] Thus, the first stage in any enquiry is to define the sampling unit and then to ascertain whether an appropriate sampling frame is available. For example, in the survey of savings carried out by the Oxford Institute of Statistics, the sampling unit was any group of people who could be expected to pool their incomes and their assets and who could agree on the use of them. This so-called 'income unit' was the sampling unit for this survey, but clearly there was no ready-made sampling frame of income units. In the event, the organisers drew a sample of households from the local authority's rating list and from such households ascertained the existence of income units appropriate for their survey In the 1953/54 Household Expenditure Enquiry too, households formed the sampling unit and, since no list of households was available, the Technical Committee had earlier[2] recommended the sample be drawn from local rating lists, rather than from either the Electoral Registers or the then available National Register.

For sampling the human population of this country there used to be three lists, *i.e.* sampling frames: the national register, the local authority rating lists, and the electoral roll. The first was the only complete list, but the second is especially useful for sampling households. The Government Social Survey was for many years favoured by enjoying access to the National Register; other survey organisations used the Register of Electors. When, however, the National Register was discontinued in 1952, the Social Survey was also compelled to adopt the Register of Electors as its sampling frame but it

[1] *Op. cit.*

[2] Interim Report of Cost of Living Advisory Committee Aug. 1951. Cmd 8238.

continued to use the Local Authority rating lists for such surveys as required them. Prior to this, however, the Social Survey had undertaken an enquiry into the value of the Register of Electors as a sampling frame[1] and the change was made without the difficulties which would have arisen without such knowledge of the Register's value. The following account is based upon that report.

The Electoral Register is estimated to include some 96% of the resident civilian population of England and Wales aged twenty-one years and over. Since the Register is used in both parliamentary and local government elections it is possible for sampling purposes to distinguish between parliamentary constituencies and local government wards. The former are broken down into polling districts which constitute the smallest sampling unit from this frame. Although the Register is revised each year so that it is reasonably up to date, any given Register is already four months old by the time it is published. It appears in March, the lists being based upon the electors' residence in the preceding November. It is effectively sixteen months old, of course, by the time the new edition appears. In other words, the Register is not continuously revised, but merely at yearly intervals, *i.e.* the November census of electors. Herein lies its main defect. Its other defect lies in the fact that a proportion of the population entitled to inclusion in the Register has not in fact been enumerated.

Unfortunately there are no accurate data available to measure the size of the error in the sampling frame arising from these omissions and unrecorded changes. The Social Survey report, however, contains an account of an investigation into this problem. It appears that about 4% of the loss is due to non-registration, while there is a further loss of $\frac{1}{2}$% monthly arising from removal. Thus, whereas at the date of publication, 94% of the eligible population are included, 12 months after, *i.e.* immediately before the new Register is due to appear, only 87% of the eligible population is correctly registered. If this short fall in numbers were evenly distributed throughout the population by reference to sex, class and income, the sampling frame would not be seriously biased. The authors of this report estimate that the 4% initial loss, *i.e.* due to non-registration is not merely relatively small in relation to the whole, but is probably unbiased in so far as it is spread over all groups of the population. The $\frac{1}{2}$% monthly loss by removal is more serious, since it appears that a high proportion of the removals are accounted for by the under-thirties, so that the age distribution of the population remaining within the sampling frame is slightly distorted. The report concludes that the current Register of Electors 'can be used with confidence as a

[1] The Register of Electors as a Sampling Frame, by P. J. Gray, T. Corlett, and Pamela Frankland, C.O.I., November 1950.

sampling frame if some procedure to deal with "moves" can be evolved'.

Random Sampling

Stress has been laid on the need to ensure that all samples should be random. To ensure random selection, various methods of selection of units from the population have been devised. The simplest is to number all the population units and then to draw, in lottery fashion, counters bearing the corresponding numbers of the selected numbers determining the choice of sampling units. This method is known as *simple random sampling*, in which every unit in the population has an equal chance of selection. This method is suitable where the population is relatively small and where the sampling frame is complete. As a matter of interest, it should be noted that even such an apparently foolproof method may give biased results. Professor Kendall has instanced the case where a particular colour of counter proved to be more slippery than those made from other colours, with the result that biased results were obtained. Instead of using this lottery method of selecting numbers from a hat, an alternative method is to use a table of random sampling numbers. Such tables have been specially prepared and comprise a series of digits drawn up in such a way that all the numbers are produced with equal frequency. An illustration of part of such a table is given in Yule and Kendall's book[1] on page 379. The electronic machine known as E R N I E, used for selecting Premium Bond winners, is really no more than a machine which produces random numbers in such a way that every valid bond held by a member of the public at the time of the 'draw' has the same chance as any other.

Systematic Sampling

In practice, simple random sampling is not practicable unless there is a sampling frame and the population is fairly small, *e.g.* in experimental work, when a sample of a group of results is to be chosen. For most practical work it is easier to select every *n*th item in a list of the population, the first of the sample units being selected by some random process. This method is termed *systematic*, or quasi-random sampling. Thus, if the lists comprise a population of, say, 25,000 and the sample required is 500, then the selection of every fiftieth item will yield the required sample. The starting point is determined by selecting at random a number between 1 and 50. Thus, if 37 turns up, the thirty-seventh item in the list is the first, the eighty-seventh and

[1] *Introduction to the Theory of Statistics* 14th edition. pp. 376-9.

one-hundred-and-thirty-seventh, the second and third in the sample, and so on. Thus, when the Social Survey carried out its enquiry among ex-miners suffering from pneumokoniosis, the first name was selected by picking a number between 1 and 7 and thenceafter every seventh card was selected from the files of the pension authority. If the 'lists' is in the form of a card index, there is no need to count the intervening cards if their number is large. By setting a ruler across the file, the card coinciding with a predetermined interval, *e.g.* every $3\frac{1}{2}$ inches, is selected.

One of the most recent official uses of systematic sampling was in 1946 by the Statistics Committee of the Royal Commission on Population. Great care was taken over the preparation of the sample comprising over one and a half million married women. By normal standards this was a fantastically large sample; most national surveys are based on about 2,500 units. The need for so large a sample was explained by the detailed breakdown of the sample which was to be be made and in some 'cells' or 'boxes', as those small sub-groups are called, there would only be a few hundred units – quite small samples. The selection was made by extracting every tenth card from the Ministry of Food's records and rejecting all cards for males and single women. The Ministry of Food records based on the coloured cards from ration books provided a classification of the population by sex and three age groups. Women could be classified from these records as married or single and the questionaire (reproduced on pp. 24-5) was sent to those described as married. If an error was made, the unmarried recipient of a form was asked to return it uncompleted. To the extent some single women had described themselves as married when first registering for the issue of a ration book, the sampling frame was inaccurate.

Strictly speaking, systematic or quasi-random sampling is not truly random. This is because once the initial starting point has been determined, it follows that the remainder of the items selected for the sample are pre-determined by the constant interval. Thus, if we are selecting every twentieth address from a street list, the first is admittedly chosen by random methods, but the remainder are thenceafter pre-selected. For this method of sampling too, a complete sampling frame is necessary. Nevertheless, this form of sampling approximates sufficiently closely to pure random sampling to justify its widespread employment. The list or sampling frame should be checked to see whether it has been previously arranged in such a way that a particular type of unit may occur at the appropriate interval and therefore be over-represented in the sample. Generally speaking, street-lists and alphabetical lists of names are free from such bias, *i.e.* non-randomness in the arrangement of the characteristic.

Stratified Sampling

So far it has been assumed that the population to be sampled consists of a single homogeneous group, *e.g.* ex-miners disabled by pneumokoniosis. In many surveys the population is far from homogeneous, but markedly heterogeneous. This applies to the adult population of a country, which comprises men and women in different age groups, in different social circumstances, and so on. Because of these differences in background, individual members of the population being surveyed to assess opinion about, say, road accidents, will have views on the problem but from experience it is known that certain social groups into which the population may be logically divided will think differently from other groups. If the population lists are classified into groups suitable for this particular survey, a more accurate reflection of these views is more likely to be obtained by sampling from each group, in proportion to the size of that group in the whole population. Each group will then be represented in the correct proportion within the sample. Such a sample is known as a *stratified* sample; in short, we speak of population *strata* and not groups.

It was argued earlier that even if the population is not stratified, any random sample will reproduce the distribution of the characteristic within the population. Stratification of the sample is derived automatically. Stratified samples can be drawn without first stratifying the population list and selecting from each stratum. Provided the relative sizes of the strata one to another are known, the sample members can be divided among the strata as they are drawn. As soon as the quota for any stratum is complete, any further items of that type are rejected and sampling continues until each stratum has its quota. This method will probably entail sampling a larger number than would be necessary if the population had been classified into its various strata at the outset. If the population can be so classified, then a stratified sample, *i.e.* one made up of random samples from each of the 'strata', is likely to be more representative of the population than any other sample of that size.

A little thought will reveal why a sample drawn from a previously stratified population is more likely to be 'representative' than a similar-sized random sample drawn without prior stratification of the population. When the population is stratified the statistician is in effect drawing a random sample from each stratum or homogeneous sub-population. Within each stratum random sampling errors must be taken into account. But the composition of the total sample, as far as its distribution between the various strata is concerned, corresponds with that of the population – because the statistician has arranged it so. In the case of a simple random sample from that population, two sets of sampling errors must be taken into account. The first are those *within* each stratum – as in the case of the subpopulations or strata within a stratified sample. Further sampling errors arise in the random sample because the distribution of units as *between* the various strata in that sample may not correspond with that of the population. It is this risk which prior stratification eliminates. The simple random sample *may* yield the correct composition of units from the various strata; but we cannot be certain and therefore when the sampling error for such a sample is computed, it is always greater than in the case of an equal sized stratified sample.

The sample can only be stratified if the statistician is in possession of information relating to the population which determines the stratification. For example, if a survey of farms in England and Wales is to be undertaken, then it is clearly relevant to ensure that all sizes are properly represented in the sample. This is possible because the Ministry of Agriculture has a complete sampling frame of farms and their sizes. It is possible to stratify a sample used in a pre-election poll depending on the occupation of the respondent. The distribution of the population by reference to these stratification factors is known from the data collected in the decennial Census of Population. Little purpose is served by stratifying a sample by reference to a characteristic or factor which is irrelevant to the survey. For example, a survey among school children to ascertain their TV viewing habits would hardly be improved by stratifying the juvenile population by reference to the political affiliations of the areas in which the children interviewed resided. Stratification by reference to whether they lived in urban or rural communities, and the type of school they attended would be significant for such a purpose.

In recent years the Social Survey has developed two indices which are employed for purposes of stratifying populations by reference to their socio-economic status. The first of these is the now well-known 'J' Index. The 'J', or *Juror* Index is based on the proportion of the population which possesses a jury qualification. An examination of the Electoral Register for each polling district will reveal that certain names are preceded by the letter 'J', which implies that those individuals are liable for jury service. This qualification is in the main dependent upon ownership or occupation of property above a certain rateable value. In other words, the larger the proportion of 'J' names in an electoral district, the higher is the number of occupiers and owners of property of a rateable value over certain limits. Since such ownership or occupation is correlated with income and social status, a high value for the 'J' Index in any area reflects a corresponding social class.[1] The second is the *Industrialisation* index, which serves much the same purpose. In this case the rateable value of an area

[1] The Social Survey report on this index is contained in a paper by Gray, Corlett and Jones entitled 'The Proportion of Jurors as an Index of the Economic Status of a District', and published by the Central Office of Information. The jury qualification is obtained either by being a householder, *i.e.* occupier of property of £30 R.V. or over in London and Middlesex and £20 elsewhere, or by ownership of property of £10 R.V. or more anywhere.

attributable to industrial hereditaments and transport undertakings is expressed as a proportion of the total rateable value of that area. The Social Survey has ascertained that in the provinces there is a significant degree of correlation between the degree of 'industrialisation' and the proportion of the population in the highest income group. This index, however, is not suitable for the London area and an amended form of the index is therefore employed. Here the rating areas are classified according to the rateable value per head of the inhabitants within each area. The higher the *per capita* rateable value, the wealthier is the district. These indices of stratification are nowadays regularly employed by the Social Survey and other organisations in the preparation of their samples. The difficulty with stratified sampling is that it is not usual for the population lists to be stratified. In every survey, therefore, the sampling units may have to be stratified in accordance with that particular factor which is relevant to that survey. Reference has already been made to the 'J' and the 'industrialisation' indices. Stratification may be based on rateable value per head, on the population per square mile in some sparsely populated areas, or by reference to size, *e.g.* population of districts. The stratification factor will depend on what type of stratification will be most useful for the particular survey. For example, the sample of Parliamentary constituencies used in the National Readership Survey carried out for the Institute of Practitioners in Advertising was stratified by reference to the ratio of non-Labour to Labour votes cast in the Parliamentary election. The higher the proportion of the Labour vote, the lower the social class of that constituency.

When describing 'simple random' sampling the point was made that this method ensured that every unit of the population had an equal chance of selection. When 'random' sampling was first defined, it was stated that this meant that each unit had a calculable non-zero chance of selection. This merely means that all the units have a chance of selection which can be estimated, but it does not imply that all units have equal chances. This is often the case with stratified samples where a larger proportion, or what the statistician terms *sampling fraction*, is taken from one stratum than from another. For example, in the Oxford Institute's Savings survey proportionately more rateable units, *i.e.* households, were drawn from the highest stratum than from the lower rated units. This was done because there are relatively few of the former compared to the others and it was essential to ensure adequate representation of this group in the final sample. When the final results are prepared, adjustments for the difference in the representation of the two strata can be made. This type of sampling is known as sampling with *variable sampling fraction*. In the case of a simple random sample the fraction will be equal, or to use the statistical term *uniform*, for all the strata. In many samples drawn from a large heterogeneous population the sampling fraction will be variable.

Multi-stage Sampling

In most surveys in which the sample is dispersed throughout the country, simple random or systematic sampling would prove to be extremely time-consuming and expensive. A sample of the electorate drawn by either of these methods would be distributed throughout the length and breadth of the kingdom in such a way that interviewers would spend very much more time travelling from one place to another than in actually interviewing, since they would be required to contact widely scattered respondents. Clearly it would be advantageous if the interviewing could be so arranged that groups of respondents were interviewed in certain areas. For example, instead of say sixty interviews being dispersed all over Yorkshire, as might well be the case if random sampling had been used, could not these same interviews be concentrated in two towns in that area, *e.g.* Darlington and Halifax, or in any other two towns selected at random from the towns in the region.

This concentration of interviewing is the objective of what is known as multi-stage sampling which, as the name suggests, is no more than a series of samples taken at successive stages. Thus, in the case of a national sample for purposes of, say, a national pre-election opinion poll, the first stage would be to break down the sample by reference to the main geographical areas, usually the Standard Regions.[1] The allocation of the sample, *i.e.* number of interviews, between the regions would be determined by the relative populations in each region. In the second stage, a limited number of towns and rural districts in each of the Regions would be selected and then, in the third stage within the selected towns and rural districts, a sample of respondents allocated to the Region could be drawn from the electoral roll of the areas selected at the second stage.

It will be seen that the essence of this type of sampling is that a sub-sample is taken of what are in effect successive groups or strata. The selection of the sampling units at each stage may be achieved with or without stratification. For example, at the second stage when the sample of towns and rural districts is being drawn, it is customary to classify all the urban areas in the region in such a way that the members of the populations in those areas are given equal chances of inclusion. This necessitates a system of sampling referred to as sampling with *probability proportional to size*. This merely means that

[1] In 1946 England was divided up into nine regions for administrative purposes. Since then Census and other national statistical data are classified on the basis of these regions. See Registrar General's annual review for details of their present composition

if we have five towns in the region, one of which contains 1 million inhabitants, and the others 50,000 apiece, then if a simple random sample of those five towns were taken, the chances of the large city being selected would be no better than those of any one of the smaller towns. This means that the individual inhabitants of the city would have a very much smaller chance of selection than the individual people living in the smaller towns. To avoid this situation, sampling with probability proportional to size is employed. This method ensures that the inhabitants of any town, whatever its size, have equal chances of inclusion. The method in principle is somewhat similar to numbering of the inhabitants of all towns and then drawing a sample using a table of random numbers. The effect of this procedure is to ensure that all inhabitants, regardless of the size of the community in which they live, will have equal chances of being selected.

At the third stage, when individual names and addresses are drawn from an electoral register of the areas sampled at the previous stage, stratification may again be used. For example, the parliamentary wards listed in the electoral roll may be classified by reference to, say, the 'J' index, and the sample then drawn in such a way as to ensure the appropriate representation of all classes in the community.

The sample used in the Household Expenditure Enquiry 1953/54 was a two-stage sample. The first consisted of the selection from some 1,800 local authorities in Britain of a sample of London metropolitan boroughs and 350 areas outside London. The second stage comprised the selection of addresses within the local authorities selected at the first stage. The first stage sample was stratified by reference to urban and rural populations, and after including the 70 urban areas with populations exceeding 100,000 apiece the selection of 200 smaller urban outside London and 80 rural areas was made on a random basis, but with probability of selection in proportion to the respective populations of the various areas.

The sample used in the 1963 National Readership Survey provides a further illustration of the above principles. At the first sampling stage 160 Parliamentary constituencies were selected after three initial stages of stratification. The advantage of using Parliamentary constituencies as first stage sampling units, instead of administrative units such as the county boroughs and counties, is that they are more uniform in size. The Parliamentary constituencies were first grouped by reference to the standard regions, then within each standard region those constituencies in towns of 200,000 or more inhabitants were grouped separately from other constituencies. Finally, within each of the resulting groups of constituencies they were ranked by the ratio of non-Labour to Labour vote and divided into strata of approximately equal size. From each of the resulting strata, two constituencies were selected with probability proportionate to the size of their electorates. Within each of the selected constituencies two polling districts were then selected, the selection again being made with probability proportionate to the size of the electorate, the wards comprising the polling districts having first been ranked by reference to the 'J' index.

Cluster Sampling

As we have seen, simple random sampling does not yield such a high degree of precision as stratified sampling, while multi-stage sampling is almost as much the product of economic as statistical considerations. But in all these samples, what is termed the *sampling unit* is the individual unit, *e.g.* an adult or a household. The larger the number of sampling units, the larger will be the cost of a survey. Sometimes the cost factor may necessitate a different form of sampling whereby interviewers concentrate all their interviews in a relatively small number of areas or groups. Suppose that a survey is being carried out over a large area in which the population is extremely dispersed. A random sample would be quite impracticable. Alternatively, the survey may be concerned with measuring the number of homes with refrigerators in a large area for which there exists no list of these homes, *i.e.* there is no sampling frame. To carry out a census to derive a sampling frame would be very expensive indeed. This was in fact done in the United Kingdom in 1950 as a preliminary to the Census of Distribution which was later to provide a sampling frame for future periodic sample surveys. But usually, where no list or frame exists, systematic random sampling is impossible. Furthermore, where the sampling units are widely scattered, the costs of a simple random sample could be considerable. If, however, in an urban area a few blocks of dwelling houses or localities were selected at random and *every* individual in each block interviewed, then if the blocks when put together form a sample which constitutes a representative group of the population, the statistician will have achieved his objective, *i.e.* a random sample of the entire population.

To meet the problem of costs or inadequate sampling frames in the United States this method of *cluster* sampling, sometimes known as *area* sampling has been devised. By the use of map references, the entire area to be surveyed is broken down into smaller areas and a few of these areas are selected by random methods. The primary sampling unit is then no longer the individual but is a group of individuals or households to be found within the selected area. Such groups are termed 'clusters'. Within each area selected every unit, *e.g.* household, may be interviewed. Sometimes it is only a proportion, say, one in four households. Nothing need be known in advance about the area, the number or type of sampling units in it, but by following these procedures the chances of inclusion in the sample can be made the same for all individual units to be found within the area.

The basic problem with this type of sample is whether or not the units within the clusters are homogeneous. The danger undoubtedly exists that clusters often tend to comprise people with similar characteristics and since the statistician is picking out only a few clusters, he may find himself with a biased or non-representative sample. If the individual clusters are heterogeneous, *i.e.* made up of all types of individual, then the final collection of 'clusters' may well constitute a random sample. If, however, the clusters are highly homogeneous in their composition, the reverse is true. In other words, whereas the statistician who wants to be able to stratify his sample is concerned to ensure that each stratum in the population is homogeneous, the same statistician using cluster sampling would prefer the areas, *i.e.* strata from which he is sampling to be heterogeneous. In practice, the statistician using cluster sampling is well advised to take a sample consisting of a large number of small clusters rather than a similar sample containing only a few large clusters.

Cluster sampling has been evolved in the United States because it permits surveys to be undertaken with low costs. For example, cluster sampling may be used in the second stage where instead of sampling at random from a group of towns, all the interviews are concentrated in one or two, in order to save time and money. This technique is also useful where adequate sampling frames for the relevant populations are not available, *e.g.* in the under-developed countries. In this country we are not confronted with the problems of widely dispersed populations although area sampling was used in the United Kingdom for the Census of Woodlands in 1942 for which it was eminently suited.

Quota Sampling

With random sampling, the interviewer plays no part in the selection of her respondents; she is merely given a name and address. To obtain her interview she may have to call several times at the address before finding the respondent at home or willing to spare the time for the interview. In practice, such call backs are usually limited in number. Usually the interviewer will call three times, and then if unsuccessful write the respondent off as a 'no contact'. Because of deaths, removals, illness, absences from home, etc., it is inevitable that non-response through 'no contacts' of this kind may be considerable.

The process of call backs is time consuming and hence expensive.

To economise in time and cost, a method of sampling known as *quota* sampling is extensively employed by many commercial survey organisations. The essence of quota sampling is that the final choice of the respondent lies with the interviewer, although in making her selection she must ensure that the respondent satisfies certain criteria which are laid down by the survey organisation. Thus, instead of obtaining from head office a list of names and addresses, as is the case with random samples, the interviewer is instructed to carry out a number of interviews with individuals who conform to certain requirements. For example, she may be asked to interview ten men and ten women, two of them upper middle class, five from the lower middle class, and thirteen from the working class. Furthermore, some of them should be between 16 and 24 years of age, others between 25 and 44, and the balance over 45. In other words, the interviewer's choice of respondent is partly dictated by these 'controls'.

The basic controls used by the survey organisations which employ quota sampling are three in number: age, sex, and social class. The first two are obviously straight forward. The determination of social class is somewhat more difficult. Additional controls may be introduced by the survey organisation in order to ensure that the interviewers carry out their interviewing with the appropriate respondents. For example, further controls such as married or single and, in the case of women, housewife or gainfully occupied may be employed. At the foot of the schedule reproduced on page 290 there are several such classificatory questions. A difficulty arises with the multiplication of such controls since it becomes increasingly difficult for the interviewer to ensure that her respondents satisfy the various criteria laid down by the controls. Since the object of quota sampling is to simplify the interviewer's task and to save the survey organisations money, it is customary to restrict the controls to the three main characteristics cited above.

Quota sampling has been severely criticised in the past by professional statisticians because it does not satisfy the fundamental requirement of a sample, *i.e.* that it should be random. Consequently there is no justification for calculating the standard error of statistics based on quota samples. In other words, it is not possible to determine the precision of the results on any valid basis. While this is the main defect, there are other criticisms. For example, the social classification of the population used to identify respondents is based on somewhat shaky statistical foundations. The Registrar General's decennial Census of Population provides a break-down of the population both by social class, *i.e.* five groups, and since 1951 a socioeconomic classification comprising thirteen groups,[1] The classi-

---

[1] These are discussed in Chapter XVI.

fication used by survey organisations in compiling their quotas, however, is very much simpler than this. Three or four main groups are usually employed. For example, the Gallup Poll uses four groups, the upper group comprising 5 per cent of the population; the upper middle class estimated to contain 21 per cent of the population; while the lower and middle working class has 59 per cent; and the very poor 15 per cent. Other organisations employ a somewhat similar breakdown. The interviewer is given instructions as to the characteristics which are relevant to each social class. The best guide is the nature of the respondent's employment or occupation, but the interviewer may pay due regard to the appearance, accent, and other visible attributes of the respondent, *e.g.* he lives in an expensive house with large car, in order to classify him.

In the 1963 National Readership Survey the interviewers were required to classify their respondents, who were selected by random sampling methods and not by quota, by reference to their social class. To ensure a standardised and uniform classification by all the interviewers, the following classification was prepared for their guidance.

CLASSIFICATION OF RESPONDENTS BY SOCIAL GRADE

Social Grade	Social Status	Head of Household's	
		Occupation	Income likely to be
A	Upper middle class	Higher managerial administrative or professional	£1,750 or over per annum
B	Middle class	Intermediate managerial, administrative or professional	£950 – £1,750
C1	Lower middle class	Supervisory or clerical, and junior managerial administrative or professional	Under £950 per annum
C2	Skilled working class	Skilled manual workers	Between £12 and £20 per week
D	Working class	Semi and unskilled manual workers	£6 10s.– £12 per week
E	Those at lowest levels of subsistence	State pensioners or widows (no other earner), casual or lowest-grade workers	Under £6 10s. per week

Source: I.P.A. National Readership Survey 1963.

They were also told that the respondent's social class was primarily determined by his occupation, or if retired, by his former occupation. Where there was no occupation, or information about it was un-obtainable, then other characteristics had to be taken into account, *e.g.* the type of dwelling, the amenities in the home or the presence of domestic help. In particular, interviewers were reminded that the above income classification was only an indication and did not deter-mine the social grade of the respondent.

In an attempt to ascertain the extent to which quota samples were truly representative of the population, Messrs. Moser and Stuart carried out certain tests on a number of experimental quota samples interviewed by various survey organisations.[1] They found that inter-viewers tended to complete their quotas with too many better educated members of the various social groups, and that furthermore certain industrial groups, particularly in the lower and less well paid occupations, tended to be under-represented given their relative im-portance in the overall population. They recommended that a form of 'industrial' control should be introduced, *i.e.* instructing inter-viewers to select respondents who followed particular occupations. The introduction of further controls, however, as is pointed out above, complicates the interviewer's task. This criticism has been met by most of the organisations by giving their interviewers instructions as to how they can contact suitable respondents, for example, in shops, office workers, housewives at home, etc. Nevertheless, the fact has to be faced that the statistical foundations for this type of social classification of respondents are still rather limited. Another weakness of quota sampling stems from the fact that a great deal of interviewing is often carried out on the streets. While this facilitates making con-tact with respondents, it means that people who are out and about are more likely to be represented in the sample than those whose work tends to keep them at home, for example housewives with children. In some cases, where the interviewer may start work a little later in the morning, she will tend to get a disproportionate number of office workers, rather than factory workers who leave home earlier. Here again, the survey organisations have sought to remedy this potential defect by instructions to interviewers as to where they should try and select their respondents.

As against these various shortcomings of quota sampling, there are a number of advantages. The first is obviously cheapness. Each inter-view tends to cost about half that of an interview in a random sample. Second, of course, there is speed. There is no need for call-backs on absent householders. An important point made in the defence of quota sampling is that the more serious defects and mistakes in any

[1] An Experimental Study of Quota Sampling, *J.R.S.S.*, Part 4, 1953.

survey tend to be made at the interviewing stage and in the processing of the schedules. There may, of course, also be defects in the schedule itself. The sample itself is probably a smaller source of error than are these factors. Consequently it can be argued that quota sampling, when the overall results are assessed, may yield just as reliable information as a random sample. Another argument in favour of quota sampling is that the problem of non-response does not invalidate or affect the representativeness of the sample. This is incorrect. Obviously there is non-response in so far as some people, when approached by an interviewer, will refuse to be interviewed. But since the interviewer can then approach someone else who fits her control and get an interview, unless she records the fact of the refusal it will be unknown. Interviewers using quota sampling are generally required to keep a note of such refusals, but Moser[1] has pointed out that these records are of limited value in so far as nothing is then known of the characteristics of people who refuse to be interviewed.

Statisticians generally and, of course, the commercial survey organisations, recognise that quota sampling cannot in a purely statistical sense be regarded as a substitute for random sampling. The Social Survey in its early days used quota samples occasionally but has for many years used only random samples. More recently, a leading commercial survey organisation announced that its public opinion polls would be based only on random samples. Certainly the worst features of quota sampling which characterised the early American experience with it, for example, interviewers sitting at home and completing schedules based on imagined interviews, have been overcome. To avoid this particular problem, all the survey organisations carry out checks on interviewers' work, usually by contacting respondents[2] or re-interviewing them. These checks are effective, no doubt, primarily in so far as the knowledge of their existence may deter an interviewer from trying to cheat the employing organisation. Survey organisations are well aware of the potential defects and in most cases take great pains to train and instruct their interviewers, on whom in the last resort the validity of the results must depend. After an extensive analysis of the results of quota sampling, Messrs. Moser and Stuart conclude that in practised hands this method gave fairly accurate results. Certainly the past record of the pre-election opinion polls lends weight to this conclusion and it seems as though it will be many years before this speedy and economical method of carrying out surveys is replaced entirely by random sampling.

---

[1] Quota Sampling, *J.R.S.S.*, Part 3, 1952.

[2] At the end of the interview the respondent is asked for his name and address. See the Schedule reproduced on pp. 288-92.

The Sample Size

The point has already been made that the costs of a sample survey are directly related to the size of the sample used. The object of sample design, as it is called, is to maximise the degree of accuracy or precision in the sample results for any given outlay. We have seen that a stratified sample will give a greater degree of precision than a simple random sample; while both multi-stage and, to an even greater extent, cluster sampling represent compromises between statistical and economic considerations. Whichever type of sample design is used in a survey, the inevitable question arises as to the size of sample to be taken. If one asks the simple question, 'what is the appropriate sized sample for a particular survey', the answer is invariably, 'the largest practicable' since every increase in the sample size brings with it some increase in the precision of the sample estimate. The point has also been made, but it bears repetition since it often puzzles the layman, that the size of the population from which the sample is to be drawn is quite irrelevant.

The key to the question as to the appropriate size of a sample is determined by the results required. Let us assume that the leaders of a political party want to know the proportion of the electorate which approves their particular policy. The statistician may inform them after an opinion poll has been taken, that he is 95 per cent certain that between 40 and 50 per cent of the electorate support the party. This is clearly of little value; 50 per cent means victory at the polls, the figure of 40 means defeat. In the example quoted, it is quite clear that the standard error of the percentage is $2\frac{1}{2}$ per cent since the 95 per cent level of confidence sets limits of twice that error about the sample statistic of 45 per cent. To give more precise results at the same level of confidence, the statistician must take a larger sample and his clients must therefore pay more for his work. Suppose the clients will be satisfied to know with 95 per cent confidence within one per cent either way the proportion of the electorate supporting them. In other words, the sample must yield a standard error of 0·5 per cent. From the formula for the standard error of a percentage, $s.e. \% = \sqrt{\frac{pq}{n}}$ we can by substitution arrive at the required sample size. Thus:

$$0{\cdot}5 = \sqrt{\frac{45 \times 55}{x}}$$

$$0{\cdot}25 = \frac{45 \times 55}{x}$$

$$0{\cdot}25x = 2{,}475$$

$$\therefore x = 9{,}900$$

Strictly speaking, the above formula applies only to a simple random sample. As has been explained, the gains in precision from prior stratification of the population or the sample are considerable. But the formula for deriving the standard error of a stratified sample is much more complex. It consists largely of summing the standard errors within each of the strata making up the sample. Similarly, the calculation of the standard error of a multi-stage sample is complicated by the fact that at each stage a random sample is taken of the relevant sampling units and these standard errors accumulate. Generally speaking, however, the above simple formula based upon the standard error of a proportion gives a useful and easily calculated guide to the maximum sample required in a survey which is concerned to ascertain the extent to which the population possesses a particular attribute, e.g. watches I.T.V. or votes Liberal, etc. When the statistician is dealing with variables, e.g. the average income of members of a given population, then a different formula is required.

The main object of the foregoing section is to impress upon the reader that sample size has nothing to do with the size of the population and that, in a sense, the statistician works backwards from his probable results to decide upon the required sample. The important consideration is the degree of precision required in the results. The importance of costs has been much stressed, but no statistician will subordinate statistical considerations to considerations of finance. His function is to advise his clients as to the best and cheapest way of obtaining the information they require. If they are not prepared to meet the cost of what the statistician considers to be the minimum sample required to yield the information they have asked for, then he will advise them that to undertake the enquiry will merely waste their money.

Conclusions

In this chapter an attempt has been made to describe in simple terms the main types of sample which are currently employed in survey work. Reference has also been made to the considerations which may determine the sample design employed by the statistician. Great emphasis has been placed upon the need for random selection of the sampling units because only if this rule is observed can the precision of sample statistics be measured by calculating their standard error. Considerations of economy and time have led to the widespread adoption by many commercial market research agencies of quota sampling. From the statistical point of view this method is inferior to random sampling; but random sampling is itself subject to other weaknesses.[1] Provided the data are available to enable quota

[1] These are discussed in Chapter XV.

samples to be stratified in some detail, the results achieved are un-doubtedly good enough for the purposes for which quota samples are generally used.

Further Reading:
Basic Ideas of Scientific Sampling. A. Stuart, No.4 in Griffin's Statistical Monographs.

Questions

SAMPLING DESIGN

1. 'Randomness is the essence of good sample design.' Explain and discuss.
I.M.T.A. 1962

2. Define the meaning of 'random sampling' and explain precisely why randomness is important in sample design. To what extent does the use of stratification affect the randomness of a sample? Illustrate your explanation.
B.Sc. (Sociology) 1963

3. Explain the role of stratification in sample design, illustrating its use and advantages with reference to any surveys known to you. *B.Sc. (Sociology) 1961*

4. Explain what is meant by (*a*) stratified, (*b*) multi-stage sampling. Illustrate your answer by reference to the sample design in any surveys with which you are familiar. *B.Sc. (Sociology) 1958*

5. What is a stratified sample? In what ways does it differ from a simple random sample, and what are the advantages and disadvantages of using this samplying technique? *B.Sc. (Sociology) 1961, University of Ghana*

6. Explain what is meant by random sampling and quota sampling. Discuss the advantages and disadvantages of each method. *B.Sc. (Sociology) 1956*

7. In what kinds of enquiry are the various forms of sampling appropriate?
I.M.T.A. 1962

8. 'Many large-scale surveys are based on stratified random samples involving two or more stages of sampling.' Explain this statement, if possible illustrating your explanation by reference to surveys with which you are familiar.
Final D.M.A. 1961

9. Explain the respective roles of randomness and stratification in sample design, illustrating your explanation with any surveys with which you are familiar.
Final D.M.A. 1963

10. Distinguish between multi-phase and multi-stage sampling, and explain the purposes and merits of each *Final D.M.A. 1962*

11. Explain the considerations determining the distribution of a sample of a given size among a given number of strata.   *B.Sc. (Econ.) 1959*

12. It is proposed to carry out a sample enquiry in your local authority to ascertain the reactions of householders to the new valuations.

Explain: (i) how you would select your sample; and (ii) how large you consider the sample should be.   *Rating and Valuation Association 1963*

PART III
SOCIAL STATISTICS

CHAPTER XV

SAMPLE SURVEYS

Development of Surveys

The enumeration of populations by means of a census is centuries old; the Egyptians and Romans both carried out censuses for fiscal and military purposes. The sample survey, however, is of quite recent origin. It is, nevertheless, customary to start any history thereof with the great social enquiry of Charles Booth entitled 'Labour and Life of the People of London' which filled seventeen volumes and took more than a decade to complete. A few years after Booth had published his main findings, Seebohm Rowntree carried out his enquiry into poverty among the working classes of York. This was published in 1901 under the title 'Poverty: A Study in Town Life'. Neither of these pioneer enquiries could be described as sample surveys. They were virtual censuses of the relevant populations. In Booth's case the population was the working class family with children of school age. For Rowntree it was all working class families, the latter being defined as households where no resident domestic help was kept!

The first survey based upon a random sample of the population was carried out in 1912 by the late Professor Bowley in Reading. Like his great predecessors, he was concerned to measure the incidence of poverty among the working classes. He incorporated Rowntree's device of measuring the incidence of poverty by reference to a minimum living standard, but in place of a census used a one in twenty sample of addresses taken systematically from a street directory. With the growth of unemployment after the war, surveys into poverty were undertaken in many cities. The best known are those in London by Professor Bowley and his assistants from the London School of Economics and that on Merseyside prepared by Caradog Jones of the University of Liverpool. It was not until the later 1930's that the public became at all aware of the growing use of sample surveys. Their attention to this subject was attracted by the well-publicised public opinion polls which had established a considerable reputation in the United States. One or two commercial agencies were also beginning to adapt the technique for market and consumer research purposes. On the outbreak of the war, the Central Office of Information created the Wartime Social Survey.

The Social Survey, as it is now called, is still in being and is contributing much to the development of the techniques used in sample survey work. Originally it was employed by various government departments on *ad hoc* surveys to learn what the public felt about certain issues, *e.g.* clothes and fuel rationing. After the war, it carried out surveys into labour problems, *e.g.* why recruitment to the nursing profession was so poor and why miners were leaving the mines. In recent years the Social Survey has become increasingly involved in the work of improving the quality of economic statistics necessary to the formulation of a coherent economic policy, in particular, consumer expenditure surveys and consumer outlays on durable consumer goods. The annual National Food Survey enables the Ministry of Food and Agriculture to adjust its policy in the light of changes in nutritional standards and dietary habits of households. The Family Expenditure Surveys undertaken for the Ministry of Labour form the basis of the index of retail prices. Other surveys have been used to determine the use made of scientific and technically trained manpower, as well as to measure the degree of mobility of labour between jobs and between occupations. Rather less emphasis is nowadays placed on surveys concerned with social issues; it is a matter for regret that the post-war Survey of Sickness was abandoned, because it provided the first comprehensive data on the nation's state of health. A similar survey among users of the National Insurance hearing aid resulted in some improvements in design, as well as the organisation of repair facilities. The Social Survey assisted in a study of Borstal inmates and the problem of recidivism; the same organisation played a major role in the statistical surveys which formed the basis of the recommendations of the Robbins Report on Higher Education. In short, the Social Survey has become an integral part of the machinery available to the government for briefing itself on any public issue. The sample survey has become both an efficient and recognised tool for the administrator.[1]

Definition

To the layman a sample survey may appear to be an inferior substitute for a census. Reasons for not taking a census may be that the population in question is too large and the census would take too long and would be too expensive. In some cases, however, the sample survey may be preferable to the census, not merely on account of the lower cost and greater speed with which results are made available, but because it is superior to the census for the particular enquiry. The Director of the Social Survey has defined a sample survey as a *method of collecting detailed information relating to representative groups under*

[1] For an interesting exposition of this point read 'The Government Social Survey: An Aid to Policy Formation' by L. Moss, a lecture to the Royal Institute of Public Administration published in the Journal of the Institute 1960

controlled conditions.[1] This particular definition brings out all its main features. The outstanding advantage of such a survey over the census lies in the fact that it is practicable to collect much more detailed information from a relatively small number of people than from a large number. The former method permits the use of trained inter-viewers who can elicit a great deal of detailed information not merely of a factual nature, but also opinions. The census is only satisfactory for collecting factual data and even in that case some of the in-formation received must be regarded as distinctly doubtful in terms of its accuracy.

The definition emphasises that the sample is representative, a fact which is all too often taken for granted. In theory, however represen-tative it may be, it is still inferior to the 100 per cent enumeration of all the population units which is implied in the term 'census'. In practice, however, no census covering a population of any size at all, is 100 per cent complete. For example, as was cited earlier, the Register of Electors in this country appears to have a deficiency of 4 per cent due to non-registration. The U.S. Bureau of the Census carried out a series of investigations to test the reliability of data assembled from its 1950 population census which revealed quite sub-stantial 'under-counts' and deficiencies for certain groups within the population. Clearly this type of problem is much more serious in a country so great in area and so diversified in race, education and language as the United States. But in varying degrees the problem is present in any census. Lastly, the survey has the merit that the in-formation is collected under what Mr Moss has called 'controlled' conditions. It is difficult, he states, 'to exercise close control at the critical point, the point when the informant's information is put down on the questionnaire or schedule'. Even when the census authorities use enumerators (few of whom can be really well trained in the work) to collect the forms and where necessary help the respondent to com-plete them, this weakness is very serious in any large census. With a sample survey covering at most a few thousand informants it be-comes possible to employ skilled investigators. They can explain the purpose of the survey to the respondent and ensure that each question is correctly put and understood.

It is for such reasons that the sample survey has come into prom-inence for its own merits rather than as an inferior substitute for the infrequent and cumbersome census. As Mr Moss has pointed out, 'experience seems to show that it is wrong to assume that a census must automatically be more correct than a sample'. Indeed, American

---

[1] The Scope of Sample Surveys, L. Moss. Read before the Conference on Modern Sample Survey Methods, organised by the Association of Incorporated Statisticians, December 1953. This, together with the other papers read before the conference, has been published by the Association in booklet form. This body is know as the Institute of Statsticians.

experience suggests that the only method for testing the reliability of census data is a properly designed sample survey!

Some survey organisations are much better known to the public than the Social Survey, for example, the survey bodies responsible for the pre-election polls which are published at regular intervals in the daily press. Among these, the Gallup Poll and the National Opinion Poll are probably the best known. The B.B.C. Audience Research Department uses part-time interviewers to learn the views of radio and television audiences. Each day 3,000 adults in the seven B.B.C. regions are interviewed as well as a further 1,000 children under sixteen years of age. The purpose of these daily surveys is to ascertain the audience size for each programme broadcast. Audience reaction to the programmes is learned from the returns made by volunteer panels of listeners and viewers. An important and well-organised survey is that sponsored by the Institute of Practitioners in Advertising. This is concerned with media research, in particular the readership of papers and journals so that advertising agencies may be better able to advise their clients on their advertising policies. For this survey a very large sample of nearly 20,000 names and addresses is drawn, the reason for the large number being the need to ensure that minority interests are adequately represented. Illustrations and references to the work of all these bodies are made in the rest of this chapter.

Stages in the Survey

Without some understanding of the principles and problems of survey work an adequate assessment of survey data and results is hardly possible. Most surveys are designed to assess individual views on current political, economic and social issues; they have in common a carefully designed schedule of questions and intensive interviewing. Different types of schedule will be employed for various types of enquiry, just as the sample may differ. The basic pattern, however, of all surveys may be described quite briefly. As a start it is sufficient merely to detail the successive stages in such a survey.

1. The first stage may be described as providing an answer to the question 'what is the problem under review and in what way can the survey help?' This cannot be answered without a detailed study of all the facts. The organiser may have to spend a long period immersing himself in the subject and learning just what his client's needs really are. Published material must be examined to see how far it is of any use for the client and whether it will simplify the survey or even make it unnecessary. The maximum information must be derived from the survey for a given cost. Not

merely must the information be relevant to the enquiry but it is important to avoid the situation which can so easily arise at the end of a survey, when it becomes clear that it would have been helpful if only some additional data had been collected on a particular point. Too much time cannot be spent on these initial stages if time and money are to be spent to the best advantage. Such observations may appear somewhat trite and obvious; it is nevertheless surprising how often it is the obvious which is overlooked!

2. How is the information to be obtained? The two main methods are the postal enquiry and the survey employing interviewers. The merits of each are discussed below.

3. The preparation of the schedule of questions and instructions for their completion. A badly designed questionnaire may ruin an otherwise well conducted survey. If machine tabulation is employed the answers to the individual questions in the schedules will have to be coded. It may seem premature to discuss tabulation at this stage, but the main tabulations, particularly those which are to bring out the inter-relationships between different characteristics of the sample units, should be carefully prepared at the outset. This ensures that there is no danger of omitting to ask for information which will be needed. Information is sometimes required from a particular section or sections of the sample which may contain relatively few units. With so small a sub-sample it would be impossible to obtain the necessary degree of precision in the results. The size of the sample will have to be increased or special steps taken to increase the number of units in the particular sub-sample. Such a point could easily be overlooked in the initial stages of the survey unless all the analyses of the final data are considered in advance.

4. The sample selected must be of such size and composition that it will yield the most reliable results for a given expenditure. If interviewers are to be employed, where the sampling is random or systematic, 'substitutes' may sometimes be drawn in the same way. If quota sampling is to be used, suitable quota sheets and instructions for classifying respondents are necessary.

5. The preparations having proceeded so far it is advisable to pretest the schedule of questions, by a *pilot* survey, which is a survey in miniature. The number interviewed is unimportant, but the respondents are selected at random. Usually the more experienced interviewers are engaged on this pilot survey, since they are capable of assessing the weaknesses of the approach or any questions on the schedule. The results are not so important in themselves as are

the lessons learned. It is not always possible to carry out a pilot survey, although it is desirable. The reasons for its omission may be financial or, more often, simply a question of time. Usually, if the organisation has had a wide experience of similar surveys, *e.g.* the same survey was carried out, say, eighteen months ago, then clearly the pilot survey may be dispensed with.

6. Before the field work starts a briefing conference for the interviewers is normally held. Any difficulties met and the lessons learned in the pilot survey are examined and a course of action laid down for specified circumstances. The interviewers will have been issued with their instructions and care should be taken to ensure that they fully understand them.

7. Soon after the field work begins, completed schedules will begin to pour into the office. The schedules should be edited for omissions and any obvious mistakes such as inconsistencies in the replies entered. When necessary, the area organisers may check back on the respondent.

8. If the information is transferred to punched cards, the tabulation may be rapidly completed by machine. The questions should have been coded in advance for this purpose. The classification of those answers which are not of the simple 'Yes/No' or 'once a week/more often' variety, but may be expressions of opinion which cannot be easily coded in advance, will require careful consideration and supervision. The data once assembled, the report on the survey may be prepared. Usually several people will discuss the results together to ensure that all aspects of the information are brought out and correctly interpreted.

These, then, are the main stages of any survey. Because each stage has been dealt with separately, it should not be imagined that each is independent of the others. Each survey must be considered in its entirety. At every stage, what has gone before, or what is to be done later, must influence the design of the survey. For example, the type of information sought will largely decide whether the postal or personal enquiry method is used, while the type of respondent will influence the design of the questionnaire.

Problems can arise all along the line, but if a survey is well planned many difficulties may be anticipated and provision made accordingly. For example, interviewers may be required to classify their respondents by social class. Unless a method of classification suitable for the survey is determined in advance and the interviewers instructed in its application, part of the data may be valueless. Different interviewers will assess individual respondents by different standards, with obvious results. It was precisely this type of problem which has led to the loss of much useful information from the National Farm Survey carried out in 1942. Part of the schedule required an evaluation of the quality of the farm holding by layout, type of farming and condition of buildings. Unfortunately, the interviewers available were inexperienced in survey work. They were normally employed by the County War Agricultural Committees and were usually local men. Consequently, they assessed the holdings for the purposes of these questions in the light of their local knowledge. The result has been that the data are most unreliable for inter-county comparisons, although they are probably satisfactory for providing a local view of farming in the individual counties.

In brief, the quality of the data derived from any survey rests largely on the efficiency of the work at three stages. The first is the selection of the *sample*. If this is unsatisfactory, then clearly no reliable conclusions may be drawn about the population from which the sample was taken. Secondly, the design of *the schedule* of questions requires careful thought. Since few respondents appreciate the full implications of any lengthy question, the questions must be such that only one interpretation is possible. Finally, the all-important task of *interviewing*. As will be seen, more mistakes may creep into the results at this stage than any other. The remainder of this chapter will be devoted to these three basic problems in survey work.

The Sample

Generally speaking, the object of a sample survey is to learn something about the population. If the sample is random then, as has been pointed out earlier, certain conclusions regarding the population may be inferred from the evidence of the sample statistics. If the sample is, for whatever reason, unrepresentative of the population from which it was drawn, then the sample results cannot be generalised with any confidence to the population. Mr Moss in his paper on Sample Surveys, gave two illustrations of the significance of ensuring that the sample is representative. In 1946 a government committee studying the problem of shop closing hours requested a sample survey to assess public opinion on the matter. Evidence already submitted by interested parties had suggested that a certain change would affect only a minority of the public. The survey revealed that this was true, but it also revealed that this minority comprised mainly working housewives who at the time constituted an important part of the labour force. Without the survey this highly important piece of information would never have come to the notice of the committee. The second case concerned Dr Kinsey's studies in the United States of the sexual behaviour of human beings, in particular the report on the female. As Dr Kinsey himself has emphasised, the sample of informants was for obvious reasons largely self-selected, *i.e.* volunteers only. When the composition of the sample is compared with the entire American female population over fourteen years of age, it emerges that Dr Kinsey's sample is seriously over-weighted with the younger married woman who has had the benefits of a college education. In other words, the report *may* be a fair summary of the behaviour of this particular group of American women, but no inferences regarding the female population at large can safely be drawn from it. As is so often the case when a sample survey attracts publicity, the warnings of the organisers tend to be overlooked.

One of the major weaknesses of sampling is the fact that not every member of the sample can be contacted or, if contacted by the interviewer, may not be willing to answer the questions put to him. For example, in the National Readership Surveys just over three-quarters of the original sample are interviewed, the bulk of the non-reponse arising from the inability of interviewers to contact the prospective respondent. Only a relatively small proportion of those contacted refuse the interview. Likewise, the Social Survey experience is that between 80 and 90 per cent of the persons approached by interviewers give an interview, but this figure does not take into account the 'no-contacts'. This problem of 'non-response' is resolved in quota sampling by the simple expedient of seeking out additional suitable repondents, *i.e.* those who correspond to the interviewer's list or quota of respondents, until such time as the interviewer's quota is complete. This policy does not really solve the question of 'non-response'; it merely ensures that the same number of interviews are made as were intended when the sample was designed. The 'non-cooperating' members of the public whom the interviewer sought to interview may, however, form a particular group which will, as a result of non-response, be under-represented in the sample. If this is the case, then the sample is biased or incomplete and to that extent the results of the survey cannot reflect the true position.

The problem of non-response with random samples in which the interviewer is given a list of names and addresses is at all times serious. The experience of the Social Survey is that in a survey of adults the chances that an interviewer will find the respondent at home on her first call are about one in three. If the respondent is not available, then a further visit is necessary. Experience shows that a maximum of three calls is the economic limit. Admittedly, continuous attempts to establish contact will produce a larger proportion of effective interviews out of the sample, but the improved results are of disproportionate value to the efforts involved. Hence, the Social Survey instruct their interviewers to make a maximum of three calls at any one address to find the prospective respondent. The Social Survey interviewers used to be given lists of 'random' substitutes in the event that contact with any address in their first list of interviews was impossible; for example, because the person had either died or moved out of the district. This practice of substitution has long been abandoned for, as was pointed out above with quota interviewing, its only merit is to ensure that the interviewer carries out the required *number* of interviews. Nowadays the emphasis is placed on the need to obtain a satisfactory interview with the selected individual. No substitutes are provided, but the interviewer is required to make a note of all failures and unsatisfactory interviews so that these 'non-respondents' can to some extent be classified. As a result of this policy the Director of the Social Survey reports that interviewers have an average response rate of between 80 and 90 per cent, once contact has been made.

The object of such a policy is to endeavour to judge whether the non-respondents are merely a representative sub-sample of the main sample, in which case the only drawback is that the statistician will have a smaller sample than he had hoped for, and his results will be to that extent less precise. The real danger is that the non-respondents will form a particular group which, in consequence of the non-response, will be under-represented in the sample. If some means of classifying these non-respondents can be found, then this risk can be reduced. For example, suppose a sample of households has been interviewed and it appears that 40 per cent of the households have no children in them. It is known from the census data that 57 per cent of households are without children under sixteen and this sample is therefore biased, hence the views of such households will not be given their due weight in the final analysis. The probable explanation of this deficiency is that in childless households all the adults go out to work and when the interviewers called they received no reply. They may have been slack about call-backs and in consequence an insufficient number of such households have been interviewed. It is for such reasons that classificatory questions are introduced into the schedule; for example, the number of children, size of income, occupation, daily newspaper read, among others.[1] The distribution of the population in respect of certain such characteristics is known and the sample should correspond with the population.

The value of such classificatory questions is largely attributable to the wealth of information collected in the decennial census of population. These data can be supplemented and in some cases kept up to date by reference to the annual reviews of the Registrar General.[2] Thus the conventional classificatory questions relate to age, sex and region, the replies to which can then be compared with the official

---

[1] See the classification schedule on page 297 and the opening section of the schedule on p. 285.
[2] See Chapters XVI and XVII.

data for the population as a whole, sponsored by The Institute of Practitioners in Advertising. Other questions in the schedule, which are of interest in themselves can also be used to test whether or not the sample was representative. Thus, replies to a question asking at what age the respondent's full-time education ceased may be compared with the information derived from a similar question in the 1951 census of population. Additional questions relating to the ownership of a car, a TV set, telephone and refrigerator also provide further checks upon the sample. Car ownership admitted or claimed can be verified against the national figures prepared by the Ministry of Transport; while cinema attendances are compared with the data published each quarter by the Board of Trade. The check data in respect of TV ownership are provided by the results of an earlier enquiry carried out by the B.B.C. Audience Research department, and that relating to the ownership of refrigerator by similar national surveys. Obviously there is a limit to the number of classificatory questions that can be inserted in a schedule but the more 'control' questions that there are, the better. American research has shown that apparently satisfactory results sometimes emerge when a sample is compared with one type of control, but when other control data are used, the sample is deficient. In other words, the more cross-checks on the sample composition the better, and it is desirable that the checks themselves should be independent of one another.

The results of two important surveys in respect of the proportion of the sample successfully interviewed will illustrate the type of problem that the organisers have to deal with. In the 1963 National Readership Survey sponsored by The Institute of Practitioners in Advertising which was based upon a random sample of individuals whose names and addresses were taken from the Electoral Register, the proportion of successful interviews was 76·9 per cent of the original sample drawn. This consisted of 19,200 names but of this number 2,655 had either died, moved away, or the premises at the address were empty or demolished. Effectively the interviewers had to contend with 16,545 available respondents and of this number another 4,013 proved to be failures. Almost one-third of them refused to be interviewed, nearly one-fifth of them were out on each of three or more calls and about one-tenth of their number were either sick, senile or otherwise un-interviewable. The effective sample of 76·9 per cent was analysed by regions and by age, the results being weighted to adjust for any under-representation.[1] The other survey was the Household Expenditure Enquiry of 1953/4. The information required of all members of households was a detailed analysis of their expenditure over a three-week period. From experience with the annual

[1] The 1963 edition of the I.P.A. National Readership Survey contains extremely clear and detailed explanatory memoranda which set out the sample, schedule of questions and interviewer instructions.

National Food Surveys it was anticipated that the refusal rate in the sample would be high; a rate of 60 per cent was considered likely. For purposes of the Expenditure survey it was considered that a sample of 8,000 effective interviews and completed budgets would suffice and, on the basis of an expected 40 per cent response some 20,000 households were selected as the sample. In the event, about 65 per cent, *i.e.* some 13,000 households co-operated. The important point to note is that the organisers of the survey would have to check the composition of their effective sample of replies very carefully against the known make-up of the population to ensure that it was fully representative. Furthermore, those households which co-operated might have rather different expenditure and consumption habits from those who refused to cooperate. In the event, the Cost of Living Advisory Committee which was responsible for the survey declared that the sample of some 12,900 returns, which they had used for constructing the new Index of Retail Prices, could be regarded as fully representative of households in this country.

An interesting illustration of measuring the extent of non-response and making allowance for it is provided by the 1946 Family Census. This enquiry covered a sample of over 1·7 million married women. When the returns were checked there was a deficiency of some 17 per cent and it was suspected that among this group, childless women were in the majority. This was an especially important group in what was really a study of human fertility. Follow-up letters asking the 'non-respondents' to co-operate produced a proportion of replies and from these it was clear that the suspicion was fully justified. In the results, the figures for childless married women were adjusted in the light of this knowledge.

The control and measurement of non-response remains among the more intractable problems of survey organisers. The solution is to be found partly in first-class interviewing with well-designed schedules of resonable length and partly in checking, as above, on the non-respondents so that allowance for their omission from the sample can be made. This should not be read as implying that the organisers guess the facts about them. It means that the answers given by respondents who appear to be similar – as far as they can be compared by reference to certain classificatory data – can be proportionately weighted in the final analysis. Because of the danger that non-response may introduce bias into the sample, it is far better to use a smaller sample in which interviewers are expert and can ensure accurate replies as well as a very high response rate, rather than a much larger sample with poorer interviewing and a lower response rate. Even if the latter sample yields a larger number of interviews, the bias may lead to erroneous conclusions.

Designing the Questionnaire

The questionnaire or schedule of questions may be described as the keystone of any survey. The basic problem is not so much what questions to ask, but what is the best way to ask them. According to two members of the Social Survey, 'the problem is . . . to design questions that mean the same thing, a single thing, a defined thing and the intended thing, to everyone'.[1] As a general rule, questions should be short; lengthy questions tend merely to confuse the respondent. The interviewers should be instructed as to whether they may depart from the form of the question as written on the schedule. Usually, the question has been carefully considered and the final form is probably the best possible whereas the interviewer's alternative may lead to erroneous interpretation of its meaning. An obvious case is to avoid mentioning proprietary brands when engaged on a market survey. The replies to the question 'What do you consider the best transistor wireless set costing less than £15?' will be very different in the aggregate from those received if the interviewer had asked, 'Do you consider the ——— radio the best set costing less than £15?' The mere mention of a name, or the hint of the exact purpose of the enquiry, is generally sufficient to influence the respondent. Even the interviewers should not know which organisation is paying for a survey to determine consumer preferences among such products as detergents, newspapers or soft drinks.

Only questions which the respondent can answer from knowledge or experience should be asked. To ask a rural housewife who has always used either open fires or paraffin stoves if she prefers to cook by gas or electricity is pointless. Yet in one survey housewives were asked which form of heating they preferred, coal fires, gas, electricity, or central heating. Not many housewives in this country have enough experience of the last-mentioned form of heating to be able to answer the question rationally.

A particular problem in interviewing is the respondent's memory. Too much reliance should not be placed on it! In consumer surveys the interviewer, instead of enquiring about the consumption of a product over a period, *e.g.* a month, usually asks whether the housewife has a particular commodity in the house, *e.g.* a soap powder, and if so, when she bought it. Alternatively, they may be asked how much of a foodstuff, *e.g.* biscuits, they bought in the current week. In the National Readership Survey, informants were asked which newspapers and periodicals they read. To avoid the risk of any paper being overlooked, each informant was shown a booklet containing reproductions of the titles, or 'mastheads' of the various periodicals and journals and asked to indicate those that he had seen. The same

---

[1] Fothergill and Willcock, 'Interviewers and Interviewing' in Modern Sample Survey Methods.

technique is occasionally used by the interviewers of the B.B.C. Audience Research but in this case the previous day's broadcasts are listed. Even these devices are not perfect; experiments have revealed that the position of any item is important, those at the top being mentioned more frequently than others. To overcome this difficulty the lists are usually rearranged at short intervals and, in the case of the readership survey sixteen different arrangements of the order in which the periodicals appear in the booklet were used in such a way that all the different versions were used in each region. Nor is the facile assumption that the respondent will answer accurately even the simplest question justified by experience. The number of wives who do not know their husband's incomes is legion, but one consumer survey in a North London suburb revealed that quite a number either had no knowledge of their husband's occupation or described them incorrectly.

All the questions asked must be so phrased that the respondent can answer them intelligently. This implies that the respondent must understand what is being asked. For this reason, in many opinion polls, a factual question or two is usually inserted at the beginning to find out whether or not the respondent knows anything about the subject upon which he is asked to express an opinion. Every question which relates to the respondent's actions in the past must be most carefully considered, because the average person's memory is so unreliable. Great care should be taken to avoid words with 'emotional' content, e.g. 'Socialist' and 'Tory' will probably affect the respondent more than (say) 'Labour' and 'Conservative'. A good illustration of this was provided by a U.S. public opinion poll during World War II. Many more respondents were 'anti-Nazi' than were 'anti-German' when asked their views on the belligerents before December 1941 in two polls, the only difference between which was the use of the word 'Nazis' in place of 'Germans' in the relevant questions.

Apart from basic principles to be observed in the construction of schedules and questionnaires touched upon above, there is also the problem of facilitating the work of the interviewer. A good deal of study has been given to the best lay-out of the form, bearing in mind that it may have to be completed on the street, or on a doorstep and not on a table. Instructions to the interviewer must be set in bold type, e.g. 'if respondent answers "No" to this question omit next section'. It is both impracticable and undesirable that the interviewer should have to try and write down the respondent's answers each time. For many questions the answers can be anticipated, e.g. Yes/No/D.K., i.e. respondent doesn't know. Similarly, where the frequency of a particular event is concerned such as the number of weekly visits to the cinema, the answer can be pre-coded, e.g. once/twice/more often/

seldom/never. The pre-coding is not always quite so simple. For example, take the question 'Would you say that television has made your home life more interesting and happier, or do you think your family life would be better without it, or does it make no difference'. This question set in a survey on I.T.V. carried the following pre-coded answers: More interesting/better without/no difference/don't know. The weakness of this type of pre-classification is that it forces the interviewer to classify the respondent's answer on the spot. Suppose the reply is on something like the following lines 'well, it keeps the kids happy and gives the missus a chance to put her feet up while they are viewing; but sometimes we have trouble in getting the kids to do their homework and I think that the missus is too soft in letting them stay up at night after their proper bed-time; but it's O.K. in the evenings sitting with the missus after supper and not having to go out . . . etc. etc.' How would the reader classify this reply?[1]

Another point to observe when formulating opinion questions is that the respondent should not be given a 'middle course' to follow. For example, 'Do you think that unemployment in this country is likely to rise or fall during the next few months, or do you think that there will be little change'. Since this question involves making an assessment of the future, quite apart from the fact that many respondents will not have the knowledge to form an opinion, it is logical for a large number of the respondents to opt for the middle course. This is particularly the case when it comes at the end of the question and obviously offers the respondent an escape from the difficult choice posed by the earlier part of the question. Such questions which offer the respondent certain alternatives are known as *dichotomous* questions (if there are two alternatives) and multiple choice or *cafeteria* questions where there may be several alternatives. Sometimes an attempt is made to pre-code all the possible answers, the interviewer ticking that one which most closely resembles the respondent's reply. A good example of this is given in Question 11 of the schedule on Diphtheria Immunisation reproduced below. Quite often, in order to create a feeling of confidence in the respondent, the interview will start with what is termed a free-answer question. This is of the variety, 'The government has been asked by a section of the public to re-introduce flogging for crimes of violence. What do you think?' The respondent can then speak quite freely – always assuming he has some views on the subject – and he is then ready for more specific questions. Sometimes these 'free-answer questions' appear later in the schedule, but whenever they appear the interviewer is confronted with the problem of summarising accurately but concisely the gist of

---

[1] This question is taken from the schedule in a report 'Parents, Children and Television' published for the I.T.V. Authority. H.M.S.O., 1958. The schedule is reproduced on pp. 29-32.

the reply. This is not always easy and the classification of such free answers is inevitably arbitrary.

The length of the interview will vary considerably from one survey to another, but it is a good rule to try and keep it short. The closer the subject of the survey to the respondent's experience and daily life, the longer can be the interview without producing weariness or plain boredom. Nevertheless, the Social Survey appears to have little difficulty in persuading its respondents to withstand interviews of an hour or more in some cases. Some schedules contain over 60 questions, but a short questionnaire used in the Diphtheria Inquiry of 1946 is given below. This illustrates the main features discussed so far. All the questions are simple and to the point. The interviewer is helped by the pre-coding of the probable answers to some questions, *e.g.* numbers 11 and 12. With this questionnaire is published a detailed but concise set of instructions dealing with the approach to the respondent and suggestions for posing certain questions and the classification of possible answers.[1]

THE QUESTIONNAIRE

THE SOCIAL SURVEY

DIPHTHERIA INQUIRY N.S.69

TOWN or DISTRICT (as on quota)	INVESTIGATOR	DATE
Urban .. .. Y	Rural.. X	Region 1 2 3 4

Age of Mother

Economic Group (C.W.E.)

Up to 24 1	Husband at Home .. 1	Up to £3 1
25–29 2	Husband away .. 2	Over £3–£4 .. 2
30–34 3	Husband decd. .. —	,, £4–£5 10s. .. 3
35–39 4	Divorced or Separated 3	,, £5 10s. –£10 4
40–44 5		,, £10 5
45 and over .. 6	*Last Type of Education—*	N.A. 0
	Elementary 4	
Working Full-time 7	Secondary, Technical 5	*Occupation of C.W.E.*
Working Part-time 8	Others 6	
Not working .. 9		
	Substitute 1	
	Original 2	

1. Do you know what causes diphtheria?
    Don't know 1
    Infection from other children 2
    Bad sanitation, dirt, etc. 3
    Other causes 4

2. Do you know how diphtheria can be prevented?
    Don't know 6
    Immunisation, inoculation 7
    Not possible to prevent it 8
    Other ways 9

[1] The Social Survey. Diphtheria Immunisation, by K. Box. October 1945. N.S. 69.

3. Have you had your children immunised?

			SEX		Under		AGE (last birthday)										IMMUNISED	
			Boy	Girl	1		1	2	3	4	5	6 7	8 10 12 14 9 11 13 15				Yes	No
Selected																		
Child	A	1		2	Y		X	0	1	2	3	4	5	6	7	8	3	4
Child	B	1		2	Y		X	0	1	2	3	4	5	6	7	8	3	4
,,	C	1		2	Y		X	0	1	2	3	4	5	6	7	8	3	4
,,	D	1		2	Y		X	0	1	2	3	4	5	6	7	8	3	4
,,	E	1		2	Y		X	0	1	2	3	4	5	6	7	8	3	4
,,	F	1		2	Y		X	0	1	2	3	4	5	6	7	8	3	4
,,	G	1		2	Y		X	0	1	2	3	4	5	6	7	8	3	4
,,	H	1		2	Y		X	0	1	2	3	4	5	6	7	8	3	4
,,	I	1		2	Y		X	0	1	2	3	4	5	6	7	8	3	4

IF SELECTED CHILD IMMUNISED
(Ask with reference to that Child)

4. How old was the Child when immunised?

Under 1 yr.	Y	3 yrs.	1	6 or 7 yrs.	4	12 or 13	7
1 yr.	X	4 yrs.	2	8 or 9 yrs.	5	14 or 15	8
2 yrs.	0	5 yrs.	3	10 or 11 yrs.	6		

5. Who suggested that the child should be immunised?

School	1
Health Visitor	2
Private Doctor	3
Welfare Clinic	4
Own idea (from publicity, etc.)	5

6. Did you have any difficulty in getting the child immunised?

    Yes .. .. Y   No .. .. X N.A. .. 0
If Yes, what difficulty?

7. (To everyone *not* answering Code 5 to Q.5)

How long after it was suggested, was the immunisation done?

Within a week ..	1	More than 3 months ..	4
More than 1 week to 4 weeks	2	Don't remember ..	5
More than 4 weeks to 3 months ..	3		

8. Was the immunisation completed? How many visits?

    Yes .. Y   No .. X

9. Was the arm painful afterwards?

    Yes .. Y   No .. X

10. Did you pay for it (at your own Doctor), or have it done under the Council Scheme?

        Paid own Doctor ..   4   Free under scheme ..   5

IF SELECTED CHILD NOT IMMUNISED
(Ask with reference to that Child)

11. Why was the child not immunised?

Had not heard about immunisation	Y
Have not bothered, not had time, will have it done	X
Don't believe in it, not worth while	0
Husband objects	1
Bad for child, would hurt or frighten it	2
Waiting till child goes to school, not old enough yet	3
Child has just been vaccinated, is ill	4
Child has already had diphtheria	5
Waiting to consult husband	6
Tried, but unable to get it done yet	7
Other reason	8
Don't know	9

TO EVERYONE NOT ANSWERING CODE Y TO Q.11

12. Have you heard or read about immunisation in any of the following ways?

Newspapers or magazines	.. Yes	.. 1	No	.. 2
Radio	.. Yes	.. 3	No	.. 4
At the cinema	.. Yes	.. 5	No	.. 6
Posters	.. Yes	.. 7	No	.. 8
Leaflet or visit from school, health visitor, etc. ..	.. Yes	.. 9	No	.. 0

13. Do you know up to what age children should be immunised?
    Right (15 yrs. old) .. 1   Wrong .. 2   Don't know .. 3

14. Do you know what is the best age to have children immunised?
    Right (1 yr. old)   .. 5   Wrong .. 6   Don't know .. 7

TO EVERYONE

15. Do you use a children's welfare clinic?
    Regularly .. 1   Occasionally 2   Never 3   Used to .. 4

16. Have you ever been present at a school medical examination?
    Yes   6   No 7

Some surveys have utilised schedules containing many more questions than the foregoing schedule contains, but these are usually in connection with major surveys. For example, the government's survey of family expenditure necessitates the completion of several foolscap sheets of questions, quite apart from the keeping of a small exercise book to record expenditure, etc. Inevitably the response rate with such lengthy schedules tends to be relatively low. In the case of the family expenditure surveys, it is around 60 per cent.

Commercial survey organisations are not generally in a position to exploit the public's patience to such an extent, and in the case of the family expenditure survey it should not be forgotten that the members of the household who complete all the forms and schedules receive £1 apiece for their trouble. Reproduced at foot of this page is a simple schedule of questions which has been used in the past by the Gallup Poll for pre-testing potential interviewers. For obvious reasons, it is a very simple schedule. None of the questions, it will be seen, are at all difficult to complete by even the most limited of respondents. Note the pre-coded answers to a number of questions such as 2a, 3, 10, etc., among others. Note in Question 13 the instructions to the interviewer in which he is specifically asked not to prompt and also to write down the respondent's exact words. The classificatory questions at the foot of this schedule, they are at the beginning of the previous schedule, are important. These enable the answers to be classified by reference to the main social groupings in the population as well as the age and sex of the respondents. Information such as that relating to the home and number of children can be used to check the extent to which the sample of respondents corresponds in respect of these characteristics to the population.

THE GALLUP POLL
211 REGENT STREET, LONDON, W.1.

When the answer is coded please ring number like this (2)
For other answers write in contact's own words

---

1. What is your favourite radio programme?

   ....

---

2. (a) About how often do you go to the cinema?

More than twice a week	..	1	Once a month	..	5
Twice a week	..	2	Less than once a month	..	6
Once a week	..	3	Never go	..	7
Once a fortnight ..	..	4	(If 'never go' skip to Q. 3)		

(b) When did you last go to the cinema?

   .... **ago**

(c) Can you remember one film you saw? (*Write in name*)

....

Ask All

3.   What is your favourite form of entertainment?

Theatre..	..	..	1	Ballet ..	..	..	4
Cinema	..	..	2	Musical Hall	..	..	5
Concert	..	..	3				

Other (*Write in*)...

4. (a) Do you ever go to watch a football match or any other form of sport?

| Yes, regularly | .. | .. | 6 | No | .. | .. | .. | 8 |
| Yes, occasionally.. | | .. | 7 | (*If 'No' skip to Q. 5*) | | | | |

(b) If YES:

What is your favourite sport to watch?

Rugger ..	..	..	1	Tennis ..	..	..	4
Soccer ..	..	..	2	Hockey..	..	..	5
Cricket ..	..	..	3				

Other (*Write in*)...

Ask All

5. (a) Have you had a holiday away from home during the past twelve months?

| Yes | .. | .. | .. | 6 | No | .. | .. | .. | 7 |
| | | | | | (*If 'No' skip to Q. 6*) | | | | |

(b) If YES:

What sort of accommodation did you stay in?

Hotel ..	..	..	1	Holiday Camp ..	..	4
Boarding House ..		..	2	Stayed with relatives/friends		5
Private Lodgings..		..	3	Other ..	..	..	6

6.   What do you consider the ideal way of spending a summer holiday?

....

7. (a) Have you ever stayed in a Holiday Camp?

| Yes | .. | .. | .. | 1 | No | .. | .. | 2 |
| | | | | | (*Skip to Q. 7(e)*) | | | |

(b) If YES: Which one?

....

(c) What did you like most about it?

....

(d) What did you like least about it?

....

(e) **Ask all those replying 'No' to Q. 7(a)**
Do you think that you would enjoy staying at a holiday camp?
Yes .. .. .. 9 No x

(f) If No: Why not?

....

Ask All
8. (a) Do you happen to be reading a novel or other book at the moment?
Yes .. .. .. 1 No 2
(*Skip to Q. 9*)

(b) If YES: What is it?

....

Ask All
9. What is the most urgent problem facing you and your family at the present time?

....

---

10. Compared with six months ago are you finding it harder, easier or about the same to make both ends meet?
Harder .. .. .. 3 The same 5
Easier .. .. .. 4 Don't know 6

---

11. Will you please tell me about how much you/your husband earns each week? (*Show card A and ring appropriate number*)
1 *2* *3* *4*

---

12. (a) Do you have a TV set at home?
If YES: Can it receive ITV commercial programmes or only BBC?
No TV Set .. .. 1 Set receiving ITV and BBC 2
BBC only 3

(b) Do you happen to have seen any TV advertisements during the past week?
Yes .. .. .. 4 No (*skip to keys*) .. 5

(c) Ask all replying 'Yes' to Q. 12 (b)
Which ones can you recall? (*Write in*)

....

....

Ask about the first advertisement mentioned in Q. 12 (c). Encourage the contact to answer as fully as possible, but do not prompt. Write down exactly what (s)he says very carefully.

13. I would like you to describe the .... advertisement to me.
   (a) Who or what was in it?

....

(b) What happened in the advertisement?

....

(c) What did the advertisement say?

....

Man   .. 1   Woman.. 2   Married   3   Single   .. 4

Date of birth..

Date of interview:..

Home:
Owner-occupier   5   Tenant   .. 6   Other..   .. 7
Age:
21 – 29 .. 8   30 – 49 .. 9   50 – 64 .. X   65 and over V

Contact's Name and Address (Please PRINT)

Mr./Mrs./Miss ....

Address:..

Town:..

No. of children and adults in household:
Children 0 – 4 years ..   ..
Children 5 – 15 years   ..   ..
Adults (16 and over including contact)..   ..   ..

                                  Total

Present/former occupation: (Husband's if housewife not working):

....

....

Manual:		Non-manual:		
Factory; mine; transport; bldg.	1	Professional	.. ..	4
Farmworker	2	Director; propr; mngr. ..		5
Other manual	3	Shop; personal service etc.		6
		Office and others	..	7

I hereby certify that these interviews have been conducted in accordance with your directions, with persons who were previously unknown to me.

Signature: ....

Many of the problems arising from the schedule or questionnaire can fortunately be settled before the field work starts, and where a pilot survey is undertaken, unexpected weaknesses may be revealed. One last point is worth mentioning, only if because it is so frequently asked. Do people tell the truth when interviewed? Individuals do, of course, vary in this respect, but since anyone may refuse to be inter-viwed, it seems pointless for him to accept the invitation and then proceed to tell lies. In any case, consistent lying or exaggeration is extremely difficult. There is some truth, however, in the contention that not every question is truthfully answered. For example, the Customs and Excise data on tobacco and alcohol duty make it clear that respondents seriously understate their expenditure on tobacco and alcoholic drink as given on the schedules of the Family Expenditure Surveys. During the war, housewives were exhorted to salvage waste material. One war-time inquiry which asked whether the respondent did salvage waste material revealed that a far higher proportion of housewives stated that they participated in the campaign than the actual collections indicated. The significance of this point is that most people probably tend to state that they conform to accepted social requirements rather than admit that they do not. In a survey among housewives questions regarding pocket money and punishment of children were put. There was universal agreement with the proposition that 'children nowadays have too much pocket money' although interviewer after interviewer commented on the fact that according to their replies their particular respondents never made this mistake! Similarly with the punishment of children. Apparently this was hardly ever necessary, if the answers are to be believed. As several interviewers commented: 'if the replies are to be believed, never were children so well behaved as today'. Clearly the respondents in many cases were giving the socially acceptable answer. It was noted that occasionally the respondent took the line 'what business is it of yours', although up to that point in the interview she had been most co-operative.

To overcome the risk of bias, conscious or otherwise, from the respondent, the schedule designer uses various devices. Thus in the Readership Survey, the following method was used to overcome the 'prestige' factor. By this is meant that respondents might claim to have read certain periodicals for prestige purposes, when in fact they never read them. It is well known, for example, that many more people 'admit' to having read *The Times* than its circulation would justify, even when allowance is made for each copy being read by several people. The following question was asked: 'When was the last time you looked at a copy of . . .?' This clearly gives the respondent a loophole to admit that he has not read a particular periodical without admitting that in fact he never does read it. Only if the respondent answered this question in the affirmative were further questions as to place and time asked of him. Much the same reasons explain the phrasing of the following question 'Do you have a T.V. set at home YET?' rather than the same question without the final word. The inference is that the respondent can afford it, but has not decided so far to buy a T.V. set.

While the drafting of a schedule of questions may at first sight appear quite a simple task – after all, it may be asked, what is difficult about asking questions? – in practice it is a highly skilled task. The reader should examine the schedules reproduced in various reports and assess them in the light of the above comments.

The Problem of Interviewing

It has been stated that the 'representativeness of a sample depends on the ability of field workers to trace their subjects and persuade them to co-operate in the completion of a questionnaire; and the accuracy of the results depends on that of the information recorded. Much hinges on the address, skill and tact of the interviewer, who thus becomes a possible source of serious bias in the enquiry'.[1]

The majority of sample surveys are conducted by interviewers. A good interviewer can persuade his subject to reply to almost any question and it is fortunate, if somewhat surprising, that there appears to be no limit to the variety of questions which the average person is prepared to answer. Such willing co-operation can only be attributed to a general human weakness of being flattered by others seeking one's views. Occasionally, the respondent may be interested in the subject of the survey and be especially willing to participate, particularly if he feels that his opinions may influence the attitude of others. The non-response will, of course, vary from enquiry to enquiry, depending on what is expected of the respondent, but it is seldom large. One exception is the high refusal rate – approx. 60% of the respondents contacted – in the National Food Survey. The reason for this lies in the fact that the housewife is expected to keep a detailed account of food purchases for a week and allow the interviewer to check the contents of her larder.

Just how significant training of interviewers may be in affecting the

---

[1] J. Durbin and A. Stuart 'Difference in Responses Rate of Experienced and Inexperienced Interviewers'. *J.R.S.S.* 1951, Part II.

response of informants and the quality of that response, is difficult to judge. The results of a detailed investigation to test the relative efficiencies of two groups of professional investigators from the Social Survey and the British Institute of Public Opinion on the one hand and a group of University students on the other suggest at first sight that training may not be all that important.[1] It appeared from this enquiry that while the professionals enjoyed a higher success rate in establishing contact with the respondents than the students, the relative differences between the three groups of interviewers with regard to the quality of response in terms of completed and accurate schedules were not such as to warrant the inference that the students were much inferior to the professionals. There was some evidence, however, to suggest that on the more difficult questions the students were not as effective as the professionals. In view of the present trend towards longer schedules and intensive interviewing of the respondent, *i.e.* 'depth' interviewing to seek out causes and reasons for attitudes, the importance of training becomes more evident. In the early stages of survey work, the main quality required of an interviewer was 'personality' in the sense that she could easily establish 'rapport' with her informant and persuade him or her to talk freely on the survey topic. Nowadays much more attention is being paid to sequence of questions, the form in which they are posed, accurate recording, hence more skill and concentration is required of the interviewer.[2]

Apart from the difficulties already discussed in connection with the schedule, the actual interview is attended by even greater problems. The simplest is the risk that the interviewer may misunderstand a reply, or merely mark off the wrong code number for any pre-coded answer. The risk of misinterpretation is greater with opinion questions, irrespective whether the main replies have been classified in advance on the schedule or not, than with questions of fact. More important, however, is the actual conduct of the interview itself. There is always the danger of prompting the hesitant respondent or even putting the answer to him. To avoid this risk according to the official B.B.C. handbook the B.B.C. interviewer is instructed to help the respondent recall his listening or viewing the day before by getting him to recall when 'he first put the wireless on'; what did you do when you came in after work, etc. In other words, she uses a technique based upon 'association of ideas'. If this methods fails, and many respondents are extraordinarily hazy on such matters, she may show the respondent the 'log' or programme list to remind him what was on.

[1] Durbin and Stuart *op. cit.*
[2] Fothergill and Willcock, *op. cit.*

A particular cause of concern is the extent of what is termed 'interviewer bias'. 'Bias', in the normal sense of the word, is a more serious danger with the public opinion polls than with social or market surveys. The Princeton Office of Opinion Research has carried out many tests in this connection and has revealed that even with professional interviewers some bias is unavoidable. In one American pre-election poll the interviewers were divided into groups of opposite political faiths. Their returns revealed quite clearly the effects of their subconscious sympathies on their respondents. Nor is this a new problem; it has been known to exist from the earliest days of sample surveys. For example, as long ago as 1914 an American sociologist found that the replies of 2,000 destitute men explaining their distress were markedly influenced by the interviewers' sympathies and views. Thus a Prohibitionist interviewer's results revealed a strong tendency among his respondents to attribute their destituion to drink; an interviewer with Socialist leanings recorded many who ascribed their position to industrial causes. According to the author, 'quantitative measures of interviewer bias in this particular survey turned out to be amazingly large. The men may have been glad to please anyone that showed an interest in them'.[1] The same authority comments that interviewers will influence their respondents' replies by the mood into which the latter are put. For example, 'the interviewer may make the respondent gay or despairing, garrulous or clammish. Some interviewers unconsciously cause respondents to take sides with them, some against them'. As Dr Deming points out, training can do much to overcome the more obvious causes of interviwer bias, but even with a well trained corps of interviewers it may arise, even quite unconsciously. The experience of the Social Survey in the course of collecting data on the probable response of ex-Service men to the offer of 1939-45 Campaign Stars is noteworthy in this respect. It was found that the age of the female interviewer had a definite influence on the attitude of the male respondent. The younger the interviewer, the more likely was the man to disavow any intention of applying for the awards. A recent survey among adolescents on the subject of juvenile delinquency revealed that respondents were markedly influenced in their answers where male interviewers had the physical characteristics of policemen, i.e. tall, etc.

Despite the intensive training given to interviewers, the holding of briefing conferences and the issue of detailed instructions with the schedule setting out the considerations which have prompted the various questions and the best way of putting them to the respondent, mistakes are inevitable. It is not the obvious mistakes and glaring inconsistencies between answers that are troublesome. These can

[1] 'On Errors in Surveys', W. E. Deming in *American Sociological Review*, August 1944.

usually be detected in the editing of the schedule. It is the minor slips, such as incorrect interpretations of an answer or poor classification of the respondent which are so difficult to eliminate, since they cannot be detected and the results are thereby distorted. When all the problems involved in the organisation and conduct of a survey are taken into account, the errors occurring at the interview remain the most serious. In the words of two members of the Social Survey, 'sampling errors are the least serious, it is the human errors such as errors in classification and memory errors on the part of the respondent . . . that are less easily detected'.[1] A more recent enquiry into this problem carried out by the Social Survey revealed that over three-quarters of the mistakes made by investigators during the course of the interviews (and noted by observers present at the interview) could not have been detected from a scrutiny of their schedules when returned to the head office.[2]

Some idea of the task imposed upon the investigator is given by the form below headed 'Classification'. This is actually a supplementary questionnaire to the main questionnaires which were concerned with surveys of the public's knowledge of and attitude towards tuberculosis, on reading habits and on savings. The purpose of this supplementary list of questions was to provide information concerning the informant's living conditions, social class, household composition, etc., independently of the three main surveys. The questions are quite clear and it will be noted how the investigator seeks to ascertain the respondent's income group. The respondent is not actually asked what he earns, but is asked to indicate to which particular income group, as given in Question VIII, he belongs. The letters S.W.E. stand for senior wage earner, usually the male head of the household. Since it may not be possible to interview him if the investigator calls during working hours, the 'subject', usually his wife, will provide the

CLASSIFICATION † *

(i) Interviewer's name ....
Interviewer's number........./........./......

(ii) **RING DATE OF INTERVIEW.**

	Sun.	Mon.	Tues.	Wed.	Thur.	Fri.	Sat.
April	16	17	18	19	20	21	22
	23	24	25	26	27	28	29
	30	—	—	—	—	—	—
May	1	2	3	4	5	5	6

(iii) Subject. Where living.

At home................................... Y
In institution, hotel................... X
As a boarder 0
In rooms 1
As resident servant.....................2

(iv) Type of dwelling.

Detached house 3
Semi-detached house 4
Terraced house 5
Self- contained flat 6
Part of house 7
Other (*specify*) 8
....

[1] Gray and Corlett, *Sampling for the Social Survey*, J.R.S.S., 1948, Part II.
[2] Fothergill and Willcock, *op. cit.*

(v) HOUSEHOLD COMPOSITION

Relationship to Subject	Age	Sex M F	Status M S W	Working F P N
A. Subject		1 2	3 4 5	6 7 8
B.		1 2	3 4 5	6 7 8
C.		1 2	3 4 5	6 7 8
D.		1 2	3 4 5	6 7 8
E.		1 2	3 4 5	6 7 8
F.		1 2	3 4 5	6 7 8
G.		1 2	3 4 5	6 7 8
H.		1 2	3 4 5	6 7 8
I.		1 2	3 4 5	6 7 8
J.		1 2	3 4 5	6 7 8
K.		1 2	3 4 5	6 7 8

Total 0—4 ....

Total 5—15 ....

Total 16 and over...............

Total in household............

Housewife is
(give letter) ....

S.W.E. is
(give letter) ....

(vi) Number of rooms ....

(vii) Subject occupation (full description)

....
....

Subject industry, trade or profession .....

....

Self employed 1
Employee.................................2

S.W.E. occupation (full description)

....
....

S.W.E. industry, trade or profession......

....

Self employed 1
Employee...............................2

(viii) Income per week less deductions plus
bonuses.

	Sjt.	S.W.E.
Nil 0		0
Up to £3 1		1
Over £3 to £5 2		2
Over £5 to £7 10s. 3		4
SHOW Over £7 10s. to £10 4		4
CARD Over £10 to £20 5		5
Over £20 6		6
Don't know 7		7
Refusal, not asked........8		8

If SJT. D.K., REFUSAL, NOT ASKED
Why? ....
....

If S.W.E. D.K., REFUSAL, NOT ASKED
Why ? ....
....

(ix) Interview situation.
CODE Informant alone ....
ALL Spouse present 1
THAT Other adult(s) present 2
APPLY Children present 3

(x) Interviewer's assessment of success of
interview.
Above average (give reason) 4
Average 5
Below average (give reason)........6
Very poor (give reason)...............7
....

(xi) Serial number on record sheet..............

(xii) Subject education.
Age left school ....
Type of last school.
Elementary ....Y
Central, Technical, Commercial...X
Secondary, Public 0
University...............................1

---

† Reproduced by permission of Director of Research Techniques, London School
of Economics and the Editor of the *Journal of the Royal Statistical Society.*

* *Source:* Durbin and Stuart, *op. cit.* This particular paper contains three
schedules of questions which will repay study by the student. The discussion
following the paper is also useful.

information. The interviewer will need briefing on the appropriate method of answering Question X; assessments by the individual directly concerned of his success or otherwise are not usually very satisfactory for comparative purposes, e.g. comparing the performance of interviewers. The question does give some indication of the tenor of the interview and indicates the degree of co-operation which the investigator received from the informant.

In the National Readership Survey 1963, interviewers were required to classify their respondents, who were a random sample of names and addresses, by social grade. To ensure that all interviewers used the same standards, the survey organisers drew up a six-fold classification of the respondents.[1] The highest social group was defined as 'upper middle class', the occupation of the head of the household in this group being described as higher managerial, administrative or professional with an income of £1,750 or more. Another group was referred to as the 'skilled working class', comprising skilled manual workers with an income of between £12 and £20 per week. The lowest group were defined as those living at the lowest levels of subsistence, with an income of under £6 10s. per week. This group comprised State pensioners or widows, casual and lowest grade workers. The interviewers were instructed that such a classification was at best a general guide; the income figures were to be regarded as at best approximate. Where no information concerning the occupation of the respondent was available, the interviewer was to base her assessment 'on environmental factors such as the type of dwelling, the amenities in the home, the presence of domestic help, and so on'.

Obviously, the consistency with which individual interviewers would follow such a classification would vary with their experience. There would also be some variability between interviewers. The solution is clearly to give the interviewers as detailed instructions as is possible and adequate training.

Postal Enquiries

Quite apart from the problems and weaknesses of personal interviewing as a survey method, an important consideration is cost. A cheaper method is to use a postal questionnaire. At first sight the postal questionnaire has several advantages. It can be sent to a very large number of people at low cost, so that the sample size may be increased considerably, relatively to the survey sample for the same cost. Further, the risk of bias or mistakes on the part of the interviewer is absent. All these apparent advantages, however, prove on examination to be fictitious. The fundamental weakness of the postal method is the low proportion of returns. On the average, a 20 per cent

[1] Reproduced on p. 262.

response is considered good. It may be argued that if 100,000 schedules can be sent out for the same cost as interviewing 5,000 people, the final sample still contains 20,000 and is therefore preferable. Unfortunately, there is no means of ascertaining whether or not the 20 per cent return constitutes a representative sample of those people to whom the schedule was sent. It almost certainly does not. It is probably true to say that the cost per completed schedule by interviewing is ultimately little above that for the postal enquiry. There is, too, the greater reliability of the former, since despite the risk of interviewer bias and mistakes, the respondents will themselves make mistakes in completing the forms.

The experience of the Social Survey with postal schedules is worthy of note. Some years ago 16,000 members of a profession were circularised and 38 per cent returned the forms immediately. A further 32 per cent replied, after a reminder had been sent. Ultimately, the response was just over 80 per cent. Such an experience was quite exceptional for the following reasons. The subject of the enquiry was connected with the future of the profession, a matter of considerable interest to the members. They would in any case return a higher proportion than would be received in an enquiry on any subject covering the general population for, as professional people, their reaction to forms would not be that of the average member of the public. Finally, a pilot survey had provided many useful lessons and the schedule of questions was devised with very great care. The last points should not be exceptional, but the fact that the organisers of the survey comment on them suggests that they are not always accorded the same degree of attention.

More recently the Social Survey has been experimenting with further postal enquiries. In a paper read to the Royal Statistical Society Mr Christopher Scott argued, on the evidence of several highly successful postal surveys, that response rates of 80 per cent and more could be achieved, provided certain obvious conditions were observed.[1] It is perhaps worth stressing again that no survey, whether it be carried out by interviewer or by post, will achieve 100 per cent response. One would be as suspicious of the results of such a survey as we tend to be of elections in 'people's democracies' where 99·9 per cent of the voters vote for the ruling party. One of the postal surveys discussed by Mr Scott in his paper concerned road safety, and the schedule used is illustrated on pages 300-1. Two points emerge very clearly from a scrutiny of that schedule. The first are the brief and simple questions, with underlining, italicised print, and large capitals to get the point of the questions across to the respondent. Secondly, the letter itself. Note the stress on confidentiality, the statement that a

[1] Research on Mail Surveys. *J.R.S.S.* 1961. Part II.

Central Office of Information,
Social Survey Division,
Montagu Mansions,
Baker Street,
LONDON, W.1.

Dear Sir or Madam,

Road Safety

I am writing to ask for your help in an enquiry we are making for the Government Road Research Laboratory.

In carrying out its work on road safety, the Laboratory needs to know who are the chief road users, how much they use their vehicles, and how much experience they have with different kinds of vehicle. At present we are asking only about motor-cyclists, including drivers of motor scooters and mopeds.

We can only get this information by writing to motor-cycle owners. To save expense we are not writing to all, but only to a small number, chosen at random from the registration records. In order to be sure that all points of view are taken into account we are anxious to get a reply from every person we write to.

The questions are on the back of this letter. Would you please fill in the answers and post the sheet back to us, using the enclosed label and envelope. There is no need for a stamp.

Please note that it is *your* reply we want, even if you do not at present ride any machine. Do not ask any one else to fill in the form instead of you, or we will not have a true cross-section of owners.

Your reply will be kept strictly confidential, and will *only* be used for counting how many people give each different answer. We shall pass these total figures to the road safety authorities, but we shall not mention any names.

I would be most grateful for your help.

Yours faithfully,

C. SCOTT.

PLEASE TURN OVER

Vehicle registration No.........................

1

1. Is the above machine   a motor-cycle <u>with</u> side-car?   ————

   a motor-cycle <u>without</u> sidecar?   ————

   a motor scooter?..........   ————

   WRITE YES
   AGAINST
   ONE OF
   THESE

   or a moped or auto-cycle? ....   ————
   (A moped or auto-cycle means anything
   which has a motor <u>and pedals</u>.)

2. Are you yourself still the owner <u>of the above machine</u>?..............
   (If your answer is NO, please state <u>when</u> you sold it, then answer
   questions 3 and 4 for the period before you sold it.) 1958
   date when sold by you

3. Have <u>you yourself</u> driven <u>the above machine</u> during the
   last 4 weeks? If so, roughly how far?

   (If you did not drive it, write 0)   ....miles   Give rough mileage for
   in last 4 weeks   the last 4 weeks. You
   Please note that the figure you write above should be the   may explain below if
   mileage <u>you yourself</u> have done, <u>as driver</u>, on the above   that period was <u>very</u>
   machine.   different from your
   normal.

4. When did you <u>first</u> take out a PROVISIONAL LICENCE to drive any vehicle?

   Month..................... **Year: 19**.........

5. Since that time, how much driving experience have you had?
   Please answer below for each type of vehicle.

   <u>Motor-cycle</u>   *When did you first drive a motor-cycle?* ....19...... **WRITE NEVER** if
   month   year   never driven.

   *Have you driven more than 1,000 miles
   on a motor-cycle? (Write YES or NO)*............

   <u>Motor scooter</u>   *When did you first drive a motor-scooter?* ....19...... **WRITE NEVER** if
   month   year   never driven.

   *Have you driven more than 500 miles on
   a motor scooter? (Write YES or NO)*............

   <u>Moped or</u>   *When did you first drive a moped or auto-*
   <u>auto-cycle</u>   *cycle?*   ....19...... **WRITE NEVER** if
   (with motor   month   year   never driven.
   <u>and</u> pedals)   *Have you driven more than 250 miles
   on these?*   *(Write YES or NO)* ....

   <u>PLEASE DO NOT LEAVE BLANKS.</u> If not sure of the month, give the year only.

6. What is your date of birth?   Year: 19......   Month:.................   Day:............

PLEASE RETURN THIS FORM AS SOON AS POSSIBLE USING THE ENVELOPE AND LABEL
PROVIDED. PLEASE DO NOT DELAY EVEN IF YOU CANNOT ANSWER ALL THE QUESTIONS

WE WILL BE VERY GRATEFUL FOR YOUR CO-OPERATION

wide response is needed. The enclosure of an envelope and stamped label is in itself important and helps the response. In this survey the response rate was equal to 84·4 per cent of the original sample and, when account was taken of those in the sample who no longer possessed a motor cycle, the effective response was equal to 91·4 per cent.

With a postal survey, it is obviously essential not merely to invoke the prospective respondent's co-operation, but where possible to convey the idea that the results of the survey are likely to be useful and beneficial to him. This approach is very obvious in the Spring 1964 enquiry carried out by the Southern Region of British Railways among its suburban passengers. The forms in this case were issued to passengers on a particular day at the various stations and to encourage their completion, the front page of the schedule contained a note which, after explaining that the enquiry was concerned to build a new pattern of services in the area, stated that the object was 'to do the best we can for you in the years ahead', and later again, 'in order to ensure that your own travel needs are taken into account, please complete this enquiry, etc.' While it is arguable that not every respondent passenger was entirely convinced by such assurances, the effect of this approach was surely to increase the response.

For all the undoubted success that has accompanied some of the Social Survey postal enquiries, generally speaking, most organisations find that the response to such forms is insufficient to justify generalising the results for the population as a whole. The plain truth of the matter is that it is, generally speaking, a self-selecting class which returns the schedules, usually because they have a particular interest therein and as such are not representative of the entire population. The obvious method of overcoming non-response with such enquiries is to make the return of the questionnaire compulsory. In practice the only body with statutory powers to compel a return is the government. For this reason they are the main users of this method and employ it for both samples and censuses.

Nevertheless, it would be idle to pretend that because the respondent is compelled to return the schedule, it is necessarily completed accurately. One of the features of recent censuses, particularly the Censuses of Production and of Distribution, has been the reduction in the size and coverage of the questionnaires sent to shops and industrial undertakings listed in the Board of Trade sampling frames for the purposes of these Censuses. When these Censuses were first introduced after 1947, the schedules were sent to most firms and were in many cases small volumes of closely printed notes and questions. Inevitably the refusal rate was high, and although statutory powers are available to compel the return, once the number of non-respondents becomes at all large this threat ceases to be effective. In consequence, the statisticians responsible for the Censuses of Production and of Distribution have gradually reduced the size of their forms for smaller firms and simplified them in the case of the larger firms. In the case of the Census of Production, whereas originally all firms with under eleven employees were approached, more recently the practice is to approach firms with 25 or more workpeople. In other words, the majority of industrial undertakings may not be approached at all, particularly in the case of the sample surveys between the main Censuses.[1] The contribution to the national output of these small firms is estimated on the basis of other data.

Lastly, it should be borne in mind that postal surveys are suitable only for the collection of limited factual data; in other words, only questions which require a simple yes/no answer, or a figure or a date, are really suited to this method of enquiry. Once the survey concerns itself with attitudes and opinions, it becomes necessary to use interviewers. One qualification of this last statement is perhaps necessary. There is nothing to prevent long questionnaires being sent out with questions on attitudes, opinions, etc., incorporated in them. For example, the basis of one well-known sociological study[2] is the collection of schedules returned by readers of a Sunday newspaper. These schedules were four foolscap pages in size, with detailed questions on all aspects of the respondent's life, ranging from his material possessions to his views on sex, marriage, etc. It could, of course, be argued that a greater degree of frankness in the answers would be achieved by this method. This may well be true, but the basic defect remains that only a limited number of people, *i.e.* a non-random sample, would take the trouble to complete the questionnaire. However interesting their collective responses might be, and the above study does make interesting reading, the fact is that there would be no statistical justification for describing the attitudes reflected in the completed schedules as fully representative of the British way of life, although it may reflect, as the author suggests, the views of the readers of this particular Sunday newspaper.

Conclusions

The opening chapters of this book stressed the need to verify the source and quality of any published statistical data. This warning applies with especial force to economic and social investigations where the existence of a few figures sometimes tempts the user to make statements the validity of which rests entirely on some highly dubious and often scanty data. If such data are to be used, their source

---

[1] See Chapter XVIII for details.

[2] Geoffrey Gorer, *Exploring the English Character*. The structure of the sample and its method of compilation are clearly described.

should be checked. Any good survey report will contain a copy of the schedule used, a summary of the sample design and the instructions to interviewers on matters such as classification, non-contacts, etc. In other words, all the main points made in the foregoing chapter should find some mention.

For the student reader interested in this branch of statistics, an ounce of practice is infinitely more instructive than a ton of theory. Participation in a survey teaches quite a lot, even if the participation is restricted to acting as an interviewer. The student who would like to try his hand at such work can sometimes find occasional work with survey organisations. These often advertise in the press for temporary staff; while a letter of enquiry to commercial organisations may sometimes produce a response. Failing such opportunities for practice, the student should get hold of as many reports as possible and study them in the light of the above comments. The list below offers some suggestions for further reading in this field.

Selected Survey Reports:

Higher Education. Appendices One and Three. Cmnd. 2154.

Family Census. Part 1 of Vol. VI of the Papers of the Royal Commission on Population.

Early Leaving. Ministry of Education. H.M.S.O. 1955.

British Income & Savings. H. Lydall. Cambridge U.P.

Television and the Child. H. Himmelweit and others. Oxford U.P.

15 to 18. Volume 11 of the Crowther Report. H.M.S.O. 1960.

Enquiry into Household Expenditure in 1953-54. H.M.S.O. 1957.

Family Expenditure Survey 1962. H.M.S.O.

Birth Control in Britain. Rowntree & Pierce. Two articles in *Population Studies*. 1961.

Royal Commission on Taxation of Profits and Income. Second Report. Cmd. 9105. Appendix I on P.A.Y.E. and Incentives Survey. 1954.

Women and Teaching. R. K. Kelsall. Ministry of Education. H.M.S.O. 1963.

Royal Commission on Doctors' and Dentists' Remuneration. Cmnd. 939. Appendix A on postal enquiry into professional earnings.

Family Limitation. Vol. I of the Papers of the Royal Commission on Population.

The Pre-Election Polls of 1948. Mosteller, Hyman and others.

Reasons Given for Retiring or Continuing at Work. Ministry of Pensions and N.I. Enquiry, H.M.S.O. 1954.

Further Reading on Survey Techniques:

Survey Methods in Social Investigation. C. A. Moser. This is the standard British work on the subject.

Research Methods in Social Relations. Jahoda, Deutsch, Cook, edited by Sellitz. An American book issued in the U.K. by Methuen.

Questions

SAMPLE SURVEYS

1. Give an account of the main principles to be borne in mind when drafting questionnaires, illustrating your answer by reference to any survey with which you are familiar.   *B.Sc. (Sociology) 1962 (University of Ghana)*

2. Give examples of the various types of non-sampling error which may affect the results of a large social survey using interviewers.   *I.M.T.A. 1962*

3. What is meant by the 'non-response problem' in the field of social surveys? Distinguish different types of non-response, and indicate possible solutions to the problem in each case.   *Final D.M.A., 1961*

4. Many social survey questions relate to the past and thus rely on the respondent's memory. Consider the possible effect of memory errors on the accuracy of survey results, paying particular attention to ways of dealing with such errors in practice.

    Final D.M.A. 1961

5. The type of interviewing used in many social surveys allows little discretion to the investigator in the way in which the questions are put. Discuss the advantages and disadvantages of this kind of approach.   *Final D.M.A., 1962*

6. Discuss the various sources of error and bias which may arise in any sample survey using interviewers.   *I.M.T.A., 1958*

7. 'No survey can be better than its schedule of questions.' Comment.

    I.M.T.A., 1960

8. 'The basic problem of questionnaire and schedule design is not so much what questions to ask, but what is the best way of phrasing them.' Explain and comment on this statement.   *B.Sc. (Sociology), 1958*

9. What is the rôle of a pilot study in the planning of a survey by a postal questionnaire?   *B.Sc. (Econ.), 1958*

10. Survey findings may be misleading because of ambiguity in the wording of questions. Give some examples of such questions and indicate how these ambiguities can be avoided.   *B.Sc. (Econ.), 1958*

11

11. Explain the different ways in which bias may arise in sample surveys.

B.Sc. (Econ.), 1956

12. For the purposes of a housing survey, a local authority needs to know the extent to which families move from its houses and flats, either into other Council accommodation or into privately owned property, and the causes of such movements. Draft the enquiry form and indicate how you would summarize the information. *I.M.T.A., 1962*

CHAPTER XVI

SOCIAL STATISTICS

The administrative needs of the Government extend beyond the fields of industry and economic planning. A major part of the Government's machine is concerned with the provision of social services and, as is the case with economic statistics, the administrative processes yield a large group of statistical data relating to social affairs. A large proportion of these are available in the Annual Reports of Ministries concerned with social affairs, for example, the Ministry of Health, the National Assistance Board, or the Ministry of National Insurance. In addition to the Annual Reports there are a large number of publications based upon periodic enquiries, as well as *ad hoc* enquiries undertaken by the Ministry or government committees. The annual sample survey of food consumption provides the Government with important information relating to the dietary habits of the population, while the survey conducted by the Ministry of Labour into reasons for retirement is a good illustration of an *ad hoc* enquiry. Anyone who has read even part of the Robbins Report on *Higher Education*, or the Newsom Report entitled *Half our Future* dealing with the state of secondary education, will be aware of the fact that a prerequisite of sound recommendations for the formulation of future policy is a firm statistical foundation. As stated above, a great deal of information relating to social affairs, such as the extent of poverty or the incidence of crime, can be derived from regular reports from the administrative bodies concerned with these matters, *e.g.* the National Assistance Board and the Home Office. Nevertheless, to an increasing extent it is being realised that the statistics generated by the administrative processes are generally inadequate for purposes of research and further investigation. This is especially true, for example, of crime statistics and the Government has appointed an expert committee to examine the material available and to make recommendations. The Social Survey in particular is kept busy year in year out undertaking particular enquiries into subjects such as when people prefer to take their holidays, or the extent to which noise at London Airport is a major problem for the residents in the vicinity of the aerodrome. Some indication of the scope and coverage of social statistics in particular fields is given in the next few pages. For the reader who would like a wider and less specialised coverage, a book entitled *A Survey of Social Conditions in England and Wales as Illustrated by Statistics* will make interesting reading.[1]

Population Statistics

The most important single source of statistical information relating to the social scene is the Census of Population. The census consists of a complete enumeration of persons in the country on a given night. In addition to counting the number of people, information relating to their sex, age, civil status, *i.e.* married, single, etc., their occupations, and the industry in which they work, together with further information relating to housing conditions, educational attainment, are usually collected. The first Census in Great Britain was taken in 1801 and there has been one at ten-yearly intervals ever since, apart from 1941. It is customary to use the national registration figures prepared in September 1939 as the best measure of the population between the 1931 and 1951 Censuses. The Census data for England and Wales are published separately from those for Scotland and Northern Ireland. Although the Census is taken at decennial intervals, since the passage of the Census Act 1920 power has existed to hold a census at five-yearly intervals. It is proposed to exercise these powers for the first time ever in 1966. The importance of the Census cannot be over-estimated. Quite apart from providing information relating to the population, its age, sex, geographical and occupation distribution, all of which analyses are relevant to the effective administration of the country, the Census provides a great deal of information relating to social conditions, particularly in respect of housing, occupational mortality, and fertility. Recent questions on the educational attainment of citizens and the number of persons with scientific qualifications are relevant to the formulation of educational policies and manpower distribution in industry. It is proposed in 1966 to collect information relating to the mobility of the population, *i.e.* the movement of persons and households from one part of the country to another, and also some information relating to private cars, garage facilities and the journey to work.

The average Census contains about two dozen questions most of which remain unchanged from census to census, *i.e.* those relating to the age and sex composition of households, etc., and the occupations of their members.[2] However, the Census is also used to collect information relating to special problems of the time. For example, the 1911 Census is often referred to as the Fertility Census, since this was the first occasion on which questions relating to fertility were asked.

[1] By Carr-Saunders, Caradog Jones and Moser, O.U.P., 1958.

[2] Guides to Official Sources, No. 2, Census of population 1801-1951, provides comparative tabulations of all the questions asked at each census since 1801, as well as providing detailed information about each census.

The 1921 Census is known as the Dependency Census, because special questions were asked relating to widows and orphans in view of the manpower losses sustained during World War I. In the 1951 Census new questions were asked relating to fertility, and also some new questions relating to housing conditions. A feature of the 1961 Census was the introduction for the first time of multi-phase sampling, whereby 10 per cent of the population were asked to complete an additional schedule, which asked questions relating to their qualifications in science or technology.

The accuracy and reliability of the information collected in a census depends on a number of factors. The first and most important are the enumerators themselves. These are part-time workers whose task is to distribute the census schedules to every household within their allotted enumeration district, of which there are approximately 70,000 in England and Wales alone. In other words, the completeness of the coverage depends on the knowledge possessed by the enumerator of his enumeration distrct. To the extent that he misses any households, there is under-counting. Having distributed the schedules, he must ensure that immediately after the census night he collects the completed schedules and checks that there are no omissions or obvious mistakes. This is by no means a simple task. In 1951, for example, the weather on the day after the census was very bad, and the scrutiny of the schedules undoubtedly suffered, since in most cases the schedules are merely handed back over the doorstep. The choice of date needs perhaps some explanation. The Census in England and Wales normally takes place on a Sunday in April. In 1951 it was April 8, in 1961 April 23. The choice is dictated partly by weather conditions, which affect the enumerators, but above all by the need to avoid periods in which there is extensive population movement; the Easter or August Bank Holidays would obviously be unsatisfactory in so far as the population of the seaside resorts would be grossly inflated at the expense of that of the inland industrial and urban centres.

Further sources of inaccuracy and error arise in the completion of the schedule itself. This is performed by the head of the household, and since the standard of education and literacy varies widely among the 14 million or so heads of households in this country, mistakes are likely to be made. Many of these can be picked out in the editing, which is carried on during the tabulation process at the census offices. Here, too, of course, is a potentially fruitful source of error. Classification of occupations, for example, is extremely complex. There are over 200 groups in the occupational classification, while industrial or commercial activities under which the employer's trade or business are classified total some 280. Errors may arise here, just as errors are possible in transferring information from the schedules on to the punched cards. In due course the information on the punched cards is transferred to magnetic tape since in 1961, for the first time ever, an electronic computer has been used in the preparation of the published data.

In view of what has been said about the likelihood of error in the census, it is hardly surprising to learn that the census office undertakes what are known as small-scale post-enumeration surveys. The object of these enquiries is to see how well the questions and instructions have been understood, and how accurate are the replies to the questions. In addition, of course, they also provide some indication as to the completeness of the count, *i.e.* the degree of under-counting or omission. In the 1951 Census, post-enumeration surveys were made into the information relating to household arrangements, *i.e.* the number of households sharing or without certain domestic facilities such as piped water, w.c.'s, bath, etc. These surveys reveal substantial errors in the information provided. This was also true of the questions relating to education in the same Census, as described on p. 345. Misstatements of age and occupation were detected by comparing the information relating to a sample of persons who had died within a few weeks of the census with that contained on their death certificates. Nevertheless, despite such mistakes, and it would be foolish to pretend that in a survey of some 14 million households containing nearly 50 million persons some mistakes are not inevitable, the census authorities believe, and it seems with ample justification, that a high degree of accuracy can be claimed for most of the information provided by the Census. Furthermore, in the case of age, special techniques for adjusting misstatements and errors in age are available to the statisticians.[1]

The volume of information produced by the Census of Population takes several years to process and publish. The pattern of publications is more or less as follows. The first report, appearing within a few months of the census itself, is the *Preliminary Report*, which states the populations of the administrative areas of England and Wales at the last two census dates, with the absolute and proportionate change therein. This is then followed by a series of County Reports, although in 1951 two invaluable volumes entitled *One Per Cent Sample Tables* were produced within less than two years of the census being held. This experiment was unfortunately not repeated in 1961, since it was hoped that the use of an electronic computer would expedite publication of the main data. For various reasons this has not proved practicable. The *County Reports* appear in rapid succession, and these contain comprehensive data based on the entire schedule of questions for the population resident in the county at the time of the

[1] See General Report 1951, page 35.

Census. They are especially useful for information relating to housing conditions and household composition.[1] Then follow the major special 'subject' volumes on occupation, housing, industry and, in 1951 the Fertility Report; and then the General Report. For most purposes the reader can obtain all the information he requires as far as his own administrative areas is concerned from the County Report and the General Report. The latter is especially valuable in drawing attention to any defects in the data that the census authorities have come across in the processing thereof. Details of the census publications are normally provided in any of the individual publications. Accompanying the census volumes are the so-called *Decennial Supplements*, which in 1951 comprised a two-volume study of occupational mortality, a separate study of area mortality, and the official English Life Table No.11.[2]

Supplementing the Census of Population data are the annual returns of the Registrar General, known as the Registrar General's *Statistical Review*. This is divided into three parts. Part I, sub-titled *Medical*, comprises tables giving detailed information relating to causes of death at different ages of life for various groups of the population, as well as their location. Part II, sub-titled *Population*, consists solely of tables and these contain information relating to the number of marriages and births for all the administrative areas of the country, classified according to the age of the parties to the marriage, or the age of the mother at the birth of the child. In addition, further breakdowns by reference to the administrative areas are given in respect of legitimate and illegitimate births, and stillbirths, as well as for marriages and electors. Information relating to divorce is also given in this volume. Part III, the final volume, is known as the Commentary. This last contains quite a number of tables, but in contrast to the two preceding volumes, it includes the official review of vital events during the past year, or any trends which appear to have emerged, *e.g.* trends towards lower age of marriage and higher fertility. The three parts of the Statistical Review are published some time after the year to which they relate, Parts I and II preceding Part III by many months.

In addition to the Annual Review the Registrar General produces a *Quarterly Return*. This contains information relating to births, deaths, marriages, etc., for that quarter of the year, classified by reference to the administrative areas of the country. In addition to this standard information, each quarterly return contains selected pieces of information. For example, the 4th quarter's return contains population projections by sex and age, as well as the regional break-

---

See section on Housing, page 35.

[1] This is discussed in the next chapter.

down of the population by sex and age, while the June quarter's return publishes the Life Table based on the deaths of the preceding three years and the estimated home population. This information is in due course brought together in the annual *Statistical Review*. The sources of these data are the local registration offices, distributed throughout the U.K. As is generally known, any vital event, *i.e.* birth or death, must be registered with the Registrar of Births, Deaths and Marriages. In the case of a marriage, if it is held at the local register office, the registration is automatic. If held in an Anglican church, then the officiating clergyman is responsible for forwarding a duplicate return of the entry in the church register. In the United Kingdom, such statistics are highly reliable.

As will be seen from the sections which follow, the Census of Population is also the source of certain important data relating to social class, housing, and infant mortality. Apart from the factual content of the publications prepared by the General Register Office, the statistician has a special interest in the analysis of these data. The analysis and interpretation of vital statistics forms the subject matter of demography, *i.e.* the study of population. Some of the simpler statistical techniques used in interpreting population statistics are discussed in the next chapter.

Housing

There would be little disagreement with the assertion that the major social problem in post-war Britain has been the shortage of housing. For this reason it is rather remarkable, to put it no more strongly, that the available statistics are totally inadequate to form the basis of a rational housing policy. Quite apart from the complications arising from the system of controlled and subsidised rents, which make a true assessment of the adequacy of existing housing accommodation virtually impossible, the statistical shortcomings are very obvious. First, there are no complete and up-to-date data relating to the supply and adequacy of existing housing accommodation, *e.g.* age and amenities; second, the statistics to measure either 'need' or 'demand' for housing, either on a local or national level, are totally inadequate. The data relating to housing are derived from two main sources. The first is the Ministry of Housing which collects data relating to the volume of current building. The second is the Census of Population which, more especially in the Housing volume, provides a great deal of interesting information relating to housing conditions and over-crowding. These two sets of data are best dealt with separately.

The statistics of house building date back to the end of the First World War when in 1919 the government first made itself responsible for the provision of subsidised dwellings. Before then, the number of dwellings could only be estimated from the latest available census returns. Once public money was at stake, however, the administration annually produced statements showing the number of houses built which were entitled to the subsidies. From 1922 on, the published data became virtually comprehensive since it included all non-subsidised accommodation with a rateable value of £78 and under.[1] With the tremendous boom in house construction of the 1930's the Ministry of Health (which from 1919 was responsible for this matter) published, as from September, 1934. a half-yearly return giving the number of houses built by both private enterprise and local authorities, as well as the volume of 'slum-clearance'. A particularly useful and informative publication of this period is the official *Survey of Over-crowding* of 1936 which covered Scotland, as well as England and Wales.

The end of the Second World War witnessed a revival of government housing activity but on a much larger scale than in 1919. Between 1946-48 a monthly return of house construction and war damage repair in Great Britain was produced by the Ministry of Health. Thenceafter it was converted into quarterly returns for England and Wales (separate returns for Scotland) analysing the number of houses built by reference to the region, and the authority responsible for the construction, *e.g.* local authority or private builder. From 1951 these *Housing Returns* were prepared by the new Ministry of Housing and Local Government. In addition, these returns included data relating to the number of tenders approved, construction begun, and completions. In addition to the data on permanent houses, in recent years similar data are provided relating to conversions and adaptations of old and existing property, as well as the progress in slum-clearance. More recently there has been a sample survey of housing in Great Britain undertaken by the Social Survey which gives details of the age and condition of domestic properties, the existence of garage accommodation, together with details of the ownership or tenure of the property.[2]

Table 35 illustrates the information given in the Annual Reports of the Ministry of Housing and Local Government. These data are by any statistical standards reasonably accurate for the year in question, and subject to minor changes in definition it is possible to construct an approximate series dating back to the 1920s. The fact that statistics of housing construction have been officially prepared since 1919 should not delude the reader into thinking that the data form a consecutive series which reflect the pattern and fluctuations in house

---

[1] In the Metropolitan Police District the limit was £105 but houses with Rateable values above £78 in the provinces and over £105 in the M.P.D. were very few in number.

[2] *The Housing Situation in 1960*, by P. G. Gray and P. Russell, S.S. 319, C.O.I. May 1962.

TABLE 35
PROVISION OF NEW PERMANENT DWELLINGS*
(England and Wales)

	1960	1961	1962	1963
New permanent houses and flats built for:				
Public authorities:				
Local authorities† ..	103,235	92,880	105,302	97,015
Housing associations ..	1,650	1,585	1,561	1,925
Government departments	2,241	4,001	4,788	3,473
Total	107,126	98,466	111,651	102,413
Private owners	162,100	170,366	167,016	168,242
Grand total	269,226	268,832	278,667	270,655

* *Source: Annual Report of the Ministry of Housing and Local Government 1963. Cmnd. 2338.*

† Including the Commission for the New Towns and the new town development corporations.

building over the past four decades. Dr Marion Bowley has emphasised in her study of housing statistics that these data abound with snags such as changing definitions, varying coverage of series, etc., and therefore require the greatest care in extraction if any long period comparisons are required.[1] For all practical purposes the various Housing Returns serve as an indication of the scale of housing construction at various periods in England and Wales and little else.[2]

The second source of housing statistics is the Census of Population. Each census since the very first in 1801 has included questions – not always the same – about housing and households. Table 36, illustrates some of the available data from successive censuses. The Housing volume of the Census of Population 1951, pending the publication of the 1961 Census housing data, is the main current source, but details relating to housing conditions are given in the County volumes as well as in the 1 per cent Sample Tables. The experiment of producing the 1% Sample Tables was not repeated for the 1961 Census. At the time of writing, mid 1964, the County volumes are the only source on housing conditions in 1961. The 1951 Census was especially important in this particular respect for it collected new information relating to what are called 'household arrangements', *i.e.* the existence of piped water supply, cooking stove, water-closet and fixed bath, as well as providing more detailed and new classifications of households not given in the 1931 volume. For example, households are analysed by the number of earners and number of children they

[1] Housing Statistics of Great Britain, *J.R.S.S.*, 1950, Part III.
[2] The most detailed official account of housing statistics, *i.e.* building, is given in an article in *Economic Trends*, June 1958.

TABLE 36
POPULATION, HABITATIONS AND HOUSEHOLDS, 1861–1951* ENGLAND AND WALES

Census Year	Population 000's	Houses 000's	Families 000's	Persons per House
1861	20,066	3,924	4,492	5·10
1871	22,712	4,520	5,049	5·02
1881	25,974	5,218	5,633	4·98
1891	29,003	5,824	6,131	4·98
1901	32,529	6,710	7,037	4·85
	Population in private families 000's	Private dwellings 000's	Private families or households 000's	Persons in private families per dwelling
1911†	34,606	7.691	7,943	4·50
1921	36,180	7,979	8,739	4·53
1931	38,042	9,400	10,233	4·05
1951	41,840	12,389	13,118	3·38

* *Source: Economic Trends.* June 1958.

† In the 1911 and subsequent censuses non-private households were excluded from the tabulations which in previous censuses had included them.

contain, as well as the proportion with the 'household arrangements' specified above.

The term 'housing' in the Census volumes does not merely cover houses as such; it is primarily concerned with available accommodation, the basic unit of which is defined as a 'structurally separate dwelling'. This has been defined since the 1921 census as follows: 'any set of rooms, intended or used for habitation, having access either to the street or to a common landing or staircase'. According to the 1951 Census there were 12·4 million 'structurally separate dwellings' occupied by 13·1 million households. The preliminary report of the 1961 Census discloses that the number of private households rose during the intervening decade to 14·7 million, an increase of 12. per cent. This compares with an increase in the number of private dwellings from 12·4 million to 14·6 million, representing an increase of 21·3 per cent.

The term 'household' is also carefully defined in the Census. To start with they are divided into two main groupings, private and non-private. The former consist of what is generally known as a household, *i.e.* any group of individuals such as a family and servants and lodger with board. The non-private households cover all the people living in institutions and hotels, etc. The private households are in their turn further sub-divided into 'primary family units' and what are termed 'composite' households. The former, which in 1951 accounted for 80% of all households, consists of the head of the household, his wife, their children (of any age if unmarried) and any immediate relatives such as brothers and sisters of the head of the household and his wife. Resident domestic servants are also included. In other words, the primary family unit is what most of us regard as the 'family'. The 'composite' household consists of a private family unit plus one or more of the following, a family nucleus; a married brother or child of the head of the household; and any others not related to the head of the household, *e.g.* a boarder. The term 'family nucleus' signifies any group in a composite household such as a married couple, with or without children, or a lone parent, *e.g.* widowed or divorced. The importance of the concept 'family nucleus' arises from the need to make some estimate of housing needs. Such 'nuclei' can be regarded as potential occupiers of their own establishment; they may only be living with the primary family unit until such time as they can acquire separate accommodation. About four-fifths of such nuclei were in fact married sons or daughters of the head of the households, awaiting suitable accommodation for themselves.

The importance of the census information relating to housing lies not so much in the actual enumeration of the numbers of households and private dwellings, as in the indication they provide of what are termed 'housing conditions'; in particular, what is termed by the Census authorities as 'density of occupation'. Tables 37 and 38 illustrate the sort of information given in the County and Housing Volumes relating to density of occupation, or what is more generally referred to as 'overcrowding'. Table 37 illustrates the three measures of occupation density used in the reports. The first is the number of persons per household, the second the number of persons per dwelling, and lastly, the ratio of households to dwellings. The three sets of figures shown refer to Cardiff County Borough, the urban areas with over 100,000 population apiece, where it may be assumed the most overcrowded conditions exist, and lastly the corresponding figures for England and Wales as a whole. In the County Volumes, these figures are given for all the administrative areas, *i.e.* the minor authorities such as the urban district councils and the various wards

TABLE 37*

	Cardiff C.B.	Urban areas with over 100,000 popn.	England and Wales
No. of persons per household	3·23	3·15	3·13
No. of persons per dwelling ..	3·56	3·19	3·14
Ratio of households to dwellings	1·102	1·012	1·004

* *Source: Preliminary Report 1961 Census.*

in the towns. In addition to the foregoing figures the County Reports also provide a density measure defined as 'persons per room ratio', and it is this latter measure which indicates the degree of overcrowding in the wards and other administrative areas. Table 38 illustrates the detailed data available in the County Volumes. Note the breakdown into the four classes of occupant; this was introduced with the 1961 Census for the first time. Nevertheless, too much should not be

TABLE 38
HOUSING CONDITIONS IN CARDIFF C.B.*

	Number of:			Persons per room
	House-holds	Persons	Rooms	
Owners	33,804 (3,327)	106,603 (8,637)	182,737 (13,532)	·58 (·64)
Renting from L.A. or New Town Corp.	17,861 (315)	72,914 (797)	79,327 (915)	·92 (·87)
Unfurnished renting from Private Landlord	19,850 (7,347)	57,151 (18,046)	85,085 (20,902)	·67 (·86)
Furnished renting from Private Landlord	3,039 (2,075)	6,711 (4,121)	8,055 (4,342)	·83 (·95)

* *Source: County Volume. Glamorgan, 1961 Census.*
Note: The figures in brackets relate to the numbers in each category who are sharing their accommodation.

read into the density statistics, on the grounds that much depends on the composition of the household as to whether overcrowding exists or not. For example, a married couple obviously need less accommodation than do two single adults of the opposite sex to achieve comparable living standards. A weakness of the persons per room ratio, quite apart from the fact that rooms vary greatly in size, ventilation, etc., is that such figures are merely averages. While they are prepared for the smallest administrative units, *e.g.* wards in the urban boroughs, or rural district councils in the counties, these areas are usually large enough to contain a wide variety of housing conditions. Unless it is known that a particular area is unusually homogeneous in respect of housing conditions, inter-area comparisons of such published density ratios as a reflection of 'overcrowding' need to be interpreted with care.

'Overcrowded' is a purely subjective term, but for statistical purposes it is essential to have some fixed standard, even though such a standard is quite arbitrary.[1] The standard used in the 1931 and 1951

[1] Compare for example the detailed standard used in the 'New Survey of London Life and Labour 1928' with the standard employed in the 1936 Overcrowding Survey.

Census reports provided for a maximum of two persons per 'habitable' room.[1] A density greater than this figure qualified for the description overcrowded. There can be little doubt that the 1951 Census definition was out of date and under-estimated the extent of the problem in so far as it set a standard which was unduly low. Since 'room' covers also the kitchen if used as a living room (the Census does not distinguish between bedrooms and living rooms, nor are their dimensions measured), then a three-bedroomed house with sitting-room and kitchen-living-room would not have been overcrowded until it had over ten persons. In the 1961 census the census authorities have raised the minimum standard by taking a figure of $1\frac{1}{2}$ persons per room as the density of occupation which may be considered for all practical purposes as the upper limit of tolerable housing conditions. To the extent that the 1961 Census employs this improved measure of overcrowding, it will not be possible until the publication of the 1961 Housing Report to judge how far the improved measure of overcrowding is balanced by the undoubted improvement in the ratio of dwellings to private households that has taken place during the intervening decade.

No attempt is made to take the sex or age composition of the household into account in this measure of overcrowding. A better criterion for overcrowding would be to use a 'bedroom' standard. This would aim to provide adequate sleeping accommodation to allow separation of the sexes over say 10 years of age (except of course for married couples) and smaller sleeping accommodation for young children, e.g. one room for 3 children under 5, or one child under with one over 5, or two children aged 5 or over of the same sex. This implies that a working kitchen and separate living room are essential during the day and should not be occupied at night. An even more precise standard could be calculated by designating each adult, for example, as a unit and adolescents as 3/4, children under 10 as half and infants under one 1/4 or nil units, and relating the household total to the floor space available. Then a given-sized house should not contain more than so many units per x square feet. Another problem is of course the size of the rooms; generally speaking the poorer housed the households by the foregoing standards, the smaller are the actual 'rooms'. Any survey of overcrowding or 'housing conditions' should be examined to ascertain what standards are used. For example, the Housing Act 1936 used both a 'bedroom' and 'floor area' standard to define overcrowding. This was deemed to exist when the number of persons sleeping in the house is *either* such that two or more of those persons being over ten years old, of opposite sex and not living together as

[1] The Census does not define a habitable room in precise terms, except to exclude sculleries and bathrooms from that category.

man and wife, must sleep in the same room; *or*, that there is an excess
of the permitted number of persons as ascertained in relation to the
number and floor area of the rooms as laid down in a schedule to the
Act. For this purpose an infant under one year is not counted and a
child between 1 and 10 years is reckoned as half a unit.

The information collected in the 1951 Census which relates to
'household arrangements' revealed a far from satisfactory state of
affairs. The Census form contained questions as to whether the
respondent household had exclusive or shared use of the bath, w.c.,
kitchen sink and water supply, or if they were without any of these
facilities. Just over half of all households in England and Wales in
1951 had all these four amenities and exclusive use thereof. There
were 1·4m. households without exclusive use of *both* a kitchen sink
and a water closet. As is to be expected, large numbers of households
in shared dwellings were without separate provision of these domestic
arrangements. Even so, over one-third of private households had no
fixed bath, 8 per cent had no water closet, 6 per cent had no kitchen
sink and the same proportion was without piped water.

Unfortunately, interesting as these data are as a reflection of
housing conditions in 1951, further enquiry by the census authorities
revealed that there was some serious misunderstanding by respondents of the questions relating to the household amenities and the
extent to which these were shared. A sample of six urban areas in
different parts of the country was surveyed after the census by the
local registration officers, and it transpired that one-third of the
claims to be lacking .access to water closets were incorrect. The
general impression, states the General Report (p. 57) is that there may
have been some appreciable overstatement of the lack of availability
of water closets. With regard to piped water it appears likely, states
the same report, that sharing was substantially overstated, while
entire lack of piped water was to a less extent overstated. Partly on
account of the unsatisfactory nature of the information culled from
the 1951 Census in respect of this subject, the 1961 Census repeated
the questions on piped water supply, w.c. and fixed bath. They also
asked a further question regarding the hot water supply. The 1951
questions about the cooking stove and kitchen sink were repeated in
1961 but restricted on this occasion to households sharing dwellings.
It is proposed to ask questions relating to baths, lavatories and the
supply of hot water in the 1966 Census.

This brief review of the official statistics relating to housing conditions reveals both the strength and weaknesses of the official data.
As is the case in so many other fields of official statistics, there is both
a wealth of information on particular aspects of the subject and an
inadequacy of information on other equally important aspects. For

example, more information is urgently needed about the age and condition of domestic properties. The student reader would be well advised not to attempt to read large sections of the official publications on housing conditions but rather to learn what information is available by the simple expedient of preparing a brief summary report on the housing conditions in his own town or administrative area. By extraction of relevant figures from the various reports such as the Preliminary Report, County Report, and Housing Volume, he will not only familiarise himself with the facts relating to the subject, but the deficiencies of the data will be thereby impressed upon his memory.

REFERENCES

Census 1951. 1 per cent Sample Table, Part I.
Census 1951. Housing Report, Ch. VI.
Census 1951. General Report, Ch. VII.

Social Class

The English, it has been observed, are extremely class conscious. Certainly most people are ready to classify others by reference to their social class, such as working, middle or upper class. There is, nevertheless, no single measure of 'class'; for example, different people might well classify a particular individual either as working or 'lower' middle class. The conventional basis for social classification is well understood, but it is not sufficiently precise for statistical purposes. In any case many people dislike 'class consciousness' as such, and would reasonably ask why it interests the social statistician. There are, however, a number of reasons why the subject is important. First of all, it is common knowledge that fertility among labourers is higher than among the professional community; just as is infant mortality. It is not sufficient merely to believe that a difference exists between the two extreme groups. We want more precise information as to its extent. For example, before the war, in certain districts of England and Wales the infant mortality was almost four times as high as in the wealthy areas of the South East.[1] This is criminal waste and once we have the information as to the extent thereof, social policy can be adapted to cure such blots on the community. It is a commonplace to talk of the opportunity open to the poor boy to get 'to the top'. But how many get there? The extent to which the academic achievement of grammar school boys is related to the home background, the latter being defined by the father's occupation, has been measured in a Ministry of Education report which is of particular sociological interest.[2] To what extent is there what the sociologist calls 'social

---

[1] See *Population and Poverty*, R. Titmuss, for a study of pre-war society based entirely on such data. Even now there are regional differences, see for example Table 40 below.
[2] *Early Leaving*, H.M.S.O., 1954.

mobility', *i.e.* do sons and daughters generally move into a higher or lower social class than their parents?[1] Before such enquiries are practicable, there must be a generally accepted classification of social status.

There are, of course, a number of criteria which are employed to classify people. Accent is often a good guide; but statistically it is useless. Education is a useful indicator but is not enough. Income is a better guide and lends itself to statistical treatment; but there are many people with a great deal of money these days whom no one would place in the 'upper class', just as few people would classify a parson in a poor living as 'working class'. The most generally acceptable basis for classifying people is their occupation. Furthermore, surveys have shown a high degree of unanimity among all classes as to the relative status of given trades, occupations and professions.[2] Since the 1911 Census the Registrar General has used a five-fold classification as follows: the figures in brackets giving the proportionate distribution of the population between the classes as disclosed by the 1951 Census. The obvious defect in this classification is that over half the population are in one class, *i.e.* III.

I. Professional, etc., occupations, *e.g.* doctor, lawyer. (3).

II. Intermediate occupations, *e.g.* business executive, manager of large store. (15).

III. Skilled occupations, *e.g.* draughtsman and policeman. (53).

IV. Partly skilled occupations, *e.g.* ticket collector and plumber's labourer. (16).

V. Unskilled occupations, *e.g.* dock labourer, watchman. (12).[3]

Such a classification is a good deal better than the rough and ready three-fold classification, working, middle and upper! It is precise, but for that reason it is arbitrary. For most people the members of classes II and III inevitably merge one into the other, and in fact the latter is so large that it is not really homogeneous, *i.e.* it is too mixed. For this reason in the 1951 Census publications the Registrar General introduced a new and more detailed classification by 'socio-economic' groups. There are 13 mutually exclusive groups within three main groupings, agriculture, non-manual, manual, and a single group for the armed forces (see Table 39). Like the social class classification, the socio-economic grouping is based upon occupations, but this more detailed classification ensures a greater degree of homogeneity within the individual group. Effectively, this classification, apart from distinguishing between the main occupational groups, *i.e.* agricultural, non-manual, and manual, as well as the armed forces, breaks down the social class classification roughly as follows: social class I becomes

[1] See *Social Mobility in Britain*, ed. D. V. Glass.
[2] Glass *op . cit.*
[3] Note percentages do not total 100 due to rounding.

socio-economic group 3; social class II becomes socio-economic groups 4 and 5; social class III becomes socio-economic groups 6, 7, 8, 9 and 10; social class IV is socio-economic group 11, and social class V becomes socio-economic group 12.

TABLE 39

STANDARDISED MORTALITY RATIOS FOR MEN, MARRIED WOMEN AND SINGLE WOMEN AT AGES 20-64 BY SOCIO-ECONOMIC GROUPINGS*

Socio-Economic Group†	Males	Married Women	Single Women
Agriculture			
1. Farmers	70	93	72
2. Agricultural workers	75	95	64
Non-Manual			
3. Higher administrative etc.	98	96	82
4. Other administrative etc.	84	81	70
5. Shopkeepers	100	99	97
6. Clerical workers	109	91	75
7. Shop assistants	84	79	82
8. Personal service	113	101	84
Manual			
9. Foremen	84	91	86
10. Skilled workers	102	105	109
11. Semi-skilled workers	97	108	99
12. Unskilled workers	118	111	103
All occupied and retired	100	100	85

* Group 13 of this classification – the Forces – is omitted.

† *Source: Reg. Gen. Decennial Supplement 1951. Occupational Mortality, Part II, Vol. I.*

Tables 39 and 40 illustrate the use to which the Registrar General puts these classifications. The former table, which also serves to indicate the socio-economic grouping referred to above, reveals the difference in mortality for three sections of the community at ages between 20 and 64, *i.e.* males, married and single women. The figures themselves are known as Standard Mortality Ratios. These are explained in detail in the next chapter, but they are in effect indices based upon the relative mortality experience of each group, allowance having been made for age differences between the groups. Farmers have the lowest mortality among males while unskilled workers have the highest. The married women, *i.e.* wives, have a broadly similar experience to their husbands, but note that farmers' wives do not enjoy the same favourable position of the males. In the data for single women, the figure for clerical workers is most marked in relation to the other two groups, *i.e.* men and married women.

The reliability of such figures depends on the accuracy of the information collected at the Census and registered on the occasion of the death. As part of the post-census enquiry into the extent to which the Census data were reliable, a sample of death certificates registered after the census were compared in respect of the deceased's occupation with the corresponding information on the census return. This sample check although, according to the authorities it was not large enough to justify firm assessments of the discrepancy between the two sources, nevertheless revealed that it was unlikely that the net discrepancy would exceed 10 per cent in any occupational order. It should be noted that the numerous occupations are grouped into what are known as orders and sub-orders, and within the entire census framework there were in 1951 27 main orders. The main discrepancies appear to have been in respect of some up-grading of occupation on the death certificate and some confusion in the case of skilled workers who after retirement had taken up unskilled work. The 1951 General Report concludes its survey of the findings of the sample check on these particular data with the comment that 'the general impression emerges that the level of reliability of occupational assignment at the census justifies the statistical analyses which are based thereon. In particular, the shift in socio-economic distribution is not sufficiently serious to threaten the validity of the occupational mortality analyses'. (pp. 55-56.)

It is pertinent to observe at this point that the census authorities pay great attention to the need for accurate recording of the individual respondent's occupation. The census form itself gives detailed instructions and before the 1961 census factories and offices were circularised with notices explaining to the employers how they should answer this particular question. The object was to ensure uniformity and accuracy, for it is well known from previous Censuses that the replies to the questions on occupation and the employer's business or trade have frequently been inaccurate. It will be readily understood that while for the larger occupational groupings such inaccuracies may not be serious, for the smaller groupings widespread errors could invalidate some of the comparisons which are made using these data, for example, in respect of occupational mortality and industrial morbidity, *i.e.* industrial disease, etc. The census workers are given extremely detailed instructions and voluminous lists of the various occupations so that they may adopt a uniform tabulation. Any reader who is sceptical as to the problems involved is advised to examine the census volume entitled 'Classification of Occupations'. He will be astounded at the number and variety of occupations.

Table 40 gives a breakdown by region and class of the infant mortality rate. Note the two rates; one relating to deaths under four weeks known as *neo-natal* mortality and the other to deaths between four weeks and one year referred to as *post-natal* deaths. The division is significant since with the marked fall in the infant mortality rates in recent decades, the main hope for its further reduction is to be found in reducing the *neo-natal* rate. The conclusions to be drawn from this table are quite clear. The regional differences are obvious, but are nothing like so marked as the inter-class differences. These tabulations of mortality by social class are extended to individual industrial groupings for certain major diseases. For example, cancer rates for different ages within individual occupations are calculated, as well as for many other causes of death. The reader should consult the relevant volume, 'Occupational Mortality', for details.

TABLE 40

NEONATAL AND POST-NEONATAL MORTALITY RATES PER 1,000 LIVE BIRTHS BY SOCIAL CLASS IN FOUR REGIONAL GROUPS

	Social Class					
	I	II	III	IV	V	All Classes
Aged under 4 weeks:						
North	12·5	16·9	19·7	21·8	23·5	20·2
Midland and East	14·9	15·6	18·0	19·8	22·8	18·5
South	12·0	15·9	15·9	18·7	19·2	16·5
Wales	16·2	21·3	19·1	24·7	26·9	21·6
Aged 4 weeks and under 12 months:						
North	4·9	7·6	13·3	18·1	23·2	14·8
Midland and East	5·2	6·5	10·3	12·8	18·1	10·8
South	5·0	4·6	7·8	9·3	13·2	8·0
Wales	4·3	7·3	13·4	16·9	22·2	14·1

Source: Reg. Gen. Decennial Supplement 1951. Occupational Mortality, Part I.

The Registrar General pays just as much attention to social class differences in fertility as in mortality. In some respects the former is more interesting to the sociologist. There have been several enquiries into fertility, in particular the questions asked in the 1911 and 1951 censuses, and the Family Census taken in 1946 for the benefit of the Royal Commission on Population. The latter is published as Volume VI of the papers of the Royal Commission under the title 'The Trend and Pattern of Fertility in Great Britain'. Table 41 is taken from that report. It shows the average number of children borne by a woman marrying between the ages of 20 and 24 in the quinquennium 1920-24. The data tell their own story; as we go down the social scale so fertility increases, with the exception of the salaried employees and non-manual wage earners. This is probably due to the fact that these two groups come into close contact with more fortunate economic groups and have a greater concern to restrict the size of their families in order to maintain their living standards. Note the slightly different classification based on nine occupational groups used in this enquiry. This was devised before the Registrar General's 1951 socio-economic classification to give a more detailed breakdown of social classes than the social class classification used in the 1911 census. The 1951 Census 1 per cent Sample Tables give an analysis of fertility according to the

TABLE 41

GREAT BRITAIN: FAMILY SIZE FOR WOMEN MARRYING IN 1920-4
AT 20–24 YEARS OF AGE*

Status Category	Number of live births per woman
Professions	2·02
Employers	2·13
Own account	2·28
Salaried Employees	1·90
Farmers and Farm Managers	2·78
Non-manual wage earners	2·26
Manual wage earners	2·96
Agricultural workers	3·14
Labourers	3·76

* Source: The Trend and Pattern of Fertility in G.B.

socio-economic groupings which confirm the pattern of reproduction revealed in the Family Census five years earlier. This revealed that the clerical workers' 'index' of fertility (actually a standardised ratio) was only 84 in 1951 compared with 124 for unskilled labourers, while the higher administrative, professional and managerial group had an index of 90.[1]

The importance of a standard classification of respondents by social class to the sample survey organisations is obvious. The reader may recall the discussion in Chapters XIV and XV of the technique of quota sampling, wherein the interviewers are required to interview respondents who fall into particular age, sex, and occupational groups. As was there explained, a too detailed social classification is impracticable, and most of the bodies use a 4- or 5-fold classification of the main groupings, stressing the importance to the interviewer of obtaining within each class a representative selection of occupations within any single social group. The classification reproduced on p. 262 illustrates well the basis of this method. It remains only to add that any statistics which are classified by reference to social class are subject to the errors inherent in any practicable system of classification that may be employed by the survey or census organisation.

REFERENCES

The Changing Social Structure of England and Wales 1871-1951. D. C. Marsh.
Studies in Class Structure. G. D. H. Cole.
Articles on Social class in the British Journal of Sociology during 1957-8.

[1] See 1 per cent Sample Tables, Great Britain 1951. Part II Tables X, Nos. 9 and 10.

Personal Incomes

Statistics of incomes in the U.K. are available from two main published sources. The first are the Annual Reports of H.M. Commissioners of Inland Revenue, which provide detailed information relating to income distribution by size and source of income. Table 6 on page 44 shows one basic classification of income by size. In addition to such analyses, there are analyses of incomes by region and by reference to family circumstances. The sources of this information are the returns made by taxpayers and their employers to the Inland Revenue. Their accuracy depends ultimately on the truthfulness of the returns made. These statistics also form the basis of the data relating to income distribution which are reproduced in the Annual Blue Book on National Income and Expenditure. These data relate to both earned and investment incomes of the national population.

The Inland Revenue assembles in the course of its routine work a great deal of statistical information relating to incomes, the bulk of which is classified by reference to the schedule of Income Tax under which it is charged to tax. For example, the ordinary annual statistics of income tax are classified in the various schedules, *e.g.* A.C.D. and E, while surtax is classified separately. Thus, information relating to the total incomes of taxpayers, such as that shown in Table 6, has to be compiled from special surveys. Such surveys are based upon samples taken from the records of the local tax offices at the end of the relevant fiscal years. The sampling is done with a variable sampling fraction; for example, in the 1959-60 survey a 1 in 40 sample was taken of the lower wages, pensions, and salaries taxable under PAYE, while in the case of the higher salaries, where tax is also collected under PAYE, the sampling fraction was 1 in 10. In the case of the self-employed chargeable under Schedule D, as well as income assessable under Schedule A, a 1 in 20 sampling fraction was used. In the case of surtaxpayers, however, a 1 in 4 sample was taken of all surtax cases up to £10,000 and, in the case of those above £10,000, all cases are included.[1]

Three such surveys have been undertaken since the war at intervals of five years, in respect of the tax years 1949-50, 1954-55, and 1959-60. Although the data are used primarily for budgetary purposes, *i.e.* for estimating the effect on tax revenue of changes in tax rates and allowances, a valuable by-product is the detailed regional analyses and classifications by source of income, which is useful for economic planning. All these data are reproduced in the annual reports of the Commissioners of Inland Revenue.[2] The results of the last survey, that covering the tax year 1959-60, were published in the 105th

[1] See *Economic Trends* article cited below in text for details.
[2] The 94th and 95th Reports for the results of the 1949-50 survey; the 99th for the 1954-55 enquiry.

report (pp. 81–201). A brief summary of the survey is also given in the report, but for a more detailed account with special reference to the statistical aspects thereof, an article in *Economic Trends* for August 1963 ('Survey of Personal Incomes for 1959-60') merits study. These surveys classify personal incomes by range of income, type of income and deductions, sex, status, number of dependents, as well as the regional classification referred to earlier.

Table 42 is reproduced from the article in *Economic Trends*, and shows the changing pattern of income distribution over the decade covered by the three post-war surveys. In particular, it provides an interesting illustration of the use of percentiles and quintiles. These, it will be remembered, are derived in the same way as are quartiles and deciles (described in Chapter VI). The top quintile contains, of course, the top 20 per cent of all incomes ranged in order of magnitude, and the 5th percentile contains the top 5 per cent.

TABLE 42

INCOME POINTS DIVIDING THE TOTAL NUMBERS INTO EQUAL FIFTHS
AND THE TOP FIFTH INTO TWENTIETHS

	Lower limit of range of income before tax (£)		
	1949–50	1954–55	1959–60
Income group:			
5th percentile	895	1,108	1,460
10th percentile	635	863	1,150
15th percentile	537	766	1,012
20th percentile	477	702	925
Top quintile	477	702	925
2nd quintile	346	530	704
3rd quintile	271	405	543
4th quintile	202	281	366
Starting point	125	155	180
Number of incomes (thousands):			
5 per cent. groups	1,025	1,015	1,048
20 per cent. groups	4,100	4,060	4,193
Total	20,500	20,300	20,967

The article in *Economic Trends* states that starting with 1962-3, the quinquennial surveys will be supplemented by similar but smaller annual surveys. This, it will be noted, follows the pattern of the practice with the Household and Family Expenditure Surveys.[1]

[1] See Section on Living Standards in this chapter.

Much more detailed analyses of earned incomes are published in the *Ministry of Labour Gazette*, the main feature thereof being the classification of earnings by reference to the industry, sex, and age of the earners. *The Ministry of Labour Gazette* is in fact the primary source of all industrial labour statistics, and should be consulted whenever any data in this field are required. The most important source of such published information is the half yearly enquiry conducted by the Ministry of Labour into the earnings of manual workers. The latest enquiry at the time of writing was made in October 1963 and published in the *Gazette* of February 1964. The results were obtained from the returns made by some 57,000 establishments employing 6·7 million manual workers, *i.e.* about 70 per cent of all manual workers employed by industries and services in the United Kingdom. The information returned related to persons at work during the whole or part of the second payweek in October 1963. Earnings were defined as total earnings, inclusive of bonuses, before any deductions in respect of tax or of contributions to national insurance. Information relating to average weekly earnings, the average number of hours worked, and the average hourly earnings, are provided for men, *i.e.* males over 21, youths and boys, *i.e.* males under 21, women of 18 years and over, girls under 18 years. Data relating to the adult women is further classified by reference to full-time and part-time workers. The foregoing information is provided for all the main manufacturing industries and the major public services, including extractive industries, as reproduced in Table 43.

Such information has been published for many years. For example, statistics of wage rates and hours of labour were published separately in book form from 1893 onwards. These data, however, related only to manual workers. Since 1959 the Ministry of Labour has carried out an annual enquiry into the earnings of salaried employees. Note that in the case of manual workers the enquiry takes place twice yearly, in April and October. The latest enquiry into the earnings of salaried employees was made in October 1963 and the results published in the *Ministry of Labour Gazette* of March 1964. The information derived from this enquiry was based upon voluntary returns made by firms to whom enquiry forms had been sent. Such forms were sent to all firms with 100 or more employees, and to half the firms with between 25 and 99 employees. In all, 19,000 forms were sent out and some 17,300 were returned suitable for tabulation. The information requested on the form concerned the number of administrative, technical and clerical staff employed in the last pay week of October, showing the monthly and weekly paid staffs separately. The figures were broken down as between male and female staff. Table 44 shows the form in which these data derived from both manual workers and salaried

TABLE 43

AVERAGE WEEKLY EARNINGS OF MANUAL WORKERS, OCT. 1963*

Industry group	Men (21 years over)†		Youths and boys 21 years)		Women (18 years and over)‡				Girls (under 18 years)	
					Full-time		Part-time			
	s.	d.	s.	d.	s.	d.	s.	d.	s.	d.
Food, drink and tobacco ..	318	2	154	4	164	6	87	9	112	11
Chemicals and allied industries	348	2	164	7	164	6	84	11	109	4
Metal manufacture	358	7	165	6	166	4	84	2	108	5
Engineering and electrical goods	337	9	136	11	175	11	92	5	108	4
Shipbuilding and marine engineering ..	323	6	133	7	163	7	79	8	§	
Vehicles	396	8	153	5	199	0	89	3	116	2
Metal goods not elsewhere specified	338	5	147	7	162	1	86	10	103	4
Textiles	307	2	152	4	166	8	86	4	118	3
Leather, leather goods and fur	306	7	145	8	161	9	88	6	108	0
Clothing and footwear ..	297	1	146	6	161	6	94	4	105	9
Bricks, pottery, glass, cement, etc. ..	344	2	173	8	160	2	87	5	109	6
Timber, furniture, etc. ..	330	4	144	3	184	6	92	2	105	9
Paper, printing and publishing ..	390	0	161	6	176	4	90	11	107	4
Other manufacturing industries	346	5	160	2	164	2	86	2	112	0
All manufacturing industries	345	9	149	2	168	2	89	2	109	9
Mining and quarrying (except coal)	328	0	185	3	170	6	§		§	
Construction ..	332	7	152	2	156	4	68	7	97	1
Gas, electricity and water	325	6	164	11	174	8	83	7	§	
Transport and communication (except railways, London Transport and British Road Services)	332	2	175	4	231	5	80	7	102	1
Certain miscellaneous services	284	11	121	2	144	4	76	6	97	0
Public administration ..	257	8	148	7	176	5	71	9	109	4
All the above, including manufacturing industries	334	11	148	8	168	3	87	3	109	2

* Source: Ministry of Labour Gazette, Feb. 1964.

employees enquiries are published in the Gazette.

In addition to this information, the Ministry of Labour Gazette publishes two official index numbers, the first of which is an Index of Weekly Wage Rates, and the second is the Index of Normal Weekly

TABLE 44

Average Earnings of Administrative, Technical and Clerical Male Employees by Industry Group: October 1963

Industry group	Number of employees covered			Average earnings				All males	
				Monthly-paid		Weekly-paid	Monthly-paid and weekly-paid combined on weekly basis		
	Monthly-paid	Weekly-paid	Total	Month of October 1963	Equivalent amount per week	Last pay-week in October		October 1963 compared with October 1962 % increase	October 1963 compared with October 1959 % increase
				£ s. d.	£ s. d.	£ s. d.	£ s. d.	Per cent.	Per cent.
Manufacturing industries:									
Food, drink and tobacco	65,619	24,699	90,318	109 14 8	25 6 6	16 5 8	22 17 0	5·0	21·8
Chemicals and allied industries	85,192	19,687	104,879	119 2 5	27 9 9	14 6 5	25 0 4	6·6	22·0
Metal manufacture	46,461	39,520	85,981	108 2 0	24 18 11	14 6 6	20 19 6	3·1	18·1
Engineering and electrical goods	187,986	192,988	380,974	114 16 11	26 10 1	16 16 4	21 11 11	4·6	21·7
Shipbuilding and marine engineering	5,812	16,057	21,869	128 8 2	29 12 8	16 18 0	20 5 8	2·8	16·2
Vehicles	54,644	90,879	145,523	124 4 6	28 13 4	17 17 9	21 18 9	5·8	23·7
Metal goods not elsewhere specified	25,944	22,239	48,183	117 8 3	27 1 11	16 16 0	22 6 10	3·8	19·8
Textiles	38,260	28,967	67,227	119 10 0	27 11 6	16 4 1	22 13 6	3·7	20·7
Clothing and footwear	10,258	13,760	24,018	125 10 1	28 10 1	17 16 10	22 11 4	4·3	24·5
Bricks, pottery, glass, cement, etc.	22,959	14,453	37,412	110 10 1	25 9 0	15 6 3	21 9 11	4·4	20·4
Timber, furniture, etc.	10,258	10,617	20,875	114 1 2	26 9 0	16 4 0	21 18 11	2·4	22·0
Paper, printing and publishing	40,779	36,196	76,975	128 1 7	29 11 0	17 12 7	23 18 11	4·2	23·4
Other manufacturing industries‡	23,759	13,681	37,440	114 3 7	26 7 0	16 2 7	22 12 4	5·2	22·7
All manufacturing industries	617,931	523,743	1,141,674	116 11 11	26 18 2	16 16 8	22 5 9	4·7	21·7
Other industries and services									
Mining and quarrying	27,161	14,302	41,463	109 10 6	25 5 6	13 13 11	21 5 8	6·4	27·6
Construction	54,378	44,778	99,156	110 18 10	25 12 0	16 6 2	21 8 1	4·7	23·8
Gas, electricity and water	70,790	22,002	92,792	99 15 2	23 0 5	14 11 7	21 0 5	5·9	27·6
All industries covered by enquiry	770,260	604,825	1,375,085	114 8 0	26 8 0	16 12 9	22 2 2	4·8	22·4

Source: Ministry of Labour Gazette, March 1964

Hours. The former, *i.e.* the Index of Weekly Wage Rates, is designed to measure the average movement from month to month in the level of full time wages in the particular industries and services in the U.K. for each of the main groups of workers, *i.e.* adult males and females, as well as juveniles. These three indices are combined for purposes of producing an 'all-workers' index. The index has been revised six times since it was first introduced at the beginning of the century. The present base is the 31st January 1956, the previous base being January 1947, so that there is in fact a single continuous series between those two dates. It is possible to link the current series based on 31st January 1956 with the series between 1947 and 1956 by specially prepared factors.[1] It should be noted that these indices are based upon the weekly *rates* and not weekly *earnings*. Thus the index will not change if the workers' earnings are increased by extensive overtime, or decline if there is extensive short-time working. The indices are based upon the rates agreed as at 31st January 1956, and the weighting of the index for each class of worker in each industry is determined by the relative size of the total wage bill in 1955. With the revision of the Standard Industrial Classification in 1958, there was some regrouping of the wages for each industrial class, but it does not seem to have affected the comparability of the published indices before or after that date.[2]

The second index, known as the Index of Normal Weekly Hours, does not reflect changes in the actual number of hours worked per week, *i.e.* it is unaffected by changes in the amount of overtime, short-time, and absences for any other reasons. It merely reflects the length of the basic working week agreed within the industry. Details of the method of calculation of this rather specialist index are given in the *Ministry of Labour Gazette* for September 1957 (pp. 330-331) and details of the weights are given in the *Gazette* for February 1959 (p. 56). By dividing the Index of Weekly Wage Rates by this Index of Normal Weekly Hours, the Ministry of Labour produces what it calls an Index of *Hourly Rates of Wages*. This index is intended to reflect any improvement in the manual worker's terms of work, since any reduction in the length of the basic week without any change in the rate for the week represents an improvement in the rate of earning.

Readers will be familiar with the controversy surrounding the question of an incomes policy. One of the by-products of the universal concern over this issue has been the production of a new statistical publication. This is entitled *Statistics on Incomes, Prices, Employment and Production*, the first issue of which appeared in April 1962, since which date three issues have appeared annually.

[1] These are reproduced in Part B of the Technical Appendix to *Statistics on Incomes, Prices, Employment and Production*. This publication is discussed below.

[2] Details of the weights are given in the Technical Appendix, *op. cit.*

This particular publication is not intended to supersede the *Ministry of Labour Gazette* as the primary source of labour statistics, not least because the *Gazette* appears monthly, and the new bulletin at roughly four-monthly intervals. The bulletin does not include all the statistical information currently available on labour and related matters, but it does include some information that is not published elsewhere. The main merit of the new publication is that it brings together in convenient form all the relevant statistics relating to manpower, earnings, prices, and profits. In addition to publishing such information as appears in the *Gazette* on earnings and hours of work, the new publication also gives this information for earlier years and there are numerous series worked back for a number of years to ensure comparability. The new publication also contains a useful appendix giving definitions of the main statistical series.

The foregoing summary can give only the merest indication of the range and complexity of the published statistical material on personal incomes and conditions of work. Even if the student has no especial interest in the field of labour statistics, a few hours spent browsing through the dozen monthly issues of the *Gazette* which make up the year, together with a recent issue of the new bulletin, would bring home much more clearly the volume of data available. As with all other published statistics, especial care is required in extraction. Most of the tables in the *Gazette* and bulletin carry footnotes, many of them copious. Thus, no warning could be clearer than the following, which accompanies the tables of manual workers' earnings. 'In view of the wide variations, as between different industries, in the proportions of skilled and unskilled workers, in the opportunities for extra earnings from overtime, night-work and payments-by-results schemes, . . . the differences in average earnings shown should not be taken as evidence of, and a measure of, disparities in the ordinary rate of wages prevailing in different industries for comparable classes of workpeople employed under similar conditions.' In other words, for each of the four categories of employees, the earnings shown are *averages*, and as the student has already learnt, the mean is not always a reliable guide to the data. In this case, the differences in earnings between various industries are not completely explained by the differing *rates* of pay in these industries. If statistical accuracy is to be achieved, such warnings as these cannot be ignored.

REFERENCES

Labour Statistics. Guides to Official Sources, No. 1. H.M.S.O. Revised Edition 1958.

Statistics on Incomes, Prices, Employment and Production.

Ministry of Labour Gazette.

Living Standards

Politicians are fond of demanding a 'decent' standard of living for all sections of the community, but seldom trouble to define just what they understand by 'decent'. For most people, living standards are measured largely in terms of their consumption levels and ownership of consumer durables such as washing machines and cars. While it is generally agreed that as the national product rises so living standards improve, there is no such unanimity as to how living standards may best be measured, or what constitutes a minimum standard below which no section of the community, *e.g.* old age pensioners dependent entirely on State benefit, should fall.

The classical approach to this problem was provided by Seebohm Rowntree's famous study of poverty in York at the turn of the century.[1] He prepared, in consultation with the leading authorities of his time, a diet which would provide each man, woman and child with sufficient calories to maintain physical efficiency. He then converted the calories into the cheapest possible foods; it was in fact predominantly starch with some carbohydrate. The diet was even more uninteresting than the workhouse diets of the time. He allowed a minimum of clothing, a basic rent appropriate to working class accommodation in York, plus fuel and light. There were no extras for tobacco, alcohol, newspapers, fares, etc. The cost of this minimum needs or bare subsistence standard was then compared with the average level of wages. Those whose earnings fell short of that sum were deemed to be in primary poverty. In his famous Five Towns survey of 1912 and repeated in 1923-24 the late Professor Sir Arthur Bowley estimated a similar needs standard, but on a slightly more generous scale, by introducing meat into the weekly diet.[2]

The difficulty with this method of defining living standards is to obtain agreement on what should be included once the level is raised above the bare subsistence level. Some years ago an Oxford research worker was ridiculed by one of the popular newspapers for an article in which she employed the Rowntree technique of a minimum subsistence lowest cost diet which any old age pensioner could have afforded. Nowadays, a 'modest' consumption of alcohol, tobacco and certainly the weekly rental of a TV set, seems to be automatically included in any 'reasonable' standard and those in the community unable to afford these things are usually regarded as 'deprived'. Discussions as to the adequacy of the State welfare benefits in unemployment and old age are clearly influenced by the standards it is felt the community should provide for these sections of the community. Nevertheless, unless the standards are clearly defined, any such dis-

[1] *Poverty: A Study of Town Life.*
[2] *Livelihood and Poverty.* A. L. Bowley and A. R. Burnett Hurst, The later study was entitled *Has Poverty Diminished*, Bowley and Hogg.

cussion is a complete waste of time except to illustrate the classic dictum of the Red Queen in *Alice in Wonderland* who always used words to mean what she meant them to mean.

The most suitable method of measuring changes in living standards at the present time is to use a cost of living index. In Britain the index is termed the Index of Retail Prices and was introduced in 1947, since when it has been subject to several revisions. It replaced the old Cost of Living index introduced in 1914, which for all its limitations, reflected quite fairly the fluctuations in living costs up to the second World War. Table 45 shows the main classes of expenditure incorporated in the 1914 cost of living index as it was known until its abandonment in 1947, together with those used in the *interim* index of retail prices introduced in that year and amended in 1952. The weights shown reflect the distribution of the weekly household income among the constituent groups of related goods. What is obvious from the table is the decline in the proportion of income spent on food as living standards have risen between the pre- and post-war period. Note that the 1947 weighting for food is distorted due to the effects of food subsidies. This was one of the major reasons for the 1952 revision.

TABLE 45

COMPARISON OF WEIGHTING IN SUCCESSIVE RETAIL PRICE INDICES

Group	Weights Interim Index of Retail Prices		Cost of Living Index 1914
	1952	1947	
i. Food	399	348	60
ii. Rent and Rates	72	88	16
iii. Clothing	98	97	12
iv. Fuel and Light	66	65	8
v. Household Durable Goods	62	71	
vi. Miscellaneous Goods	44	35	
vii. Services	91	79	4
viii. Alcoholic Drink	78	101 ⎫ 217	
ix. Tobacco	90	116 ⎭	
	1,000	1,000	100

It is important for the reader to realise that such an index is an average which measures the change, month by month, in the retail cost of a particular or standard collection of goods and services. In this case it is the goods and services normally bought by the average household. Since this index is used as the basis for wage and salary negotiations it must be acceptable to all parties to such negotiations. The present index of retail prices originated in the 1953-54 Household Expenditure Enquiry undertaken jointly by the Ministry of Labour and the Social Survey. This produced a three-weekly account of expenditure, household and personal, from each member over 16 years of age in nearly 13,000 households. Because the expenditure patterns of households subsisting in the main on State pensions and those where the head of the household earned more than £20 per week differed markedly from the bulk of households, the sample of usable returns was cut down to some 11,600 household budgets. The Cost of Living Advisory Committee, which had been formed in August 1946 to advise the government on the desirability and form of a new post-war index, were of the opinion that this sample of budgets gave an adequate reflection of spending habits of over 90 per cent of the households in Britain.[1] It recommended the introduction of a new Index (note the word 'interim' was then dropped since wartime conditions had passed) of Retail Prices with a base 100 as at mid-January 1956.

In March 1962 the Advisory Committee presented a report entitled the 'Revision of the Index of Retail Prices' (Cmnd. 1657). The proposals for revision were based not upon a further major sample enquiry on the lines of the Houshold Expenditure Enquiry, which had sampled 20,000 households and got a 65 per cent response, but on smaller national sample surveys covering 5,000 households each year. Of these samples approximately 3,000 households cooperated and returned usable information on their expenditure. The Family Expenditure Surveys, as they are known, were continuous throughout the year and all members of the household were required to keep an account of their spending for a fortnight. Since 1958 private households in N. Ireland are included in the sample. In exchange for this work they were each paid £1. In their report the Advisory Committee recommended that revised weightings should be introduced based upon the average of the three preceding years' expenditure patterns as disclosed by the surveys. It is not practicable to use the results of a single year's survey for the construction of a new index, since there is a substantial sampling error in respect of certain large items in household expenditure, e.g. furniture, motor cars, etc., which are purchased only at intervals. Thus the weighting used for the index during 1963 was based upon the results of the three years surveys to June 1962. In other respects the current index is similar to that first introduced in January 1956.

It will be self-evident that the surveys indicate the goods and services to be included in the index and also provide the basis for the individual weights. The index is constructed each month and pub-

[1] This Committee first reported in March 1947 (Cmd. 7077) recommending a new index based on the 1937-8 enquiry into working-class expenditure. This formed the basis of the 1947 'interim' index of retail prices which was used with some modifications in 1952 until 1956 when the results of the 1953-4 survey became available.

lished in some detail in the *Ministry of Labour Gazette*. The compilation of the index to ensure its representativeness is interesting. Information on individual prices is collected by visiting several shops selling the same kind of goods, such shops being those which normally handle the bulk of ordinary households' spending. There is also a classification of areas in which the shops are located, ranging from Greater London to small townships with less than 5,000 inhabitants. This ensures both adequate geographical representation and type of retail outlet in the prices collected. The prices are collected on the Tuesday nearest the 15th of .the month. The official account explains the construction of the index as follows:

1. The price relative for each item in each town is calculated and the resulting figures combined as an unweighted average for all the towns in *each* population group.
2. The separate indices for the various population groups are averaged to give indices for *each item* for the country as a whole.
3. The national indices of the items, *e.g.* bread, are next combined to arrive at indices for each group, *e.g.* food. In the construction of the group index, the percentage increase in each item in the group is weighted by reference to its proportionate share in the aggregate outlay on all items in that group. Thus the group index is the weighted arithmetic mean of the percentage changes in its constituent prices.
4. The indices for the various sections are then combined, being weighted as shown in Table 31 on page 192 which gives the index for each of the main expenditure groups, and the final all-items figure. It will be noted from Table 31 that all indices for sections are given to one place of decimals. This has the advantage that quite minor changes in prices will be reflected in the index.

It may reasonably be concluded that the present index is an adequate method of measuring changes in the cost of living as far as the average household is concerned. It has been suggested that a separate index should be prepared for the professional and middle classes, whose expenditure probably differs in some major respects from that of the 'average' household, *e.g.* heavy outlays on private education, less on entertainment, etc. Since the latter in recent years have demonstrated their capacity for maintaining their real income despite the inflation, this particular proposal seems to have been dropped. There is, however, a very good case for preparing a similar index for old age pensioner households, where at least half or even more of the total income comes from a State pension and national assistance. Its construction would pose several problems, not least whether the index should measure the price changes of the goods and services which they can actually afford at the present time, or whether it should incorporate prices of those goods which the public, not to mention the pensioners, feel that they should enjoy in the affluent society. There is some precedent for a separate index for the pensioner group in so far as the National Food Survey, which is undertaken annually, publishes details of the food consumption of O.A.P. households separately from the average households. Even here some critical comment has been evoked since only about 60 per cent of the households approached by the survey organisers are prepared to cooperate. It is almost certain that those who do are on average more intelligent, etc., than those who do not. In other words, it is inevitable that the food survey results should reflect, not the average O.A.P. household's conditions, but the better section more able to meet its daily problems. To this extent it can be argued that the official data understate the extent of poverty in O.A.P. households.

Quite apart from the index of retail prices and the Family Expenditure Surveys upon which the latest indices are based, the annual Blue Book on the National Income and Expenditure contains some additional data relating to spending habits of the public. Table 7 on p. 46 shows a summary of the main Table 19 in the 1963 edition of the Blue Book. The analysis of consumer expenditure is given both in terms of actual prices for each year and, as in Table 7, the expenditure adjusted to a common basis, *i.e.* in terms of 1958 prices These data are used by the Central Statistical Office for the compilation of an index of consumer prices. This index differs quite substantially from the index of retail prices. First, there is the coverage of the two indices. The Blue Book index covers all sections of the community and their expenditure on all goods and services. The retail prices index covers only some 90 per cent of households and only selected goods and services. The method of construction also differs slightly since while the retail price index is base 'year' weighted, the Blue Book index uses current year weights. The different type of weighting used will affect the relative movements of the two indices.

From the purely statistical point of view both indices have their limitations. The breakdown of consumer expenditure given in the Blue Book is at best based upon estimates which in some cases, *e.g.* private motoring, travel, are rather tentative, to put it no more strongly. Where the goods are subject to tax, *e.g.* tobacco, or purchase tax, some effective check is provided on the estimates. In the case of the retail price index the validity of the weighting and the choice of items depends on the accuracy and truthfulness of the respondent household's replies. There is certainly considerable understatement of expenditure on alcohol and tobacco, also on sweetstuffs including ice cream, and in respect of meals eaten outside the home. Some of the figures can be estimated with a fair degree of reliability from other sources, *e.g.* tobacco duty receipts, but even so there remains the problem of apportioning the duty in terms of consumption between the households at different income levels.

While there is a good deal of published information available on both the retail price index and the Blue Book index of consumer prices, it should never be forgotten that statisticians, no more than other people, cannot make silk purses out of sow's ears. The data are probably uneven in quality and reliability and whenever such information is used, it is essential that the usual caution be exercised in inferring conclusions from small changes in the published data.

REFERENCES

A detailed account of the construction of the Index of Retail Prices is given in *Index of Retail Prices. Method of Construction and Calculation*, published by H.M. Stationery Office.

The latest review by the Advisory Committee is the *Report on Revision of the Index of Retail Prices (Cmnd.) 1657*, March 1962. Earlier reports from the same body appeared in March 1952 (Cmd. 8481) and in March 1956 (Cmd. 9710).

The *Report of an Enquiry into Household Expenditure 1953-4* appeared in 1957 and contains a good account of the sample and methods of conducting the enquiry. Its successors, the sample surveys known as the Family Expenditure Surveys, three in number up to mid-1964, have appeared relating to 1957-59, 1960 and 1961, as well as 1962. Each contains a summary account of the survey methods.

Economic Trends, June 1962 contains a special analysis of retail price movements during the previous six years.

The main source of information on Blue Book data relating to consumer expenditure is the C.S.O. study *National Income Statistics Sources and Methods*. H.M.S.O., 1956, and this is brought up to date by the notes in the current annual edition of the Blue Book.

Crime

Interest in crime has grown in recent years, not least because of the fact that certain types of crime which have attracted a great deal of publicity, *e.g.* robbery with violence, have tended to increase. Furthermore, there is growing concern with the prevalence of juvenile crime, particularly by young people under the age of 14 years. A prerequisite for the formulation of policies intended to improve upon the present situation is a clearer understanding of the nature of crimes and their causes. The existing statistics on crime are singularly unsatisfactory for such an investigation.

The main source of statistics relating to crime is the annual Home Office Blue Book entitled *Criminal Statistics: England and Wales*. A separate report is published for Scotland. The content of the Blue Book is more clearly defined by the sub-title 'Statistics Relating to Crime and Criminal Proceedings for the Year 1962'. The first and most important limitation of the published statistics of crime is that only crimes which are reported to the police are ever recorded. In other words, if the police are not aware of the crime, it does not exist statistically. For this reason tables showing the number of offences committed over a period of years (in the 1962 Blue Book the period covered was 1938–1962) are headed 'Offences known to the police'. It is possible only to guess at the extent of the crime not disclosed to the police. It is probable that it is fairly extensive, for example, in the case of shoplifting many shopkeepers do not institute proceedings against those they catch and, in the case of larceny and housebreaking, the persons affected may not report the matter to the police in view of the small losses sustained or the fact that they have little confidence that the culprit will be found.

Some crimes are less likely to be reported than are others. Blackmail and sexual offences are likely to be under-reported, since the victim in each case will often be reluctant to court the publicity which generally attaches to such cases. Another important factor which has to be taken into account is the willingness of the public to cooperate with the police in reporting crime. Recent years have seen a deterioration in relations between the public and police, and this has almost certainly affected the willingness of the former to cooperate. In short, the official statistics of crime understate the total amount in varying degrees from year to year, so that comparisons over the period are not always reliable as an indication of the scale of crime. The under-reporting of crime is usually referred to as the 'dark figure', *i.e.* the difference between the crimes reported and those actually committed. To the extent that the police forces in England and Wales are responsible to their individual local authorities, police practice in respect of some crimes may tend to vary. For example, in some areas prosecutions may more readily be undertaken in respect of driving offences than in others, likewise in respect of drink. Furthermore, the attitude of the magistrates in different areas will also affect the charges which the police are willing to bring against offenders. The expansion in juvenile crime, however, which is in part attributable to the larger number of young people in the population, is also due to changes in the law and an increasing tendency to charge young offenders before the juvenile courts. In the past, particularly with very young offenders, justice was dealt out in more summary fashion either by the police or an adult who had sometimes witnessed the offence, and the courts were not troubled. Considerations such as these must be taken into account in assessing the long period changes in the apparent increase of juvenile crime.

Apart from one or two tables and graphs indicating the number of offences reported to the police, by far and away the bulk of the Blue Book is devoted to a classification of crimes which have been cleared up. Table 46 provides a summary of the number and main types of crime cleared up during 1962. It may be thought that the inclusion of motoring offences in the grand total tends to give a misleading picture of the scale of crime when the total alone is quoted, as it often is. There is an obvious case for eliminating minor driving offences from this annual aggregate. In Table 47 these data are classified by reference to the type of indictable offence and by reference to the age of the offender, these data being given for several years for comparative purposes. The age groups used for these classifications are 8 and under 14, 14 and under 17, 17 and under 21, 21 and under 30, 30 and over. A measure of the rate of crime for each age group and each sex is derived by relating to the number of persons found guilty of the main indictable offences per 100,000 of the population under the age group in England and Wales.

TABLE 46

TOTAL NUMBER PERSONS GUILTY OF ALL KINDS OF OFFENCES,
ENGLAND AND WALES, 1962

Offence	Number of persons found guilty	Per-centage of total
Traffic offences (dealt with summarily)	785,816	62·0
Larceny	119,034	9·4
Drunkenness and other offences against intoxicating liquor laws	88,484	7·0
Breaking and entering	42,760	3·4
Revenue law offences (mainly faiure to take out licences for dogs or motor cars) (dealt with summarily) ..	34,259	2·7
Breach of local and other regulations	25,498	2·0
Railway offences	20,425	1·6
Malicious injuries to property and malicious damage..	18,952	1·5
Wireless Telegraphy Acts	14,776	1·2
Violence against the person	11,986	0·9
Assaults (non-indictable)	10,645	0·8
Receiving	9,946	0·8
Vagrancy Acts offences (dealt with summarily) ..	7,468	0·6
National Insurance Acts offences	7,389	0·6
Sexual offences (indictable)	6,068	0·5
Betting and gaming offences (dealt with summarily)..	2,556	0·2
Offences by prostitutes	2,482	0·2
All other offences	58,052	4·6
Total	1,266,596	100·0

Source: Criminal Statistics 1962.

The main classification of crimes is twofold. The two classes are 'non-indictable' and 'indictable offences', and the former are very much more numerous than the latter; the non-indictable offences are those which are dealt with summarily by the magistrates. In the case of indictable offences there is usually a jury. The classification of the crimes in the Home Office annual report is extremely detailed, there being almost 100 classes of non-indictable, and over 100 indictable, offences. The distinction between the two types of crime, apart from that indicated above, is also somewhat blurred, since this particular classification was introduced over 60 years ago and there are some overlaps between the main classes. It is fair to assert that the present classification under the main headings of offences against the person, offences against property with violence, are more justified on grounds of tradition rather than any special usefulness for criminal or socio-logical research.

TABLE 47

AGE DISTRIBUTION OF PERSONS FOUND GUILTY OF ALL KINDS OF OFFENCES, ENGLAND AND WALES, 1962

Age Group	Persons found guilty of indictable offences	Persons found guilty of non-indictable offences	Total
Age 8 and under 14*	31,889	11,406	43,295
Age 14 and under 17	34,333	43,319	77,652
Age 17 and under 21	35,353	139,639	174,992
Age 21 and over	102,200	868,457	970,657
Total ..	203,775	1,062,821	1,266,596

Source: Criminal Statistics 1962.
* No child under the age of 8 can be guilty of any offence.

For each main category of offence the statistics show both the number of offences known to the police and the proportion 'cleared up'. The overall percentage in recent years has been about 40–45 per cent, which at first sight does not suggest a very high rate of detection. In view of the fact that nearly two-thirds of the offences known to the police are relatively minor cases of larceny, the proportion described as cleared up of the total of reported cases is clearly dependent on the size of that figure. For cases of 'violence against the person' about 90 per cent are cleared up, and in some years more than 90 per cent of fraud cases are solved. Thus, particular care has to be taken when using the figure of the proportion of total offences cleared up as a measure of the efficiency of the police force.

Comparisons of the crime totals year by year are, for reasons already stated, somewhat unreliable. The Blue Book gives compara-tive figures based on the annual average of short periods of years rather than single years in the past. For recent years and, in par-ticular, for the year covered by the Blue Book, the figures are for single years. A very large proportion of the tables in the Blue Book are devoted to the work of the courts, their decisions and sentences. These are given in rather more detail than are the figures for the crimes known to the police. Table 48 illustrates the type of information relating to the sentences on persons aged 21 and over which have been passed by higher courts, classified according to the main type of sentence.

TABLE 48

SENTENCES ON PERSONS AGED 21 AND OVER FOUND GUILTY OF
INDICTABLE OFFENCES BY HIGHER COURTS*
(England and Wales)

	1938		1954		1961		1962	
	Number	Per-centage	Number	Per-centage	Number	Per-centage	Number	Per-centage
Conditional discharge..	982	15·4	902	7·1	1,398	6·8	1,259	6·3
Probation	827	13·0	1,530	12·0	3,042	14·8	3,069	15·4
Fine	—	—	1,400	11·0	3,449	16·8	3,410	17·1
Imprisonment ..	4,222	66·3	7,664	59·9	11,394	55·4	11,135	55·7
Corrective training ..	—	—	439	3·4	358	1·7	271	1·4
Preventive detention ..	—	—	217	1·7	191	0·9	123	0·6
Otherwise dealt with ..	336	5·3	633	4·9	751	3·6	706	3·5
Total†	6,367	100·0	12,785	100·0	20,583	100·0	19,973	100·0

Source: Criminal Statistics 1962,
* This table does not include persons sentenced by Quarter Sessions after having been found guilty by a magistrates' court.
† Includes fines in 1938 only.

The adequacy of the official statistics relating to crime can be judged from the fact that in late 1963 the government appointed an expert committee to examine the official statistics and to make recommendations. The committee comprises criminologists and sociologists, as well as statisticians. Certainly there is plenty of scope for improvement, as may be judged from a study of the existing Blue Book. Quite apart from the basic deficiencies in the data, i.e. the over-all knowledge relating to crime, the presentation of the material in the Blue Book leaves much to be desired.

The foregoing comments relate to the Home Office Annual Report, but it should be remembered that there are well over 100 police forces under the jurisdiction of their own local authorities and chief constables. Additional information relating to crime is given in the annual reports of the chief constables to their local Watch Committees or Joint Standing Committees. in the case of counties. In the case of the Metropolitan Police Area, the annual report of the Commissioner of Police is a substantial document which is published by the Stationery Office. In addition to this source material, there are the annual reports of the Prison Commissioners, which provide information relating not so much to crime as to the state and condition of H.M. Prisons, and those inside them.

For the reader who is interested to learn what information can be derived from statistics on crime, as well as studying the employment of statistical methods, the publications of the Home Office research unit published by H.M. Stationery Office deserve close study. The first of these is a rather specialised and statistically sophisticated analysis of prediction methods which is likely to be beyond the scope of the average reader of this book.[1] The three phamphlets, entitled *Time spent awaiting Trial, Delinquent Generations* and *Murder* respectively, are primarily factual statements of the relevant material together with a commentary on the lessons to be drawn from them. The last-mentioned pamphlet is an especially interesting example of the use of elementary statistical analysis in clarifying an issue which, in recent times, has been clouded by prejudice and ignorance.

REFERENCES

Annual Reports on Criminal Statistics for England and Wales, H.M.S.O.
Social Aspects of Crime in England between the Wars. H. Mannheim.
Statistics in Criminology, M. Grünhut, *J.R.S.S.*, 1951.
Criminal Statistics, T. S. Lodge, *J.R.S.S.*, 1953.

Education Statistics

Nothing is so effective in stimulating the production of statistical information as the fact that government money is involved. Thus, with the rising expenditure of both the national Exchequer and local authorities on education, so the volume of statistical information has increased and its presentation improved. Even so, the statistical unit attached to the Robbins Committee on Higher Education demonstrated the statistical vacuum in which that Committee was compelled to work in order to assess the potential demand for higher education and the capacity of the existing institutions to meet it.

The bulk of educational statistics are, as is usually the case with official data, the by-product of the administration of the educational system. Since 1962 this information has been considerably extended, and each year a three-volume study of educational statistics is produced. Part I of the annual *Statistics of Education* covers the school population, the type of schools in which school population is taught, together with information relating to their teachers and the finance of the system generally. This part is published annually each winter and covers the foregoing subjects, for which information is made available during the second half of the calendar year. Parts II and III of the Annual Statistical Report are published later each year and cover

[1] Prediction Methods in Relation to Borstal Training. H. Mannheim and L. T. Wilkins.

educational events, such as examination successes, scholarship awards, as well as educational building and information relating to school leavers. All this information is published in very considerable detail, with brief explanatory notes. Apart from the national statistics published by the Ministry of Education, the local education authorities produce similar statistics in the annual reports of their Education Departments. In this case, of course, the data are restricted to the area for which the authority is responsible, but by virtue of the smaller area such reports often contain more detailed information than is available in the national reports.

Until the publication of the Robbins Report entitled *Higher Education*, together with its four statistical appendices,[1] the main statistical information relating to university education was contained in the quinquennial Reports of the University Grants Committee, as well as the White Paper known as the 'Annual Return from Universities and Colleges in receipt of Treasury grant'. The information contained in these documents relates not only to financial expenditure incurred by the various activities such as student grants, salaries, administration, etc., but also the number of students and their distribution between the faculties and departments and their examination successes.

The Robbins Report on Higher Education constitutes the latest and largest source book of material on higher education in its widest connotation. The Report collected a great deal of data by means of six major sample surveys, specially undertaken to provide information which was not elsewhere available. Thus, surveys were made among undergraduates, post-graduates and advanced students. A special enquiry was made of university teachers regarding their qualifications, teaching load and method of teaching. One survey investigated teacher training facilities while another survey examined the social background of 21 year olds, in which the educational career after leaving school of boys and girls who had attended different types of school was ascertained. An especially interesting aspect of the statistical appendices to the Robbins Report is their incursion into the field of educational statistics with special reference to educational opportunity and attainment. In the past this has been a field which has been primarily the concern of the educational psychologist and more recently that of the educational sociologist. It is a field in which further rapid development may be expected, not least since the establishment of the Statistical Unit for Research into Education that has been set up at the London School of Economics as the direct outcome of the work of the Robbins Committee statistical unit. Published shortly before the Robbins Report, is the

[1] Cmnd. 2154

Report of the Central Advisory Council for Education (England) better known as the Newsom Report[1] which gives a certain amount of information on the social and educational background of children attending secondary schools in this country. While part of the statistics contained in this report are derived from published material, an important section thereof is based on a survey carried out by the Council in 1961 in modern and comprehensive schools. The object of the survey was to throw light on the conditions affecting the education of pupils aged 13–16. The Central Advisory Council some years earlier had sponsored the three sample surveys which formed the statistical background to the Crowther Report.[2] The first two of this trio of surveys was concerned to examine the sample of respondents, the type of school attended, the home background, age at leaving school and reasons for leaving school. The first of these two enquiries was based upon a sample of boys who had left school three years before the survey; the second interviewed a sample of National Servicemen. The final survey concerned a sample of students working part-time for National Certificate courses at technical colleges. Like the Newsom Report, the Crowther report was of especial interest to the educational sociologist; in contrast, the basis of the Robbins Committee recommendations is their statistical projections of demand for higher education.

Statistical data relating to the administration and structure of the present-day educational system in England and Wales is published in substantial detail in the above-mentioned annual statistical report of the Ministry of Education. The main statistical problems arise in attempts to compare the educational structure of different local authorities, where the types of school differ, and some definitional problems with inter-local comparisons can arise. In the field of educational sociology, the Robbins and Newsom Reports may be regarded as a turning point, although a certain amount of work by individual research workers in varying parts of the country has been published in the past. Other data are published by the National Foundation for Educational Research in England and Wales, as well as by the Scottish Council of Research in Education.

An attempt was made to fill some of the gaps in the existing information on the educational standards attained by the population in the 1951 Census. Two questions relating to education were asked; the first asking of those persons no longer receiving full-time education the date at which full-time education had ceased, while the second sought to ascertain the extent to which persons attending an educational institution were receiving either full or part-time education.

---

[1] Half Our Future, H.M.S.O.
[2] The surveys and their findings are set out in 15 to 18, vol. 11, *Surveys*.

According to the General Report of the 1951 Census, there was substantial non-response to the first of the above questions, *i.e.* on the terminal age for full-time education. The other questions on part-time education created some confusion among respondents, and the information derived from them is described as 'seriously defective'.[1] In the 1961 Census, an attempt was made to determine the proportion of the population in possession of scientific qualifications. A system of multi-phase sampling was used whereby one in ten of the households were given in addition to the standard census form a further schedule requesting information on scientific qualifications that they possessed. At the time of writing this information has not yet been published. In the prospective 1966 Census of Population it is proposed to ask for particulars of higher educational qualifications held by the public.

REFERENCES

Statistics of Education, Annual. Parts I, II and III.
Higher Education and 4 appendices, H.M.S.O., 1964.
Half Our Future, H.M.S.O., 1963.
The Statistics of Education, Doris M. Lee, *J.R.S.S.*, Part III, 1954.

Accident Statistics

From the viewpoint of the community the importance of accident statistics arises from the fact that each accident results in economic waste. Whether the accident results in damage, personal injury or death it causes the expenditure of material, time or life for no return whatsoever. Consequently close watch is kept upon statistics of accidents as a measure of the success or otherwise of any move to prevent accidents. The interest is hardly surprising when it is estimated that in 1961 in Great Britain road accidents alone cost £230m.

The basis of classification of accidents is, at present, by the place of occurrence of the accident as distinct from classification by the type of accident. The reason for this is that there is no single authority responsible for collecting and analysing statistics of all types of accident. Accident statistics can therefore be divided into the particular spheres of influence of the interested bodies:

(1) Industrial accidents covered mainly by the Ministries of Labour and Pensions and National Insurance;

(2) Transport accidents covered mainly by the Ministry of Transport; and

---

[1] General Report, p. 111.

(3) Home accidents covered partly by the Royal Society for Prevention of Accidents. This organisation is a voluntary body and has no authority to enforce the collection of statistics on a uniform basis.

In dealing with accidents it must be remembered that as all deaths must be reported, the analysis of accidental deaths is far more reliable than that for any other accident statistics. Only in some situations do accidents resulting in personal injury have to be reported, e.g. on the roads and in the factories. However, while in 1960 there were over 20,000 accidental deaths (about $3\frac{1}{2}\%$ of total deaths) there were some $2\frac{3}{4}$m. cases of accidental injury of which 200,000 were classified as serious. The important point is that the proportion of deaths to injuries is by no means constant. Thus, on the road at least one death is recorded for every fifty injuries while in the coal mines the proportion falls to one death for every six hundred and fifty injuries. This does not mean that the mines are safer than the roads (the proportions of injured to those exposed to injury is totally different). It does mean that where only the number of accidents resulting in death are known (e.g. in the home) it is very difficult to estimate from experience in other spheres the number of accidents resulting in non-fatal injuries.

A further difficulty arises in attempting to estimate the seriousness of injury arising from accidents. The only reasonably uniform basis of classifying injury is that provided by the Ministry of Pensions and National Insurance, where an accident results in a claim for benefit under the National Insurance Acts. In these cases the number of days absent from work can be used as a measure of seriousness of the injury arising from the accident. While this is hardly a perfect basis of assessing the injury, it does at least provide the basis for a uniform classification.

Where an accident does not involve personal injury such statistics that are produced, and there is no regular series available, are reduced to the level of intelligent guesses. The reason for this is the lack of necessity to report such accidents. In some spheres, e.g. industrial accidents and collisions between cars that result in a claim for damage under insurances, there is some factual guidance but the supporting evidence, even here, is incomplete.

As stated above the sphere of accident statistics can be divided roughly into three main groups discussed individually below.

Industrial Accidents

The two main agents in the collection of statistics in this field are the Ministry of Labour and National Insurance, and the Ministry of Pensions. The interest of these two bodies is somewhat different. The Ministry of Labour operates under the Factories Acts and is interested in the incidence of accidents primarily from the point of view of accident prevention. Analyses are published in the Annual Reports of H.M. Chief Inspector of Factories. The Ministry of Pensions and National Insurance is mainly interested in accidents from the point of view of benefit entitlement under the National Insurance (Industrial Injuries) Acts. Summarised information is published in the Ministry's Annual Report, although more detailed analyses are available to research workers in the 'Digest of Statistics Analysing Certificates of Incapacity for Work' which is available from the Ministry upon request.

The scope of the two sets of statistics produced by the Ministries differs somewhat. Thus, an accident in a factory may be reportable under the Factories Acts but not give rise to a claim for industrial injury benefit, and *vice versa*. Further there are many accidents of a minor nature that are not reportable under either Act. The Factory Inspectorate statistics relate to a large but restricted sector of industry, whereas the Ministry of Pensions and National Insurance statistics cover the whole field of employment.

The Factory Inspectorate statistics are based upon reports that must be made when 'an unexpected happening' which occurred at a particular instant in time to a person employed in factories, as defined in Section 151 of the Factories Act 1937, or employed on construction work as defined by Section 152 of that Act, or in docks, ships in port, or warehouses as covered by Sections 105 and 106 or at electricity stations covered by Section 103 of the Act, and resulting in death or disablement from work for more than three days. The unit of enumeration is the killed or injured person and the recorded accidents are classified according to causation, nature and site of injury and by industrial classification. Further details are published in the 'Guide to Statistics Collected by H.M. Factory Inspectorate', H.M.S.O.

The Ministry of Pensions and National Insurance statistics arise from successful claims to benefit arising from an unexpected happening resulting in personal injury that occurred out of, and in the course of, insurable employment. Detailed analyses are published in the Ministry's Annual Report of the type of injury and the cause of the accident in a 5% sample of the total claims received. A table is also given of the estimated average population at risk in each industry, the number and estimated duration of spells of injury benefit, together with the total number of incapacity days recorded. The 'Guide to Official Sources No.5, Social Security Statistics' gives further information on the make-up of these statistics.

Apart from the main coverage of industrial accidents mentioned above, farm accidents are reportable to the Ministry of Agriculture and coal mine accidents to the Ministry of Power; while accidents involving explosives or petroleum are reportable to the Chief Inspector of Explosives at the Home Office. All these bodies publish summaries of their work in their respective annual reports.

Transport Accidents

The most important source of transport accident statistics is provided by the authority of the Road Traffic Acts. These provide that any road accident resulting in personal injury must be reported to the police. The resulting information is published, primarily by the Ministry of Transport at monthly intervals. The Ministry of Transport is trying in 1964 to obtain further detailed analyses of road accidents from the police in an attempt to draw some conclusions as to steps that would reduce the number of accidents. A good example of the periodic analyses of road accidents is given in the study *Road Accidents 1959* (H.M.S.O. 1960).

The statistics collected by the Ministry of Transport are supplemented and analysed further in a monthly publication of the Royal Society for the Prevention of Accidents entitled 'Road Accident Statistical Review'. An interesting feature of the Review is the definitions of terms used in the booklet listed on the inside front cover. Much additional road accident information is given in the Review, but the basic table is reproduced overleaf.

The Ministry of Transport also analyses accidents between those occurring at night and day; and those on roads subject to the speed limit and those not. From time to time insurance companies publish figures relating to claims for damage to cars caused by accidents.

Railway accidents are analysed in detail in an Annual Report to the Minister of Transport upon Accidents which occurred on the Railways of Great Britain. In general the classification is into:

(*a*) Train Accidents;
(*b*) Movement Accidents;
(*c*) Non-Movement Accidents.

and into:

(i) Passengers;
(ii) Servants;
(iii) Others.

A number of accident categories attributable to certain particular circumstances, *e.g.* at level-crossings, are subject to a more detailed analysis.

The numbers of aircraft accidents are comparitively low, less than $\frac{1}{3}$rd of 1% of all deaths from accidents in Great Britain in 1960, but because of the seriousness of any one accident the statistical analysis

TABLE 49

NUMBERS OF PERSONS KILLED AND INJURED IN ROAD ACCIDENTS
IN GREAT BRITAIN DURING DECEMBER 1963

Figures issued by the Ministry of Transport

Class of Road User	Persons Killed		Seriously Injured		Slightly Injured		Total Casualties	
	1962	1963	*1962*	1963	*1962*	1963	*1962*	1963
Pedestrians								
Child	26	39	396	468	1,125	1,313	1,547	1,820
Adult	294	314	1,389	1,503	2,881	2,864	4,564	4,681
Total	320	353	1,785	1,971	4,006	4,177	6,111	6,501
Pedal Cyclists								
Child	2	9	105	107	335	376	442	492
Adult	52	57	518	462	1,747	1,591	2,317	2,110
Total	54	66	623	569	2,082	1,967	2,759	2,602
Mopeds								
Riders	7	6	106	153	320	385	433	544
Motor Scooters								
Riders	23	27	394	394	1,109	1,172	1,526	1,593
Child Passengers	—	—	—	1	6	11	6	12
Adult Passengers	1	3	50	48	203	190	254	241
Total	24	30	444	443	1,318	1,373	1,786	1,846
Motor Cycles								
Riders	51	47	797	917	1,656	1,759	2,504	2,723
Child Passengers	—	—	13	6	31	28	44	34
Adult Passengers	7	10	153	152	373	377	533	539
Total	58	57	963	1,075	2,060	2,164	3,081	3,296
Other Vehicles								
Drivers:								
Car and Taxi	82	120	1,259	1,548	4,181	4,599	5,522	6,267
Other Vehicle	23	21	414	349	1,415	1,245	1,852	1,615
P.S.V. Passengers:								
Child	1	—	14	17	159	157	174	174
Adult	4	7	196	176	1,377	1,241	1,577	1,424
Other Passengers:								
Child	6	7	131	118	629	691	766	816
Adult	96	99	1,437	1,677	4,758	5,264	6,291	7,040
Total	212	254	3,451	3,885	12,519	13,197	16,182	17,336
Total, Children under 15	35	55	659	721	2,289	2,581	2,983	3,357
Total, Adults	640	711	6,713	7,375	20,016	20,682	27,369	28,768
TOTAL ALL ROAD USERS	675	766	7,372	8,096	22,305	23,263	30,352	32,125
On Restricted roads	430	508	4,790	5,482	16,743	18,047	21,963	24,037
Unrestricted Roads	245	258	2,582	2,614	5,562	5,216	8,389	8,088

Index of Motor Traffic (average month in 1958 = 100)

Index	*1961*	*1962*	1963
December	107	113	128

* *Road Research Laboratory estimate*

of accidents is important. Statistics are primarily used to search for preventive measures or to assess the effectiveness of such measures as have been taken. However, statistics are also used to measure the liabilities of airline operators, in insurance studies, in planning rescue services and in designing accident investigations themselves. The two basic units of measurement in this field are fatalities per aircraft kilometres flown and fatalities per hundred million passenger kilometres. Some care in the interpretation of these statistics is needed. A comparison between two air-lines, one of which has long routes and the other short-distance routes, is difficult. The latter may well appear to have a higher accident rate per hundred million passenger kilometres, but since a high proportion of aircraft accidents occur either on take-off or on landing, the aggregate distances flown may be misleading if used as the sole guide. The central body for the collection and publication of these statistics is the International Civil Aviation Organisation. Their statistical digests are published through the H.M.S.O. and are summarised in the United Nations statistical publications, *i.e. Monthly & Annual Digest of Statistics.*

Accidents in the Home

Nearly one half of all accidents that enter into statistics occur within the home. The field is obviously important but it is badly documented. The only accurate but limited measure of home accidents is the analyses of the causes of death published by the Registrar-General. Some estimates are made of injuries arising from home accidents by sampling hospital admission lists, but the results cannot be considered as more than rough indications of the order of magnitude. The most important source of statistics in this field is the Royal Society for the Prevention of Accidents. Their annual report is a valuable source of information in which they collate and analyse

TABLE 50

ANALYSIS OF HOME ACCIDENT DEATHS IN GREAT BRITAIN IN 1962

| Cause of Death | Age-group (years) | | | | | Sex | | Total Deaths in 1962 | Total for the year 1961 |
	0-4	5-14	15-44	45-64	65+	Male	Female		
Poisoning	42	13	350	535	1,067	902	1,105	2,007	1,600
Falls	58	14	67	318	4,247	1,405	3,299	4,704	4,689
Burns and Scalds	119	54	77	135	539	329	595	924	749
Suffocation and Choking	594	11	75	71	84	490	345	835	692
Others	98	32	75	70	116	233	158	391	363
Total	911	124	644	1,129	6,053	3,359	5,502	8,861	8,093

Source: The Registrar General's Statistical Review of England and Wales for the Year 1962 (Part 1 – Medical Tables), and the Annual Report of the Registrar-General for Scotland – 1962 – No.108.

all available published data. They also publish an early preliminary analysis of deaths from domestic accidents. The type of information published therein is shown on page 351.

The study of accident statistics is most important if the waste resulting from accidents is to be eliminated. While some fields are quite well documented there are serious gaps and some whole areas that are barely touched upon. Currently the only attempts at comprehensive statistical coverage of accidents are made by the Royal Society for the Prevention of Accidents which is largely dependent for its statistical material upon outside sources.

Conclusion

The importance of social statistics lies in the fact that they provide information relating to matters on which there is often widespread public interest and even concern, *e.g.* juvenile crime, housing, etc. Since the government is primarily responsible for the social services, the bulk of the data is to be derived from official publications, in particular the annual reports of the relevant ministries and the reports of special enquiries on particular problems. While none of the social problems which attract so much publicity at the present time is new – it is merely that from time to time public attention is focused much more sharply upon such matters – the statistical information collected in the past has often been somewhat slender in quantity and limited in coverage. As with economic statistics, there has been a very real expansion in both the quantity and quality of information relating to social conditions in Britain in recent years.

The reader will appreciate that it is quite unnecessary to seek to memorise large sections of the information contained in the various reports referred to in this chapter. The sole object of making reference to them is to indicate where statistical information relating to the problems discussed may be found. Generally speaking, the only occasion on which these source materials are consulted is when a report or speech is being prepared on the subject. It is then that some statistical training comes in useful, not least the awareness that definitions are important, that the coverage of the data or various series may change in the middle of the period under review. Above all, one of the first lessons that will be learned when extracting information is that very often the data which the reader requires is not provided in the form in which it meets his needs, either because the information is insufficiently detailed, or the coverage of the data is too wide. In such cases one must learn to make the best use possible of the information that is available. On such occasions, always assuming that the available data are relevant to the topic under discussion, great care needs to be taken in drawing any conclusions from such data.

It is essential to remember that the extraction of information from statistical reports on subjects such as these is not a simple clerical task. It requires great care, attention to detail, and very often considerable judgement in deciding whether or not a particular series of figures is appropriate for a special purpose. The only way in which the reader can familiarise himself with the contents of these reports and the limitations of the statistics published therein is by actually trying to extract information relevant to a particular problem, for example, the industrial distribution of unemployment or the number of slum houses in his particular area. Perhaps the most surprising lesson to be learned in this sort of work is that, despite the massive volume of statistics published, many gaps still remain, particularly in the field of regional as distinct from national statistics.

CHAPTER XVII

VITAL STATISTICS

Introduction

In recent years, the term 'vital statistics' has been defined as the use of three figures to describe a single figure. Popular usage apart, vital statistics are derived from the enumeration of the human population and the registration of births, marriages and deaths. The statistician specialising in this branch of statistics is usually described as a demographer, the subject of demography concerning the analysis of population data. The collection of information relating to human populations by governments dates back to biblical times. The Romans used such information for raising both armies and tax revenues, and as early as the 8th century there appears to have been a system of registration of births, deaths and marriages in some parts of Japan. In Europe, systematic vital registration appears to have begun in Spain in the 15th century, and in 1538, the clergy of England were required by Henry VIII to record baptisms, marriages and burials. Around the same time several European countries adopted the practice of registration of vital events by the ecclesiastical authorities, but the first instance of registration by a secular authority was that introduced in the colonies of Massachusetts and New Plymouth in 1639. The advent of Napoleon and the adoption of the Napoleonic Code influenced strongly the development of vital registration systems throughout Western Europe and Latin America, wherever French influence prevailed.

The first Census of Population in England and Wales was undertaken in 1801, and in 1837 the General Register Office was created. Two years later Dr William Farr became 'Compiler of Abstracts' in that office, and, according to a biographer, 'the next forty years of his life were almost exclusively devoted to the, to him, congenial task of creating and developing a national system of vital statistics, which has not only popularised sanitary questions in England in such a manner as to render health progress an accomplished fact but which has, practically, been adopted in all the civilised countries of the world'.[1] Apart from underlining the tremendous work carried out by Farr in the analysis of vital statistics during the 19th century, this quotation also reflects the change in emphasis which has taken place in vital statistics since the days of Farr. In the past, the analysis of vital statistics was predominantly a concern with public health, hence

[1] N. A. Humphreys, quoted in the *Handbook of Vital Statistics Methods*, U.N. 1955.

the stress on birth and death registration. More recently, however, more importance has been attached to marriage registration statistics, as well as those relating to divorce, adoption, legitimation and separation, etc. According to a United Nations study, 'the vital statistics system includes the local registration, the statistical recording and reporting of the occurrence of, and the collection, compilation, presentation, analysis, and distribution of statistics pertaining to vital events, *i.e.* live births, deaths, foetal deaths, marriages, divorces, adoptions, legitimations, recognitions, annulments, and legal separations'.[1]

Census and Registration Data

Reference has already been made in the preceding chapter to the two basic methods of collecting vital statistics. The first is the decennial census of population, which provides in effect a snapshot of the population at a given time. It also provides a benchmark against which changes can be measured, and is a necessary adjunct to the second, *i.e.* a system of continuous compulsory registration. Both methods of collecting vital statistics are essential, but in some cases, *e.g.* in underdeveloped territories, a limited census or even a sample survey, may be the only means of obtaining information about the population and its current condition. It should be noted that the mere setting up of a registration system does not by itself automatically ensure the notification to the authority of vital events. For example, in some countries, the birth of female babies is frequently unnotified, but this deficiency can be picked up by comparing the ratio of registered male to female births. Since this ratio is more or less constant, any variation is immediately evident. Likewise, undernotification of deaths can also be deduced by comparing the death rates derived from the available data with those of other countries where living conditions are almost comparable and where the registration system is better developed.

The practice of registering vital events is well established in this country, and it is unlikely that the figures are subject to any significant degree of inaccuracy.[2] There are both legal and other kinds of pressures to ensure registration. For example, a corpse cannot be buried until the Registrar's certificate of registration has been produced. The birth of infants is notified by the hospital authorities and medical practitioner to ensure that the parents of the child ultimately register all the requisite details. A marriage is automatically registered whether it takes place in church or register office. On the occasion of registration, a good deal more information is required than the mere

---

[1] U.N., *op. cit.*
[2] See General Report, 1951 Census for a discussion of such inaccuracies.

fact of the vital events. In particular, the Population (Statistics) Acts of 1938 and 1960 require the informant to give not only the obvious facts concerning the vital event, *e.g.* birth of a child, but also information relating to the duration of marriage, age of parents, occupation of father, etc. The later Act also collects further information about still-births, on the basis of which the statistician can carry out detailed analyses which may ultimately form the basis for future health policies. In brief, then, it can be said that the system of vital registration in Britain is efficient and the data collected thereby highly reliable. Once this has been proved to the statistician's satisfaction, it then remains to consider to what purposes these statistics are put.

TABLE 51
POPULATION CENSUS OF ENGLAND AND WALES 1841–1961

Census Year	Population 000's	Absolute Increase 000's	Decennial Percentage increase %
1841	15,914	—	—
1851	17,928	2,014	12·7
1861	20,066	2,138	11·9
1871	22,712	2,646	13·2
1881	25,974	3,262	14·4
1891	29,003	3,029	11·7
1901	32,528	3,525	12·2
1911	36,070	3,542	10·9
1921	37,887	1,817	5·0
1931	39,952	2,065	5·5
1941*	41,748	1,796	4·5
1951	43,758	2,010	4·8
1961	46,072	2,314	5·1

Source: Reg. Gen. Statistical Review 1962, Part II. Table A.1.
* Mid-year estimate; no census held in this year.

Table 51 shows the growth in the population of England and Wales between 1841 and 1961. The figure shown for each year is the census count, with the exception of 1941 when the mid-year estimate, *i.e.* as at June 30, is used. The two adjacent columns illustrate an exercise in very simple statistical analysis, a comparison of the absolute increase in population numbers with the proportionate or percentage increase. Thus it will be seen that although the absolute increase between 1841 and 1851 was almost the same as that between 1941 and 1951, the relative rate of increase in the earlier decade was two and a half times as high. Table 52 provides a reminder of the importance of definitions in all statistical work, but particularly in this field. It gives three definitions of the population. The first is the *home* population, which is the number of people of all types actually in England and Wales on the night of the census. This is sometimes referred to as the *de facto*

TABLE 52

ESTIMATED POPULATION OF ENGLAND AND WALES, 30TH JUNE, 1962, 000's*

	Persons	Males	Females
Home ..	46,669	22,651	24,018
Total ..	46,768	22,748	24,020
Civilian ..	46,379	22,374	24,005

* *Source: Reg. Gen. Statistical Review 1962, Part II, Table A.2.*

population. The second definition, *i.e. total*, is the home population plus members of H.M. Forces belonging to England and Wales but serving overseas, minus the forces of other countries temporarily in England and Wales. The final definition, *civilian*, is the total population minus members of H.M. Forces belonging to England and Wales at home or overseas. It will come as no surprise to the reader to learn that pre-war population estimates were defined on a slightly different basis, and that the *home* population is the present figure most nearly comparable with pre-war totals.

Table 53 illustrates the manner of publication of a part of the information relating to marriages and births in the Registrar General's annual *Statistical Review*. It will be noticed that instead of using data for single years, the comparative periods are inter-census decennia, with the exception of the last four figures quoted, which are single

TABLE 53

MARRIAGE AND BIRTH RATES, ENGLAND AND WALES, 1841-1962

Period	Marriages 000's	Marriage Rate[1]	Live Births 000's	Crude birth rate[2]	Legitimate fertility rate[3]	Illegitimate rate[4]	Infant Mortality[5]	Male births per 1,000 female births
1841–50	1,355	16·1	5,489	32·6	—	—	153	1,049
1851–60	1,602	16·9	6,472	34·1	281·0	65	154	1,046
1861–70	1,770	16·6	7,500	35·2	287·3	61	154	1,042
1871–80	1,961	16·2	8,589	35·4	295·5	50	149	1,038
1881–90	2,047	14·9	8,890	32·4	274·6	47	142	1,037
1891–1900	2,394	15·6	9,155	29·9	250·3	42	153	1,036
1901–10	2,641	15·5	9,298	27·2	221·6	40	128	1,038
1911–20	3,076	16·6	8,096	21·8	173·5	48	100	1,044
1921–30	3,025	15·5	7,129	18·3	143·6	44	72	1,045
1931–40	3,615	17·7	6,065	14·9	111·1	43	59	1,053
1941–50	3,673	17·2	7,251	16·9	114·0	61	43	1,061
1951–55	1,755	15·8	3,377	15·2	105·0	48	27	1,059
1956–60	1,724	15·2	3,698	16·3	113·5	50	23	1,060
1961	347	15·0	811	17·5	122·1	60	21	1,056
1962	348	14·9	839	17·9	124·1	60	22	1,051

Source: Reg. General Annual Reviews. Parts I and II.

[1] Persons marrying per 1,000 population of all ages.
[2] Total live births per 1,000 population of all ages.
[3] Number of legitimate births per 1,000 married women aged 15–44.
[4] Number of illegitimate births per 1,000 total live births.
[5] Deaths of infants under 1 year of age per 1,000 live births.

years and two quinquennia. This mode of presentation has two obvious advantages. First, a period of more than a century can be compressed into a relatively short table, and second, while the annual fluctuations are eliminated the figures for the successive decennia facilitate the measurement of trends and changes. The meaning of the figures in each of the columns is defined in footnotes, but other measures are possible. For example, the *crude marriage rate* shown, *i.e.* 14·9 per thousand in 1962, is the result of expressing the number of men and women marrying in that year per thousand of the population of all ages. Over long periods of time, this may prove to be an unsatisfactory measure for comparative purposes, since not everyone in the population is eligible to marry, *i.e.* what proportion of the population is already married, what proportion of the population is too young? Thus, the Registrar General also calculates marriage rates per thousand of unmarried men of 15 years and over, a similar rate for women aged 15 and over. In addition to these there is a further rate for unmarried men aged 20-44, and one for unmarried women aged 15-39. The choice of the last two age groups is reasonably self-evident. These are the age limits between which most people marry.

The *crude birth rate*, like the marriage rate, is a simple measure derived by expressing the number of births in each year as a rate per 1,000 of the mid-year population of that year – in England and Wales the population as estimated on the 30th June. The reader should note the important statistical point that the date of registration is seldom the date of the vital event. This means that children born at Christmas may not be registered until the New Year, and thus might be counted in to the births of the following year for purposes of calculating these birth rates. At present it is customary to take the date of occurrence into account, not dates of registration.[1] But this was not always the practice. The crude birth rate as a measure of fertility suffers from the defect that a large section of the total population has nothing to do with child-bearing, *i.e.* the very young or aged, and the majority of unmarried people. A better measure of changes in fertility is provided by the *legitimate fertility rate*, which expresses the number of children born to married women as a rate per thousand of married women who are aged between 15 and 44. The illegitimate rate shown in the above table is merely a form of percentage of all births. Thus, in 1962 the rate of 60 per thousand merely indicates that 6 per cent of all live births were illegitimate.[2] The *infant mortality rate*, to which reference was made in Chapter XVI (social class) is interesting not least because of the tremendously rapid fall that has taken place within this century. It is derived by expressing

---

[1] The registration of a birth may be made any time up to 6 weeks after the event.

[2] An illegitimate fertility rate can be calculated by expressing the number of such births as a rate per 1,000 unmarried; divorced and widowed women aged 15-44.

the number of deaths in the year among babies under one year of age per thousand of all live births recorded in that year. Note that, strictly speaking, the sum of the deaths of such infants occurring in the year may, in fact, relate to children born in the previous year, but no adjustment is normally made for this fact.

One of the most interesting features disclosed by successive censuses is the changing pattern in the age distribution that has emerged over the past century. Even in the past fifty years there have been marked changes. Whereas at the 1911 census children under 15 constituted 30·6 per cent of the entire population and only 5·2 per cent were over 65, in mid-1961, the proportion of under 15s had fallen to 22·9 per cent, while those who had passed their 65th birthday comprised 11·9 per cent of the population. One of the consequences of the postwar boom in births (there were 881,000 live births in 1947) is the current demand for places at institutes of higher education. The Registrar General's *Reviews* devote considerable space to analyses of the changing age structure. The population is classified by reference to age usually by single years up to age 20, although there are subgroupings for each quinquennial age group, and thenceafter, *i.e.* from 20 onwards, quinquennial age groups, *e.g.* 20–24, 25–29, etc., are used. This is done for both sexes and for each of the three definitions of population to which attention was drawn earlier. The age structure of the population is important, not merely in terms of knowing how

Figure 21

POPULATION PYRAMIDS. ENGLAND AND WALES

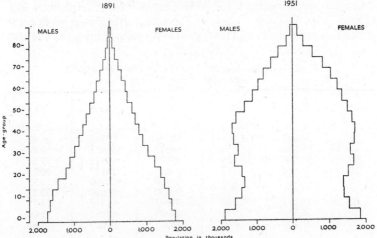

Source: 1951 Census, General Report, p. 90.

many young working and old people there are, but also for purposes of making population projections. Since the Royal Commission on Population reported at the end of the last war, the Registrar General has prepared a number of population projections which appear in the Annual Review and the fourth quarter's issue of the *Quarterly Return*.

A highly practical method of bringing home the effects of the changing age structure is illustrated in Figure 21. Basically the diagram is akin to the bar diagram illustrated in Fig. 7 (p. 70), the bars to the left of the central ordinate measuring off the number of males at each age (this is usually done in quinquennial age groups), while those to the right measure the number of females. The age structure of the population in 1891 with its predominantly young population and very small proportion aged over 60 is in marked contrast to that of 1951, which shows in every sense of the term a middle-aged spread. These two illustrations are derived from the *General Report* of the 1951 Census.

Life Tables

One of the most useful methods of presenting information relating to a given population in such a way that it can be immediately compared with a comparable data for other populations, is the *life table*. The life table is also used for calculation of the probability of death or survival. An official life table, usually referred to as the English Life Table, is prepared on the occasion of each population census. The current life table is No.11, which was prepared by the Government Actuary on the basis of the 1951 Census and published as one of the decennial supplements to the 1951 Census.[1] An excerpt from this table is given in Table 54 below. The actual supplement gives comparable data for males and females at every year of life up to 104 and

TABLE 54
EXCERPT FROM ENGLISH LIFE TABLE No. 11, 1950-52
Males (Selected ages only)

Age x	l_x	d_x	p_x	q_x	$\mathring{e}_x$
0	100,000	3,266	·96734	·03266	66·42
10	95,866	50	·99948	·00052	59·24
20	95,151	123	·99871	·00129	49·64
30	93,820	147	·99843	·00157	40·27
40	91,968	267	·99710	·00290	30·98
50	87,591	745	·99150	·00850	22·23
60	75,823	1,796	·97631	·02369	14·79
70	52,350	2,958	·94349	·05651	9·00
80	21,130	2,880	·86371	·13629	4·86

Source: Registrar General's Decennial Supplement, 1951. Life Tables.
[1] H.M.S.O., 1957.

109, respectively. For illustrative purposes, ages at intervals of 10 years are reproduced above. The l_x column is the most generally quoted for demographic analysis, and it shows the number out of a hypothetical population of 100,000 new born males which would survive to successive ages 10, 20, 30, etc. Thus, by age 40, given the mortality upon which the life table is based of the 100,000 new born male infants, 91,968 would be alive. Obviously as age increases, so the number in the l_x column diminishes, until finally at age 105 there are no survivors. The l_x values are determined by the column headed q_x, which gives the probability of dying between successive ages. The other column, which is normally reproduced together with the l_x column, is the final column $\mathring{e}_x$. This shows the average length of life lived after each age. For example, the average length of life lived by each of the original 100,000 males is 66·42 years. At age 50, the average expectation of life is 22·23 years. Note that this figure is an arithmetic average and like all averages suffers from the defect that it may not be representative of any single unit in the population. For example, the figure of 68·1 years as the expectation of life at birth in 1960–62 (Table 55) is derived by aggregating the lives, short and long, of the entire hypothetical population of 100,000 males, and averaging the total over each member. This figure should not be read, however, as implying that every English male child born in 1960–62 can expect to live 68·1 years. He may; he may also live longer; he may die within the next year. The columns in Table 54 headed d_x, p_x and q_x are necessary to the calculation of the two main life table values, l_x and $\mathring{e}_x$, and are shown here because it is in this form that the official life table is published. Their derivation and their usage in the computation of a complete life table is explained in an appendix to this chapter. For the moment they may be ignored.

The life table provides a highly useful reflection of living conditions. For example, if we learn that the expectation of life of a newborn male child in India at the present time is less than 40 years, compared with 70 in the United States, then we infer from this that living conditions, including the application of medical science, etc., in India are considerably inferior to those in the United States. The extent to which living conditions in England and Wales have improved over the years is shown in Table 55. Thus, in 1841, the average expectation of life for a male child at birth was 40·2 years. Today, some 120 years later, it is 68·1. Note that the really significant improvements in the expectation of life have occurred in the earlier years of life. For the last two ages shown, *i.e.* 55 and 75, the improvement that has taken place over the past 12 decades is relatively slight. In other words, the average expectation of life has increased not because people generally are living longer, the natural life span has not been generally extended

TABLE 55
EXPECTATION OF LIFE, MALES, ENGLAND AND WALES

Age	1841	1870-2	1900-2	1910-2	1930-2	1950-2	1960-62
0	40·2	40·4	45·9	51·5	58·7	66·4	68·1
5	49·6	49·8	54·1	57·1	60·1	64·0	65·1
10	47·1	46·7	50·1	53·1	55·8	59·2	60·2
15	43·3	42·7	45·7	48·6	51·2	54·4	55·3
20	39·9	38·9	41·5	44·2	46·8	49·6	50·6
25	36·5	35·4	37·4	40·0	42·5	45·0	45·8
35	29·8	28·7	29·5	31·7	33·9	35·6	36·3
55	16·7	16·1	15·9	16·9	17·9	18·3	18·7
75	6·5	6·0	6·1	6·5	6·4	6·7	7·1

Source: Reg. General Annual Reviews, Part I

beyond the 'three score years and ten', but because many people who in the last century died at early ages now are living longer due to improved medical science, better sanitation and living conditions generally. For example, in the period covered in the above table, the infant mortality rate has fallen from 150 to 20 per 1,000.

Birth and Fertility Rates

In 1962 there were registered in England and Wales 838,736 live births. Ten years earlier, the corresponding figure was 673,651. Thus, within the decade there has been an increase of nearly one-quarter in the annual number of births. A number of factors have to be taken into account, however, in assessing the true significance of such an increase. The simplest and most obvious factor which determines the number of births is the size of the actual population. When this is taken into account we can calculate a *crude birth rate* which relates the actual number of births to the estimated population at June 30 each year. The result is then that in 1952 the crude birth rate was 15·1 and in 1962 it was 17·9 per 1,000, which suggests a somewhat smaller increase than the absolute totals. The crude birth rate suffers from the defect that the denominator comprises every member of the population, many of whom will have no effect on the birth rate, *i.e.* the young and the very old. This might not matter if these elements in the population formed a constant proportion thereof. They do not, and therefore the crude birth rate is unsatisfactory as an indicator of changes in fertility except in the relatively short run for a population within the same geographical area with little or no migration. An alternative, as was explained earlier, is the so-called *fertility rate*, whereby the number of births are related to the number of women of child-bearing age, normally taken as between 15 and 44. This is not completely satisfactory, for while nearly 95 per cent of births take place within wedlock, a small proportion do not, and therefore if the proportions of women in the population who are married changes significantly, the fertility rate may be affected, merely by the greater frequency of marriage in the population.

An improved measure of fertility may be derived by using the female *age specific* fertility rates, *i.e.* we compare the fertility rates of each age group of women of reproductive age, *i.e.* 15–19, 20–24, etc. This comparison can be simplified by aggregating the age specific fertility rates as in Cols. 2–4 in Table 56 below. This provides a hypothetical total of babies who would be born to a group of women who commenced their child-bearing period together, and neither die nor migrate until they have reached the end of that period. Such a rate is known as the *gross reproduction rate* (G.R.R.) and this has a value of unity where the number of female children born corresponds with the number of women in this hypothetical group. The G.R.R. suffers from the defect that a number of women will die during their child-bearing period, and thus the G.R.R. gives an upper limit to the number of babies that may be born given the age specific fertility rates. To take account of mortality, there is calculated a so-called *net reproduction rate* (N.R.R.). This estimates the average number of daughters that would be produced by women throughout their lifetime if they were exposed at each age to the fertility and mortality rates on which the calculation is based. As with the G.R.R., so an N.R.R. of unity represents replacement of the female generation. Table 56 below illustrates the calculation of both reproduction rates. The age specific fertility rates are based on both male and female births to the women in the various age groups. Column 5, headed l_x, shows the number of female survivors in the successive age groups, while col. 6 shows the number of total years of life lived within each 5-year period by the survivors in each age group.[1]

The reader should note that the only difference between the gross and net rates in the above calculation is that while for purposes of calculating the female G.R.R. it is assumed that the original cohort of 1,000 women remains alive throughout the reproductive period 15–49, in the case of the female N.R.R. allowance is made for some of them dying.[2] No significance attaches to the fact that in the G.R.R. the cohort of women is 1,000 and for the N.R.R. it is 10,000. In the first case this follows because all the fertility rates are already expressed per 1,000 women; in the second an adjustment from the rate per 10,000 to the rate per 1,000 is needed since the values in col. 5 are based upon a cohort of 10,000. In both cases the rates measure the

---

[1] The basis of this column is explained in the appendix to this chapter dealing with the construction of life tables. For the moment the reader is asked to accept the statement.

[2] It is possible to calculate both a male and all persons N.R.R. but the female rate is the best measure and most generally used.

average number of daughters that would be produced per woman if, in the case of the G.R.R. the women were subject to the fertility rates used in the calculation, and in the N.R.R. subject not only to the G.R.R. assumption re fertility, but also if the cohort of women were subject to the mortality experience summarised in the life table from which cols. 5 and 6 are derived.

TABLE 56

CALCULATION OF GROSS AND NET REPRODUCTION RATES

(1) Age x	(2) Total number of women '000's	(3) Total number of births M & F	(4) Live births per 1,000 women or Age-specific F/R per 1,000 (f_x)	(5) l_x	(6) $5L_x$	(7) All babies born to generation of 10,000 $5L_x \times f_x$ ('000s)
15–19	1,424	27,639	19·41	9,645	48,130	934
20–24	1,531	226,817	148·20	9,607	47,900	7,099
25–29	1,653	280,506	169·70	9,554	47,610	8,079
30–34	1,658	194,526	117·30	9,489	47,265	5,544
35–39	1,741	113,966	65·48	9,416	46,850	3,068
40–44	1,669	32,363	19·39	9,324	46,315	898
45–	1,561	2,215	1·42	9,201	45,515	65
				9,005		
		878,032	540·90			25,687

G.R.R. = 540·9 infants in one year but, since each age group covers five years, the correct figure is 540·9 × 5 = 2,704 infants per 1,000 women. These infants are both male and female and the proportion of female births in that year was 0·485. Hence the number of female births to 1,000 women is 2,705 × 0·485 = 1,312 signifying a G.R.R. of 1·31. N.R.R. is given by adjusting the total number of births in col. 7, *i.e.* 25,687,000 to female births only; *i.e.* multiply by 0·485, which yields a total of female births of 12,458. This relates to 10,000 women, *i.e.* the original hypothetical cohort which was reduced by mortality to 9,645 by the time it entered into the reproductive period (see col. 5). Thus the N.R.R. is equal to 1·25, *i.e.* 12,485 divided by 10,000. (Note that whereas with the G.R.R. the total of births in col. 4 had to be multiplied by 5, this has already been done in the N.R.R. calculation when the figures in col. 5 are converted into those in col. 6.)

When first developed by Dr Kuczyncski during the early 1930s, it was assumed the net reproduction rate provided the basis for reliable population projections. It was widely adopted, and even now when its defects are generally recognised, the United Nations demographic publications and the Registrar General's annual Reviews all give the latest indices, both gross and net, for the various populations. The basic defects of the N.R.R. are that it is determined by the rates of fertility and mortality that are adopted for purposes of its calculation. Thus in so far as the rates for any selected year are employed, the N.R.R. merely indicates what will happen on the basis of these rates. But since it is common knowledge that mortality and, more especially, fertility may change rapidly within the space of a few years, the N.R.R. as a basis for projecting population trends is quite unsatis-factory.

For purposes of evaluating trends in birth rates a number of factors have to be taken into account. Quite apart from the number of women in the population, there is the question of how many of them are already married and for what period, how many have just married, to what extent they are having small or large families, and the extent to which their family-building is being compressed into a short period or extended over most of their married lives. For some years now, since the publication of the Family Census, undertaken by Professors Glass and Grebenik on behalf of the Royal Commission on Population,[1] the Registrar General has used what is known as *cohort fertility* to study the trend of births.[2] Instead of making projections on the basis of the fertility experience of any single year, cohort fertility necessitates following through the family-building history of a cohort of married women. The cohort may be of women born in a particular year or, more usually, of women married in a particular year. The former are known as *birth cohorts*, the latter as *marriage cohorts*. By studying the trend in the cumulative numbers of children born to married women at successive points of time up to the end of their child-bearing period, the limitations of annual rates such as fertility rates or N.R.R., are avoided. Thus, if at the end of the child-bearing period each member of a marriage cohort has produced on average 2·2 children as compared with a completed family of 5·8 children of an earlier generation (*i.e.* marriage cohort), then it is clear that fertility has fallen.[3] The main limitation of cohort analysis is that definite conclusions regarding the fertility of a particular marriage cohort can be drawn only when that cohort has reached the end of its reproductive period. Since at any time a large proportion of married women are under 45 years of age, it is arguable that their future child-bearing activities may markedly affect the resultant figures. A partial solution to this problem is provided by the fact that one of the

[1] Vol. VI of the Papers of the Royal Commission on Population entitled *The Trend and Pattern of Fertility in Great Britain*, H.M.S.O. 1954.

[2] See for example the 1961 *Review*, Part III.

[3] These are the actual numbers of children in completed families for women who married durin 1870–79 and 1925 respectively. See Glass and Grebenik report *op. cit.* Table 1.

characteristics of family-building in present times is that, as well as marrying at a younger age than in the past, the majority of women tend to complete their families within the first 7–10 years of their married lives. Thus, while estimates must be made to take account of uncompleted fertility, such estimates are not liable to such margins of error as would be the case if family-building habits, as far as timing at least is concerned, were more volatile or more extended.

Standardisation

The crude death rate for Torquay in 1962 was 17·0 per thousand. In Stevenage, with a population only slightly smaller, the crude death rate in the same year was 4·1 per thousand. Likewise, the crude birth rate in Torquay for the same year was 12·1, as compared with the rate of 27·5 per thousand in Stevenage. These figures suggest that the population in Stevenage is more than twice as fertile, and that inhabitants of Torquay are more than four times as likely to die as the inhabitants of Stevenage. These contrasting figures bring home the basic defects of the crude rates which, as already explained, relate the number of events, *i.e.* either births or deaths, to the total population. In the case of births a much better denominator for the derivation of a rate is the number of married women in the child-bearing ages. While in the case of deaths, since from the age of 12 upwards mortality rates at each age increase, it follows that the higher the proportion of older people in a population, the higher will be the crude death rate. Thus some part of the explanation of the disparity between the rates for Stevenage and Torquay, in respect of both births and deaths can be explained, first, by the fact there are relatively more women of child-bearing age in Stevenage than in Torquay, and second, in Torquay a higher proportion of the population are in the older age groups.[1] Since comparisons of fertility and mortality rates are made between administrative areas in England and Wales,[2] and for that matter other areas too, some means of making effective comparisons must be devised. The method used to this end is the technique of standardisation.

Standardisation involves the adjustment of the crude rates so as to compensate for the differences in the age and sex composition of the populations to be compared. There are various means of doing this, but the two most generally employed methods are known as the direct and indirect methods. The *direct* method consists of calculating the total number of deaths which the death rates for a given population would have produced in a *standard* population. As may be seen from Table 57, towns A and B have somewhat similar crude death

[1] When these factors are taken into account, the adjusted (or standardised) rates, Torquay first, are, births 13·3 and 18·4; deaths 11·9 and 9·9.
[2] See Table E in *Review*, Part II.

TABLE 57
CALCULATION OF STANDARDISED DEATH RATES (DIRECT METHOD)

Age Group	Town A			Town B			Stan-dard Popu-lation 000's	No. of expected deaths in standard population if subject to mortality of	
	Popula-tion 000's	No. of Deaths	Crude Death Rate per 1000	Popula-tion 000's	No. of Deaths	Crude Death Rate per 1000		A	B
(1)	(2)	(3)	(4)	(5)	(6)	(7)	(8)	(9)	(10)
0—	5	40	8	7	70	10	12	96	120
5—	15	15	1	10	20	2	24	24	48
15—	15	15	1	15	60	4	20	20	80
25—	15	30	2	19	114	6	30	60	180
45—	25	375	15	25	500	20	10	150	200
65—	5	400	80	3·5	210	60	3	240	180
85—	1	300	300	0·5	125	250	1	300	250
	81	1,175	14·5	80	1,099	13·7	100	890	1,058

rates of 14·5 and 13·7 per 1,000 respectively. The crude death rate is simply the weighted arithmetic mean of the *age specific death rates*. An age specific death rate is merely the death rate of a particular age group arrived at by dividing, for example, the number of deaths of males aged 5–14 by the total number of such males in the population. By applying the age specific death rates of each town to a standard population, as is done in columns 8–10 inclusive, we can calculate the number of deaths that would have arisen from the mortality rates of towns A and B if their populations had been identical with the standard population. The result in this case is, whereas the crude rates per 1,000 are 14·5 and 13·7, the standardised rates for the two towns are 8·9 and 10·6 respectively. In other words, the mortality in town B is significantly higher than in town A, despite the initial impression given by the crude rates that the reverse was the case.

The alternative *indirect* method is slightly more complicated, and is normally used where the age specific death rates of the local populations are not available. In this case, instead of using a standard population a set of standard age specific death rates are applied to the two populations to be compared. The standard death rates are the national age specific death rate of England and Wales in 1951. The basis of the calculation is demonstrated in Table 58. In this case, we derive the number of deaths which would have occurred in the two areas had their individual populations experienced a common mortality, *i.e.* the same or standard age specific mortality rates. Expressing the total of these deaths (columns 7–8) per thousand of the actual populations, we derive what are termed *index* death rates. The index death rate must not be confused with the crude rate. The index

TABLE 58

CALCULATION OF AREA COMPARABILITY FACTORS FOR EASTERN REGION AND TYNESIDE CONURBATION, 1954

Age Group	Est. Mid-Year Popln. 1954. 000's		Number of Deaths 1954.		Standard D/R per 1,000 England & Wales 1951	Expected Deaths on basis of 1951 death rates	
	Eastern Region	Tyneside Con-urbation	Eastern Region	Tyneside Con-urbation		Eastern Region	Tyneside Con-urbation
(1)	(2)	(3)	(4)	(5)	(6)	(7)	(8)
0—	248	70	1,279	492	6·53	1,619	457
5—	491	128	189	61	0·51	250	65
15—	419	107	306	85	0·95	398	102
25—	464	131	511	197	1·45	673	190
35—	448	113	924	315	2·63	1,178	297
45—	445	117	2,367	851	6.90	2,990	807
55—	342	87	4,587	1,551	18·0	6,156	1,566
65—	255	58	9,234	2,622	46·1	11,755	2,674
75—	146	27	15,916	3,492	138·0	20,148	3,726
	3,258	838	35,313	9,666	12·5	45,167	9,884

	Eastern Region	Tyneside Conurbation
Index death rate per 1,000	$\frac{45,167}{3,258} = 13.86$	$\frac{9,884}{838} = 11.8$
Crude rates, per 1,000	$\frac{35,313}{3,258} = 10.8$	$\frac{9,666}{838} = 11.5$
A.C.F. $= \dfrac{\text{Standard death rate}}{\text{local index death rate}}$	$\frac{12.5}{13.86} = ·90$	$\frac{12.5}{11.8} = 1.06$
A.C.F. × local crude death rate = local adjusted death rate	$·90 \times 10.8 = 9.72$	$1.06 \times 11.5 = 12.19$
Ratio of $\dfrac{\text{local adjusted death rate}}{\text{standard rate}}$	$\frac{9.72}{12.5} = 0.78$	$\frac{12.19}{12.5} = 0.98$

death rate will tend to be lower than the actual national crude death rate if the local population is abnormally young. Correspondingly, it will tend to be higher than the national crude death rate if the population is older than the population of the country as a whole. By dividing the index death rate into the standard or national death rate, we derive what is known as an *Area Comparability Factor* (A.C.F.). The A.C.F. is then multiplied with the local crude death rate and the product, termed the local adjusted death rate, is then comparable *both* with the national death rate and any other local death rate calculated on the same basis. Thus, in Table 58, in place of the crude death rates of 10·8 and 11·5, the adjusted or standardised death rates as they are known, are equal to 9·72 and 12·19. In other words, the crude death rate of the Eastern region tends to over-estimate mortality because there are relatively more people in the older age groups than in the standard population; whereas in the case of the Tyneside conurbation, the crude rate under-states the true mortality by virtue of the fact that there are proportionately more young people in that population than in the national population. Finally, the ratio of the standardised death rate to the national standard rate is given in the table. These are shown merely to explain the principle upon which these rates, which are given in the Registrar General's *Statistical Review* Parts I and II in the table showing the birth and death rates for the administrative areas of England and Wales,[1] are calculated.

Standardised Mortality Ratio

In the 1951 Census report on *Occupational Mortality* (the latest available in August 1964) the mortality experience of each occupational group was summarised by means of a *Standardised Mortality Ratio* (S.M.R.). This is defined as the number of deaths registered of men in a given occupational group at ages 20-64, expressed as a percentage of the number that would have occurred if the death rates in each separate age group within the occupation had been the same as in a standard population consisting of all males in England and Wales. The method of calculating an S.M.R. (all causes of death) is illustrated in the following example for the occupational grouping known as Farmers and Farm managers taken from the 1951 report.[2] Columns 1 and 2 give the age distribution of this occupational group.

TABLE 59

CALCULATION OF STANDARDISED MORTALITY RATIO

Ages (1)	Census Population (2)	Standard death rates per million 1949–53 (3)	Expected deaths in occupation $\dfrac{5 \times (2) \times (3)}{1,000,000}$ (4)
20—	7,989	1,383	55
25—	37,030	1,594	295
35—	60,838	2,868	872
45—	68,087	8,212	2,796
55—64	55,565	22,953	6,377
Total standard deaths 20–64			10,395

Total registered deaths of farmers and farm managers aged 20–64 = 7,320.

$$\text{S.M.R.} = \frac{7,320 \times 100}{10,395} = 70 \text{ per cent.}$$

[1] See Tables 13 and E in Parts I and II respectively of the 1962 *Review*.
[2] Registrar General's Decennial Supplement 1951, Occupational Mortality, Part II, vol. I, p. 17.

13

The basis of the calculation of the S.M.R. is the expected deaths in the last column. The standard death rates are based on the registered deaths in the five year period 1949–53 inclusive, the rates shown being annual averages. The population in each age group is at risk for a period of five years so that the expected deaths will be five times the annual rate times the population at risk. On this basis the expected number of deaths will be 10,395, *i.e.* this is the number of deaths to be expected if farmers had experienced the same mortality as all males between 20–64 in the quinquennium 1949–53. In fact the number of deaths of farmers registered in the same quinquennium was only 7,320. This number expressed as a percentage of the expected number of deaths yields a percentage of 70. In other words farmers and farm managers experienced a considerably lower mortality, age for age, than the male population as a whole.

Standardised Mortality Ratios are calculated not only for 'all causes' of death but also for individual diseases such as tuberculosis and cancer, as well as various types of accident. The mortality rates for each disease are analysed by social class and by sex, married women being distinguished from single women. In addition, the S.M.R. is used to compare regional differences in respect of mortality as well as comparing class and regional experience in respect of infant mortality and still births.

Since 1958, the Registrar General has used the S.M.R. for showing the trend of mortality over many decades. In this case the S.M.R. shows the number of deaths registered in the year as a percentage of those which would have been expected in that year had the sex and age mortality of the standard period (in this case 1950-52) operated on the sex and age population of the year of experience. Until 1958, the *Statistical Review* Part I published not the S.M.R., but an index known as the C.M.I. (Comparative Mortality Index). This was derived by the method of direct standardisation, *i.e.* by using a standard population as in Table 57. The standard population was the mean of the population of 1938 and the year under review. In other words, instead of taking as the standard a population of any single year, an 'average' population was used which, in the opinion of the Registrar General, was more representative and provided a better base for long period calculations.[1] By dividing the standardised death rate for the 1938 mean population into the standardised death rate for the other year, a ratio known as the C.M.I. was derived. The base year was 1938, in which year the C.M.I. had a value of unity. This method, as already stated, has now been discontinued.[2]

It should be noted in passing that, although the illustrations of

[1] The C.M.I. was worked back to 1841.
[2] Method of calculation is given in the *Review* for 1940-45, Vol. I, pp. 6–11.

standardisation given above relate to mortality rates, precisely the same techniques are employed in connection with birth rates. For the reader who wishes to consult the method of presentation of these data, he will find the standardised rates for the year 1962 (the latest Review available at the time of writing) in Part II, Table E. The rates are shown for both live births and deaths. Apart from the actual numbers there is shown the crude rates, the area comparability factor, and the ratio of the local adjusted birth rate to the national or standard rate.

LIFE TABLES

The data required for the construction of a life table are the age distribution of the population and the age specific mortality rates, *i.e.* the mortality rates appropriate to each year of life. The accuracy of the life table depends upon the accuracy and completeness of the registration of deaths and the enumeration of the population at the census. Generally speaking, even in Britain the data are least reliable for the highest age groups and in some underdeveloped countries the data for the first few years tend also to be somewhat unreliable. The methods used to construct a life table range from the simple arithmetic process illustrated in the examples below, to highly sophisticated mathematical techniques as used by actuaries for life tables used in life assurance and the Government Actuary's English Life Table. The Census decennial supplement describes the method used for calculating this life table on the basis of yearly death rates, *i.e.* showing the number of survivors and the mean expectation of life at every age. Alternatively, an *abridged* life table may be constructed, in which quinquennial and decennial age groups are used. As will be seen in the main example below, the abridged life table gives reliable estimates of the life expectation at various ages.[1]

Before embarking on the calculation of an abridged life table, using recent data relating to the population of England and Wales, the principles underlying its construction and the relationship between the various columns may best be brought out by a simplified example, as in Table 60 below. Starting with a single year age distribution in column 1, column 2, *i.e.* p_x, is defined as the probability of

---

[1] The Registrar General calculates an abridged life table each year. See 1962, *Review*, Part I, Appendix B. The table is also reproduced in the Annual Abstract of Statistics. See Abstract No.100, Table 34.

surviving one year from x. The probability of surviving from birth to the first birthday is ·9; from the first to the second birthday, p_x is equal to ·7. This merely means that 90 per cent of the original cohort of births survive the first year of life, but only 70 per cent of those who reach their first birthday live to see their second. Persons who do

TABLE 60

SIMPLIFIED LIFE TABLE CALCULATIONS

(1)	(2)	(3)	(4)	(5)	(6)	(7)	(8)
Age	Prob- ability of surviving one year from x	Prob- ability of dying within one year of x	Number of sur- vivors at x	Number dying within year of x	Years lived between x and $x+1$	Years lived after x	Mean expec- tation of life at x
x	p_x	q_x	l_x	d_x	L_x	T_x	$\mathring{e}_x$
0	·9	·1	100	10	95	290	2·90
1	·7	·3	90	27	76·5	195·0	2·17
2	·6	·4	63	25	50·5	118·5	1·88
3	·7	·3	38	11	32·5	68	1·79
4	·6	·4	27	11	21·5	35·5	1·31
5	·4	·6	16	10	11	14	·88
6	·0	1·0	6	6	3	3	·50
7			0				

not survive are those who die, and the probability of dying between any two birthdays is represented by the symbol q_x. It is self-evident that $p_x + q_x = 1$, so that if $p_x = ·6$, then $q_x = ·4$. Given these rates, which as will be seen from the example which follows, are derived in practice from the actual mortality experience of a given population, the rest of the table is no more than simple arithmetic. The l_x column shows the number of survivors at each successive age. Starting with 100 live births, since the probability of survival during the first year of life is ·9, then the number of the original cohort who will be alive to see their first birthday is 90. Likewise, during the next year of life, i.e. between ages 1 and 2 with $p_x = 0·7$, then out of the 90 1-year-olds, 63 will reach their second birthday. It will be self-evident that the numbers dying within the successive years, i.e. d_x, is merely the dif- ferences between the successive values of l_x. For example, $l_0 - l_1 = 100 - 90; = d_0 = 10$. Column 6 is headed 'Years lived between x and $x+1$' and this is merely the sum of the years lived by the cohort during each year of life. For example, during the first year of life, only 90 survive and these therefore contribute 90 years to the sum of years lived, while it is assumed (this point is discussed in more detail below) that the 10 who died lived on average six months piece. So they contri- bute 5 years. Thus, the total years lived between age 0 and 1 by the original cohort of 100 infants is 95. $T_0 = 290$ is merely the sum of all the values in column 6. The final column, e_x, i.e. the mean expectation of life at age x, is derived by dividing T_x by the corresponding value of l_x.

The same principles underlie the construction of the very much longer and apparently more complex table set out in Table 61 which illustrates the construction by the simple method of an abridged life table. The table is described as abridged since, instead of giving each single year of age, apart from the first five years of life the ages are grouped. For example, from 5 to 25 the age groups are quinquennial and thenceafter they are decennial. This greatly reduces the amount of calculation, and despite the degree of approximation involved, the results differ only slightly from those which would be derived from a full life table. The age grouping is practicable because the mortality rates do not vary significantly throughout the age intervals used, i.e. the average death rate for the quinquennium is a good approximation of the death rate of the individual years in the age group. This does not apply to the first five years of life, hence the use of single years in the calculations. Likewise, at the other extreme, for the age group 85 and over, the mortality rate used is an approximation, but since there are so few lives at risk, for the purposes to which an abridged life table may be put, this does not matter.

While the lay-out of Table 61 (pp. 374-376) appears complex, its actual form, as will be seen, is determined primarily by the nature of the calculations. Reference to Table 54 will show that only five columns, apart from the age distribution, are reproduced in the English Life Table No.11. There are twelve in Table 61 and of them only two, l_x and $\mathring{e}_x$ appear in the English Life Table; all the other columns are working columns. The first seven are needed to provide the l_x values, i.e. the number of lives at each age x, and the remainder to give the $\mathring{e}_x$ values which are derived from the l_x values. The first three columns of Table 61 are self-explanatory. They show the number of males by age enumerated at the 1951 census, adjusted to show the number at mid-1951. The third column gives the total number of male deaths at each age recorded in England and Wales during the three year period 1950-52 inclusive. The next column (3) is the result of dividing the number of deaths at each age by three to adjust them to an annual basis and then by the population of that age given in the second column. The resultant figure is a crude rate of mortality for that age group, written m_x. Thus the rate for the first year of life is ·03402 or 34 per 1,000 of the population of that age. These are known as *central* death rates since they are based on the average population of the year as given by the mid-year estimate.

For purposes of a life table the crude 'central' mortality rates given

TABLE 61

ABRIDGED LIFE TABLE – ENGLISH MALES

Age x	1951 Population (000's) (1)	1950-2 Deaths (2)	m_x (3)	$\dfrac{2-m_x}{2+m_x}$ (4)	$\log p_x$ (5)	l_x (6)	$\log p_x + \log l_x$ (7)	$l_x + l_{x+1}$ (8)	L_x (9)	T_x (10)	$\log T_x - \log l_x$ (11)	$\overset{\circ}{e}_x$ (12)
0	338	34,494	·03402	1·9660 2·0340	0·2936 0·3083	10,000	4·0000 Ī·9853	19,668	9,834	664,525	5·8225 4·0000	
					Ī·9853						Ī·8225	66·45
1	353	2,573	·00243	1·9976 2·0024	0·3006 0·3014	9,668	3·9853 Ī·9992	19,317	9,658	654,691	5·8161 3·9853	
					Ī·9992						Ī·8308	67·74
2	373	1,592	·00142	1·9986 2·0014	0·3008 0·3013	9,649	3·9845 Ī·9995	19,287	9,644	645,033	5·8097 3·9845	
					Ī·9995						Ī·8252	66·86
3	411	1,232	·00100	1·9990 2·0010	0·3009 0·3012	9,638	3·9840 Ī·9997	19,270	9,635	635,389	5·8031 3·9840	
					Ī·9997						Ī·8191	65·94
4	429	997	·00077	1·9992 2·0008	0·3009 0·3012	9,632	3·9837 Ī·9997	19,257	9,628	625,754	5·7965 3·9837	
					Ī·9997						Ī·8128	64·98

Age												
5 –	64·03	5·7898 3·9834 / 1·8064	616,126	48,042	19,217	3·9834 1·9985	9,625	0·6013 0·6028 / 1·9985	3·9933 4·0067	·00067	3,262	1,616
10—	59·24	5·7545 3·9819 / 1·7726	568,084	47,900	19,160	3·9819 1·9989	9,592	0·6015 0·6026 / 1·9989	3·9946 4·0054	·00054	2,335	1,429
15—	54·38	5·7162 3·9808 / 1·7354	520,184	47,727	19,091	3·9806 1·9980	9,568	0·6011 0·6031 / 1·9980	3·9906 4·0094	·00094	3,745	1,335
20—	49·62	5·6744 3·9788 / 1·6956	472,457	47,460	18,984	3·9788 1·9971	9,523	0·6006 0·6035 / 1·9971	3·9864 4·0136	·00136	5,840	1,427
25—	44·93	5·6285 3·9759 / 1·6526	424,997	93,860	18,772	3·9759 1·9931	9,461	0·2976 0·3045 / 1·9931	1·9841 2·0159	·00159	14,969	3·140
35—	35·56	5·5199 3·9690 / 1·5509	331,137	91,775	18,355	3·9690 1·9874	9,311	0·2947 0·3073 / 1·9874	1·9712 2·0288	·00288	28,468	3,291
45—	26·48	5·3793 3·9564 / 1·4229	239,362	86,925	17,385	3·9564 1·9648	9,044	0·2830 0·3182 / 1·9648	1·9194 2·0806	·00806	71,444	2,955

TABLE 61 (*continued*)

ABRIDGED LIFE TABLE – ENGLISH MALES

Age x	1951 Population (000's) (1)	1950-2 Deaths (2)	m_x (3)	$\dfrac{2-m_x}{2+m_x}$ (4)	$\log p$ (5)	l_x (6)	$\log p_x +$ $\log l_x$ (7)	$l_x + l_{x+1}$ (8)	L_x (9)	T_x (10)	$\log T_x -$ $\log l_x$ (11)	$\mathring{e}_x$ (12)
55—	2,032	140,845	·02310	1·7690 2·2310	0·2477 0·3485	8,341	3·9212 Ī·8992	14,954	74,770	152,437	5·1829 3·9212	
					Ī·8992						1·2617	18·27
65—	1,373	227,322	·05519	1·4481 2·5519	0·1607 0·4067	6,613	3·8204 Ī·7540	10,366	51,830	77,667	4·8903 3·8204	
					Ī·7540						1·0699	11·74
75—	545	208,487	·1275	0·7250 3·2750	Ī·8603 0·5152	3,753	3·5744 Ī·3632	4,584	22,920	25,837	4·4123 3·5744	
					Ī·3451						0·8379	6·89
85—	62	53,031	·285	1·0000 0·285	0·0000 Ī·4548	831	2·9376	—	2,917	2,917[1]	3·4649 2·9195	
					0·5652						0·5454	3·51

m_x = central death rate of age group x to x + 1.
l_x = number of a generation surviving to age x
p_x = proportion of the generation surviving to age x + 1 having reached age x.
L_x = total years of life lived by the generation between ages x and x + 1, x + 5 or x + 10.

T_x = total years of life lived by entire generation to age x.
$\mathring{e}_x$ = average length of life of those surviving to age x.

[1] Since $l_x \times \mathring{e}_x = T_x$ then $l_{85-} \times \mathring{e}_{85-} = 831 \times 3·51 = T_{85-} = 2917$.

above require adjustment. The population figures in column (1) are the estimates of the population as at 30th June, 1951. Since the deaths occurring in any given year of life are spread over the entire twelve months, the actual population of any age which is at risk from the beginning of the year can be estimated by adding half of the annual deaths at each age ($\frac{1}{2}$D) to the mid-year population at that age (P). The probability of dying at age x, represented by q_x, can then be written $q_x = \frac{D}{P+\frac{1}{2}D}$. The chances of dying at any age are directly related to the actual death rates which were represented above as $m_x = D/P$. By multiplying both sides by P, this relationship can be re-written $Pm_x = D$. Substituting this in the formula for the probability of dying at age x, i.e. $q_x = \frac{D}{P+\frac{1}{2}D}$, we get $\frac{Pm_x}{P+\frac{1}{2}Pm_x}$ which by eliminating the P's give $\frac{m_x}{1+\frac{1}{2}m_x}$. This multiplied by 2 to remove fraction yields: $\frac{2m_x}{2+m_x}$. This, it may be repeated, is the formula for obtaining the value of q_x, i.e. the probability of dying at age x.

In any year of life x the probability of survival for the period of year x is represented by the symbol p_x just as the chances of dying during that year are represented by q_x. Thus $p_x + q_x = 1$ or certainty. In other words, the chances of survival can be re-written as $1 - q_x$ and since $q_x = \frac{2m_x}{2+m_x}$ then $p_x = 1 - \frac{2m_x}{2+m_x} = \frac{2+m_x}{2+m_x} - \frac{2m_x}{2+m_x}$ which reduces to $p_x = \frac{2-m_x}{2+m_x}$. Since the values of m_x for all relevant values of x are given in column (3) the probability of survival (p_x) between any age x to $x + 1$ can be derived.

It is the formula which must be applied to the crude death rate at each successive age. To illustrate the calculation of p_x the numerator and denominator of the fraction $\frac{2-m_x}{2+m_x}$ for each value of m_x are set out in column (4).

Take, for example, the first line where m_x is equal to 0·03402, which rounded to four figures is 0·0340. This, subtracted from 2 for the numerator and added for the denominator, gives the values shown in the next column (4), i.e. 1·9660 and 2·0340. In the next column, the logarithms of these values are set out and the one value subtracted from the other. The reader will recall that subtraction of the logarithms of two numbers provides the anti-logarithm of the quotient when those two numbers are divided one into the other. The value of the anti-log $\bar{1}$·9853 is 0·9668 which is the value of p_x at age 0. Lack of space makes it impossible to insert a column of p_x values, just as it is not possible to show the q_x values; but they can both easily be derived from the data given and the student may like to perform the necessary calculations. Strictly speaking, the p_x rate for the first year of life needs to be calculated differently, since the average life of those who die in that year is barely 2 months. But this does not alter the basic principles of constructing the life table.

Once the probability of survival during the first year of life is known, we can calculate how many of a generation of 10,000 new born males will survive to their first birthday, *i.e.* start upon their second year of life.[1] All that is needed is to multiply the number of males alive at the beginning of the year (l_x) by the probability of survival (p_x). Thus, for the first year there are 10,000 males with a probability of survival of 0·9668; so that 9,668 of the original generation of 10,000 embark upon the second year of life. This calculation is shown in column 7; the upper figure is the logarithm of the l_x value, *i.e.* the log. of 10,000 is 4·0000 and the log. of p_x which was already obtained in column (5) is $\bar{1}$·9853. The anti-log., *i.e.* their sum, is 3·9853, which is set out below the first two logs and its value 9,668 is inserted in the l_x column (6) indicating that such a number of lives are at risk at the beginning of that next year.

The entire calculation described in the preceding paragraph is repeated for the values opposite age 1. The m_x value is adjusted to give the numerator and denominator of the fraction which gives the value of p_x. The log. of p_x, *i.e.* $\bar{1}$·9992, is added to the log. of 9,668 shown in column (7) which gives the anti-log., 3·9845, of the product of p_x and l_x where x is *1*. This figure is then carried down to give the number of males alive at the beginning of the third year, *i.e.* 9,649.

When we come to the age group 5—, there starts a quinquennial as opposed to single year grouping. The fraction for deriving the value p_x has hitherto been based upon single years, but for the five year age groups it is adjusted to $\dfrac{2-5m_x}{2+5m_x}$.

This change is explained by the fact that if the mortality in any single year is given by m_x, then in a period of 5 years, each assumed to be the same as the other, the mortality will be five times as great in a population starting at the beginning of the first year of the quinquennium. To facilitate the calculation of m_x whereby we derive the logarithm of $_5p_x$, the fraction is multiplied by 2 to yield $\dfrac{4-10m_x}{4+10m_x}$. It is easier to multiply m_x by ten, than by five! The reader can check the figures in columns (3) and (4) accordingly; 4 minus 10 times m_x, *i.e.* ·00067, yields 4 — ·0067 = 3·993. The remainder of the calculations are as described above. At age 25 there is a further change; the age

---

[1] It is customary to start with a hypothetical population of male (or female) births. This is often termed a radix. By convention the actual number may be 1,000, or 10,000, or 100,000. The Registrar General's life table proper uses 100,000, while in the abridged table 10,000 is customarily adopted as the original radix. In the above illustration, the hypothetical population will consist of 10,000 males, subject to the mortality rates experienced by males in England and Wales in 1950-52.

grouping being altered from five to ten years. The formula for deriving p_x must therefore be changed from its quinquennial form to 10 year and so it becomes $2 - 10m_x$ divided by $2 + 10m_x$. Having obtained from this formula the log. of $10p_x$ it is added to the log. of l_x, i.e. l_{25}, to yield the anti-log. of the number of lives at risk at the beginning of the following period. The calculation is repeated for the remaining 10 year age groups.

For the age group '85 and over' the adjustment to the m_x rate to give the probability of survival (p_x) is different from that used hitherto. According to the full life table prepared by the Registrar General, there are survivors aged 104 years. But the number of survivors between ages of 85 and 105 is so small that detailed calculation is pointless; furthermore the assumption on which the earlier p_x rates were based, that deaths are spread evenly throughout the period is no longer valid. The full calculations for these age groups are far too complex for this book but a good approximation to the value of $\mathring{e}_x$ derived by the Government Actuary is obtained by calculating $\mathring{e}_{85-}$ directly from the formula $1/m_{85-}$. From the data we get $1/0\cdot285$ which equals $3\cdot51$ years. This corresponds very closely with the expectation of $3\cdot48$ years in the official 11th English Life Table.

At this stage the information provided by the table enables us to derive the value of p_x and q_x, i.e. the probabilities of survival and death respectively at certain ages. The column l_x shows the effect of such rates on a hypothetical generation of 10,000 new-born males, as they grow up. Thus 332 (d_x) fail to survive the first year, leaving only 9,668 (l_x) to celebrate their first birthday. The number surviving to their 15th birthday (l_{15}) is 9,568, and at 55 there are 8,341 of the original 10,000 still alive. But this is only part of the complete life tables; i.e. that part comprising the first four columns of Table 54 above, which was taken from the published official English Life Table No.11. The final column thereof is headed $\mathring{e}_x$, i.e. the expectation of life at age x and it is to the calculation of this figure that we now turn.

Column (8) is headed $(l_x + l_{x+1})$ and the first figure in it, 19,668, is obtained by adding together the first two values in the l_x column (6), i.e. 10,000 and 9,668. The next figure of 19,317 is the sum of the second and third values in the l_x column, i.e. 9,668 and 9,649. For the next value, the fourth will be added to the third and so on all the way down the column. Having done this the totals in column (8) are divided by two to give the values in the next column headed L_x. The purpose of this averaging technique must be clearly understood. Take, for example, the first two values in the l_x column, i.e. 10,000 and 9,668. From these figures it emerges that 332 lives were lost in the first year. Since it may be assumed that the deaths were spread out over the year, the *average* live population under one year of age was 10,000 less 166

$\left(\frac{332}{2}\right)$ which corresponds to the first value in the L_x column. In other words, the total period of life lived by the original 10,000 males during that first twelvemonth was approximately 9,834 years.[1] At the beginning of the next year there were 9,668 lives at risk, but only 9,649 survived and following the same principle as before, we estimate that they lived in the aggregate 9,658 years between year x and $x+1$. As was pointed out earlier, the average life of infants dying before their first birthday is rather less than six months. Hence this method of averaging l_0 and l_1 to derive l_0 tends to give an over-estimate. A better estimate is given by using the formula $0.3l_0 + 0.7l_1$. This compares with $0.5l_0 + 0.5l_1$, used in this illustration.

When we come to the 5 year age groups an adjustment to the figures in column (8) is required. This column headed $l_x + l_x +_1$ should now be described as $l_x + l_{x+5}$, for we are estimating the total years lived by the survivors during the quinquennium and not merely one year as for the age groups under 5. As before, the sum of the two successive values of $l_x + l_{x+5}$ is divided by two and then the quotient multiplied by five to give the value to be entered in the L_x column, which from this point on should be described as $5L_x$. The explanation can be illustrated by reference to the figures corresponding to age 5 —. The $l_x + l_{x+5}$ values are 9,625 and 9,592; their sum is 19,217 which divided by two gives the 'annual' population and this when multiplied by five yields 48,042. This is simply the number of years lived by the 9,625 males who entered their fifth year until they reached their tenth birthday or died. This can be checked by a simple calculation. If the entire group of 9,625 males who entered upon their fifth year had survived until their 10th birthday, they would have lived altogether $5 \times 9,625$ years, i.e. 48,125. In fact 33 of them (9,625 — 9,592) died before the 10th birthday and if we assume they lived on average $2\frac{1}{2}$ years, they lost by their premature deaths $2\frac{1}{2} \times 33$ years of life, or 83 years in all. This figure taken from 48,125 equals 48,042 years, which value is to be found in column (9). These calculations are repeated for all the quinquennial age groups to 20 —.

The next age group 25 — is a ten year grouping so that after adding successive values in column (8), dividing by two and then multiplying by five, instead we now multiply the quotient by *ten* for the reasons already stated and get the values in column (9). In this case they are $10L_8$.

When we reach the final age group there is only a solitary l_x value left, in this case 831. The L_x value can only be derived indirectly from the $\mathring{e}_x$ value of 3.51, the derivation of which was explained above. The

---

[1] Alternatively, if 9,668 survived the first year, they lived in total 9,668 years. 332 died. some at the beginning so that they may have lived only a few days, while others died at the end after 11 months, of life. On average .we may assume the 332 non-survivors lived 6 months apiece, or 166 years. Thus the original 10,000 babies lived in all 9,668 + 166 years, i.e., 9,834.

values of T_x in column (10) are obtained by aggregating all the L_x, $5L_x$ and $10L_x$ values. Without L_{85} the T_x column could not be obtained. Since $\mathring{e}_x$ is obtained by dividing T_x by l_x, it is possible to calculate the value of T_{85-} from the values of $\mathring{e}_{85-}$ and l_{85-} which are already known. Thus we get $3.51 \times 831 = 2,917$ to give the value of T_{85-} which corresponds in the final age group, as will be seen below, with the figure for L_{85}.

By adding up all the values in the L_x column we get a total of 664,658 years. This is the total duration of the lives of the entire 10,000 males from birth until death. Since we started with 10,000, the average length of life was $664,658 \div 10,000$, i.e. 66·46 years which is the first figure shown in the final column headed $\mathring{e}_x$. To obtain the expectation of those who reach their first birthday, we again add up all the years lived by the 9,668 males who started from age 1 and divide the total of 654,824 by 9,668 so that the average expectation of life for those surviving their first year is 67·6 years.[1] The student may now perhaps understand why it is possible for the expectation of life to increase after having lived one year, i.e. at birth it is 66·4 and at age 1 it is 67·6! The reason is that both values are averages and in calculating the second, the extreme low values (i.e. average life 6 months) of which there are a large number (332) are dropped. The student will recall the discussion of the arithmetic mean and how it was affected by extreme values, particularly where the corresponding frequencies were high.

It is interesting to compare some of the values for $\mathring{e}_x$ in Table 61 with those shown in Table 55 which is taken from the official life table No. 11. It will be seen that the divergence at any age is at most 0·1 of a year. Thus, it can be seen that even with simple approximate calculations, results similar to those arrived by using what are known as graduation formulae can be obtained.

Population Projections

The life table is a singularly useful device for summarising the mortality experience of an entire generation. It is possible to visualise a purely hypothetical population which is fed by a constant stream of births, e.g. 100,000 per annum, and subject to the mortality summarised by the life table. If year after year this population is fed by this stream of births, ultimately we would get a population in size and age distribution indentical with that shown in the L_x column. This is, to repeat, a purely hypothetical model of a population in which it is assumed that fertility remains unchanged, i.e. the same number of births each year, mortality remains unchanged, i.e. that

---

[1] In practice, the same result is achieved by taking away from the aggregate the total lives lived by the previous age-group, e.g., 664,658 less 9,834 = 654,824.

summarised in the life table, and lastly that there is no migration or emigration.

Such a hypothetical population is referred to as a *stationary* population. The estimated mid-year popualtion is given by the sum of L_x values, *i.e.* T_0, and the distribution by age of that population is given by the L_x column. For this population it is possible to work out a crude birth rate which is derived by expressing the annual stream of hypothetical births, *i.e.* 100,000, over the total number of people in the population, *i.e.* T_0. Since by definition the number of deaths is equal to the number of births in any year, the death rate is the same as the birth rate.

This model of a hypothetical population has limited utility, but it is useful in so far as by means thereof we can estimate the prospective numbers in the population at any age, given the mortality on which the life table is based. The L_x column, which shows the number of persons in each group, is used in making population *projections*. The demographer speaks of projections rather than forecasts, because what he is doing is to project a trend which is already established, *i.e.* the data upon which the life table is based. He can, of course, make different assumptions regarding both fertility and mortality and the result is determined by the assumptions made.

To estimate the number of persons at any particular age or in any age group who will be alive in say ten years time, we apply to the actual population at the lower age what is termed a survivorship factor. This is calculated by deriving the ratio of two L_x values. For example, using data in the abridged life table set out in Table 61, it will be seen that there are 1,429,000 males aged 10-14. On the assumption of the mortality on which that life table is based, how many of this number can be expected to be alive 10 years hence, when they will be 20-24 years old? The survivorship factor is derived by expressing L_{20-} as a ratio of L_{10-}, *i.e.* $\dfrac{L_{20}}{L_{10}} = \dfrac{47,460}{47,900} = 0.9908$. By multiplying the actual population aged 10-14, *i.e.* 1,429,000, by the survivorship factor, *i.e.* ·9908, it is estimated that there will be 1,416,000 males alive in 10 years time. It must be remembered that this is no more than an estimate based on specific assumptions, which may or may not be borne out in the event.

Conclusion

The calculation of a life table is practicable only where there is a census, since this normally provides reliable information regarding the age structure of the population. Abridged life tables are normally calculated for intercensal years. As already stated, in the last resort the reliability of the data it contains depends entirely on the accuracy of the q_x values, *i.e.* the probability of dying. These values are them-selves derived from the recorded deaths in the registered population. Although life tables are based upon the actual mortality experience of a given population at a given time, they do not reflect the mortality experience of any particular cohort of births. It is a purely hypo-thetical creation which illustrates what will happen to a population in terms of mortality and survivorship if it experiences the same mortality as that reflected in the q_x rate.

REFERENCES

Demography, P. R. Cox, Institute of Actuaries, 1959.

Introduction to Demography, M. R. Spiegelman, Society of Actuaries, U.S.A., 1955.
Both the foregoing texts contain a large amount of descriptive material, but the reader who is unfamiliar with the conventional mathematical notation will find some of the discussion of standardisation – which is designed for student actuaries – difficult to follow.

Techniques of Population Analysis, G. W. Barclay, John Wiley & Sons Inc., 1958. This is the student's standard text on life tables and related topics.

Principles of Medical Statistics, Sir Austin Bradford Hill (7th ed.). The chapters on 'Standardisation', 'Life Tables', and 'Morbidity Statistics' are especially useful.

For the more advanced student, *Elements of Vital Statistics*, B. Benjamin, Allen & Unwin, 1959, is the standard work on the current statistics produced by the Registrar General and other official sources in the U.K. The same author has contributed a useful chapter summarising the objects and contents of the Popu-lation Census in *Society Problems and Methods of Study*, published by Routledge and Kegan Paul, 1962.

An interesting application of life table principles is to be found in *The Length of Working Life of Males in Great Britain*, H.M.S.O., 1959. There is also a summary account of the method of constructing a life table which will supplement the above account.

An article by R. A. M. Case and Jean M. Davies in *The Statistician* 1964, No. 2, gives a detailed classification of the sources and records of official vital statistics.

Questions

VITAL STATISTICS

1. Explain the purpose of standardising birth and death rates. Outline the methods used by the Registrar-General in his annual review. *I.H.A. 1962*

2. Define and explain the various rates used by the General Register Office to measure fertility in England and Wales. *I.M.T.A. 1960*

3. Outline the basis of calculation of the occupational mortality statistics pub-lished by the Registrar-General. Discuss the value of these figures to the public health statistician. *Final D.M.A. 1961*

4. Describe methods of calculating gross and net reproduction rates. What are the relative merits of the net reproduction rate and the effective reproduction rate as employed by the Registrar General?

From the following data calculate the gross and net reproduction rates for 1841.

Age (x) ..	15	20	25	30	35	40
$5L_x$ ..	3,427	3,287	3,140	2,984	2,821	2,651

Annual female fertility rates (daughters per 1,000 women) 1,841

Age group	15–19	20–24	25–29	30–34	35–39	40–44
Fertility rate	12·65	73·28	108·60	107·98	92·99	51·97

What general conclusions can be drawn from a comparison of the reproduction rates for 1841 with those for 1951 *viz.*: G.R.R. $= 1·044$ and N.R.R. $= 0·996$?

B.Sc. (Soc.) 1958

5. Define and contrast some measures of fertility with which you are familiar.

Final D.M.A. 1962

6. Define the life table functions p_x, l_x, L_x, and ℓ_x. Describe *briefly* how they may be calculated given appropriate data from death registration and a population census.

B.Sc. (Soc.) 1954

7. The following l_x and L_x for females in Egypt are given:

Life Tables for Females, Egypt, 1936-38

X	Survivors to age X (l_x)	Stationary Population ($10L_x$)
0	100	745
10	63	623
20	61	598
30	58	561
40	54	515
50	49	458
60	42	375
70	32	—

Source: U.N. Demographic Year Book, 1957

(a) Given that the number of years lived by the life table generation after age 70 (*i.e.* T_{70}) is 275, compute the values of T_0 and T_{10} and the expectation of life at age 10.

(b) It is known that in many cases fairly good estimates of L_x may be obtained by averaging successive values of l_x multiplied by the appropriate number of years in the age group. In the above table, for example, an estimate for $10L_{40}$ can be calculated as $\frac{1}{2}$ (54+49) 10=515. Explain briefly why this does not apply to $10L_0$.

(c) Compute the crude birth rate and the general fertility rate of the stationary population (the general fertility rate may be calculated on the number of women in the age group 10–50).

(d) How many women aged 20–30 would survive under the mortality of the given life table, from one census to the following if the interval between two censuses is twenty years and if it is known that at the first census there were 50,000 women in that age group? *B.Sc. (Soc.) 1962*

PART IV

ECONOMIC & BUSINESS STATISTICS

CHAPTER XVIII

ECONOMIC STATISTICS

Manpower

Between mid-1951 and mid-1964 the total working population in Great Britain increased by about 7 per cent to nearly 25 million. Of this total two-thirds were men and one-third women. The importance for the economy of an adequate supply of labour hardly needs stressing, and in recent years there has been a tremendous improvement both in the quantity and quality of statistics relating to the working population. One of the results of these recent changes in labour statistics is that comparisons with earlier years, *i.e.* pre-war and post-war, are rendered difficult, and in some cases impossible.

The change in the nature of the statistics reflects the change in the attitude towards manpower problems that has taken place in the post-war era. Before the war the Ministry of Labour, which is the primary producer of labour statistics, published reliable data relating to the unemployed, but in respect of the working population estimates only could be made. Up to that time estimates of the working population could be derived partly from the Census of Production, from the Census of Population, and the Ministry of Labour annual estimates. These various figures were neither reliable nor comparable. The first steps to improvement were taken during the war as the result of the direction of labour policy and the necessity to conserve labour supplies. After the war, with the establishment of full employment, *i.e.* a very low level of unemployment, it became equally necessary to know more about the quantity and distribution of the labour force. Hence, during the war and more especially since 1945 there has been a succession of changes, amendments, revisions of classification, etc., in labour statistics, the most important of which arose from the introduction in July 1948 of the National Insurance Act. This provided that every gainfully occupied individual was legally bound to register. Unfortunately there are a number of omissions due to failure on the part of some members of the community to register, *e.g.* some married women and self-employed persons. In 1948 a Standard Industrial Classification was introduced whereby all industrial statistics were classified by industry on a common basis, and this classification was amended in 1958, with inevitable results for the comparability of certain series. The consequences of such a change in the basis of classification are well illustrated in the following table, in which the

TABLE 62
CHANGES IN COMPOSITION OF EMPLOYMENT (000's)
1951–61 (end June figures)

	1951 (a)	1959 (a)	1959 (b)	1961 (b)
Males				
Distribution	1,453	1,578	1,689	1,707
Manufacturing	5,819	6,271	5,738	6,064
All industries and services	14,859	15,308	15,308	15,682
Females:				
Distribution	1,147	1,422	1,520	1,605
Manufacturing	2,917	2,898	2,739	2,864
All industries and services	7,355	7,889	7,889	8,243

(a) S.I.C. 1948; (b) S.I.C. 1958.
Source: Treasury Bulletin for Industry, May 1962.

figures for 1959 are shown on the basis of the old and new Standard Industrial Classification.

The basic method of estimating the working population at the present time is to use the quarterly exchange of insurance cards, since each quarterly group constitutes a random selection of insured persons. The industrial classification and analysis which is carried out once yearly is based upon the June quarter figures, which are supplemented by the annual June returns from employers with five or more work people. This particular enquiry covers more than three-quarters of the employed population.

Monthly returns are also required of employers by the Ministry of Labour for certain industries, and these are used for the monthly estimates of the distribution of the employed population published in the *Ministry of Labour Gazette*. The June totals of the working population are published in the *Gazette* in considerable detail in the following February. These aggregates cover all industries in both Great Britain and the United Kingdom, the totals being analysed by sex and by age. In the case of the latter characteristic it is merely a break down as between those under and those over eighteen years. These estimates include not only any employees who may be absent owing to sickness or other cause, but also those who are unemployed. In the March issue of the *Gazette*, the same data are analysed by industries and classified according to the administrative regions of the Ministry of Labour. Here too changes have taken place, the most recent being a minor change of boundaries in April 1962 between Lincolnshire and the Midlands. The Ministry is continuing to prepare statistics on the basis of the old regions which are available upon request. This is the usual practice when such changes in the basis of classification are made, the two series being maintained for a number of years before the old is discontinued.

The importance of juvenile workers for the economy is in part reflected in a special classification published in the August issue of the *Gazette* each year, which shows the number of young persons under 18 years of age taking up employment for the first time during that year. The new entrants are classified by the industry, their age, and the type of employment taken up.

Apart from the analyses relating to the aggregate working population, the *Gazette* regularly publishes data illustrating particular aspects of the labour force. Among the more important are the following:

(1) labour turnover for employees of both sexes in each manufacturing industry in selected months;

(2) the number of operatives working short- or over-time in each industry, together with percentages and averages designed to indicate the extent of the abnormal working in various industries;

(3) the number and percentage of administrative, technical, and clerical workers in manufacturing industries is published each year, just as is

(4) the number of women in part-time employment in manufacturing industry; the information relating to this is in fact collected quarterly;

(5) the numbers employed by local authorities and in the police force analysed by age and sex is published annually; while

(6) the total number registered under the Disabled Persons (Employment) Act is analysed in part each month and in their entirety once yearly. Likewise the numbers admitted to industrial rehabilitation courses are published each month.

One of the most recent and interesting developments in the field of labour statistics has been the publication (*Gazette*, September 1962 and October 1963) of forecasts of the future working population up to 1980. Forecasts are made for each year up to 1973, for both the United Kingdom and Great Britain, covering male and female workers classified by age. The basis of these forecasts is the Government Actuary's forecasts of total population, and the statisticians then apply to each age and sex group the ratio between the working and total populations to estimate the working population in the years ahead. The relationship or ratio, between the working and total population in each age/sex group, is termed 'activity rate'. They are no more than estimates for among the factors influencing the activity rates are the rates of retirement, disablement, migration, school leaving, etc. Closely related to these data are the periodic studies of regional employment and unemployment, with special reference to the degree of labour migration between the regions.[1] An article in *Economic Trends* (September 1962) describes the basis of these forecasts of working population. According to the most recent assumptions made, the working population as a whole will grow by nearly one million, *i.e.* just under 4 per cent, between 1962 and 1972. There will be a rise in the proportion of the total working population between 15 and 25 years, and there will be a small but steady increase in those age 60 and over. Furthermore, it is expected that the over-all ratio of male to female workers will remain fairly constant, but the proportion of married women among the female labour force is expected to rise.

Apart from the statistical information relating to the working population, the *Ministry of Labour Gazette* publishes extensive information each month about the unemployed. Unemployment figures are based on the number of cards lodged mid-month at the employment exchanges by unemployed workers. The published classifications of 'registered' unemployed are extremely detailed, *e.g.* by age and sex, by industry, duration of unemployment, by region, by principal towns and development districts. In addition to these, each month there are analyses of the number of placings and unfilled vacancies given in respect of men, boys, women and girls by industry and by region.

It is evident from the foregoing account of labour statistics that this field is now very well documented. The primary source of information is the *Ministry of Labour Gazette*, and the student reader is urged to obtain a copy of the *Gazette* to see the type of information which is published regularly each month. In April 1962, the *Gazette* was supplemented by a new publication entitled *Statistics on Incomes, Prices, Employment and Production*. This has since been published three or four times a year and, while virtually all the information therein has appeared in the *Gazette*, the advantage of this publication is that it brings together in a single publication a large amount of comparative data covering fairly long periods of time which is not normally done in the *Gazette*.

REFERENCES

Guides to Official Statistics, No.1, Labour Statistics.
Economic Trends, September, 1962.

Production

The importance for the national economy of knowing the level of production and its composition needs no emphasis at the present time. Many industries publish data relating to their activities, *e.g.*, the

[1] See for example Treasury Bulletin for Industry, January, 1963.

post-war Ministry of Power publishes quarterly statements on the coal mining industry, as well as an annual digest of statistics on all forms of fuel production and consumption. *Lloyd's Register of Shipping* provides an annual return of all ships over 100 tons gross under construction in the United Kingdom. The iron and steel industry publishes a monthly bulletin of statistics containing figures relating to the level of employment, output of various products, prices, international trade, and foreign production. Invaluable as these published data undoubtedly are, they relate only to segments of the national economy. The only way to find out the total value of all production in the country is to carry out a census of production.

The Census of Production is the most valuable source of information available to the Central Statistical Office for computing the national product. It is particularly important in view of the information it gives relating to changes in stocks of both finished products and work in progress, as well as of materials and fuel. The information provided by the recent Census of Production relating to capital expenditure on plant, machinery and vehicles, as well as new building work, is a big step forward in filling a very serious gap in our national economic statistics.

For statistical purposes the term 'production' requires careful definition. Thus, from the economic point of view, any goods or services produced and exchanged for value constitute 'production'. The census, however, is restricted to the extractive, building and manufacturing industries in both private and public ownership. The first category includes mining and quarrying, but not agriculture, the last group includes firms which are engaged in repair work for the trade, *e.g.*, a ship-repairer. Despite the use of the term 'census', the enumeration of firms is far from complete. In Great Britain only those firms employing more than a given number of workers return the full schedule. For smaller firms, *i.e.* those with fewer than the given number of workers, a return giving the nature of the trade carried on and the number of employees only is required. In certain trades, however, in which small firms are believed to represent a large proportion of total output the full censuses require such firms to make a simplified return. This varied from trade to trade. Such establishments are known as 'small firms'; it should be noted that the data derived from the Northern Ireland censuses does not include any information relating to small firms.

Despite such limitations a census provides much information on other points. It reveals the division of the national industrial product between the various industries. The changes in these data over time bring out the trend and relative importance of the individual industries. Without such information, central economic planning in respect of the distribution of labour and new capital construction is virtually impossible. Estimates can also be made of labour productivity and the ratio of supervisory staff to operatives in the different industries although such figures are of limited accuracy and value. Without all these data, the index of industrial production would be unreliable and estimates of the national product subject to wide margins of error. Table 63 illustrates some of the data for one industry derived from successive enquiries.

TABLE 63

CENSUS OF PRODUCTION 1957: BREWING AND MALTING INDUSTRY

	Unit	1954	1955	1956	1957
Gross Output	£m	454·1	471·9	494·7	520·3
Net Output ..	,,	101·3	107·7	113·7	127·1
Average number employed:					
Total including working pro-	Thou-				
prietors ..	sands	68·3	69·7	67·6	68·9
Operatives..	,,	53·1	54·0	52·0	52·8
Other employees	,,	15·2	15·7	15·6	16·1
Wages and salaries:					
Operatives..	£m	21·2	22·9	24·1	25·9
Other employees	,,	10·3	10·9	11·4	12·5
Change during the year in:					
Stocks of materials and fuels..	,,	− 3·2	− 0·2	+ 0·5	+ 0·5
Work in progress and stocks of					
products	,,	− 0·3	− 0·5	+ 0·7	+ 1·1
Capital expenditure:					
Plant, machinery and vehicles					
acquired	,,	5·6	5·8	6·5	8·7
New building work ..	,,	2·3	3·0	3·2	3·7

Source: Board of Trade Journal, 21st November 1958.

The first census of production in the United Kingdom was taken in 1907, and was followed by others in 1912, 1924 and 1930. The last pre-war census, known as the fifth census, was held in 1935. Since the end of the war, following upon the Statistics of Trade Act, 1947, a partial census has been taken in respect of industry in 1946, while a full census was held in 1949 relating to industry in 1948. A census was taken for the years 1949 and 1950, but the information then required was rather less than was required in the full census covering the year 1948. The 1948 census was restricted to Great Britain, *i.e.* no census was taken in Northern Ireland, but with the passage there of an Act similar to the 1947 Act in this country, censuses were taken in Northern Ireland for the years 1949-51 inclusive and the results incorporated with those for Great Britain in the appropriate Board of Trade census reports for those years. It was intended to hold an annual census of production as from 1948 onwards. In the event, a complete census like that of 1948, in which a great deal of detailed information was required from firms, is to be carried out now only once every five years. The years to which full censuses refer are, since the War, 1948, 1951, 1954, 1958, and one was planned for 1963. In the intervening years the Board of Trade has conducted either a restricted or limited *census* which covers a wide range of industry with a few questions or a *sample survey*, which selects fewer industrial units, but asks the full census range of questions. Thus a limited census was taken for 1959 while sample surveys were taken for the years 1960–1962.

For purposes of the sample surveys, separate samples are drawn each year from base year size tabulations, the sampling unit being normally the larger firm or business unit. Normally one in every ten of the firms were included in the sample. The sample census held in 1960 had a fairly broad industrial stratification consisting of thirty-one headings. The sampling within each strata was random and all the units in the same stratum had an equal chance of being selected, except that, where possible, the smaller units selected in one year were excluded in the following year. The estimates of the larger establishments were obtained by multiplying the total figures for each item returned by the selected units in each sampled stratum of each industry, by a 'grossing-up' factor and then adding together the results for all the strata of each industry. The grossing-up factor was the denominator of the fraction used in selecting the sample. Thus, if one firm was selected in every ten, the grossing up factor was 10. Beginning with 1953 the figures obtained for each sampling stratum were adjusted by multiplying the results by a correction factor consisting of the ratio of total employment in the sampling field as recorded in the census for the base year, and the estimate of employment in the base year obtained by grossing up the base year figures for those units which were excluded from the sample. It is estimated that the sampling errors in total manufacturing industries in 1957 ranged from 0·1 % for employment to 1·0 % for capital expenditure.

Some changes were introduced in the full detailed census for 1958. Some of these were due to the revised Standard Industrial Classification which replaced in 1958 the original version published in 1948. Apart from the changes arising from the S.I.C., an important variation was to exempt firms employing less than 25 persons as opposed to the previous practice where the exemption limit was under 11 persons. There were also some changes in the instruction governing the making of returns for two or more establishments operated by the same firm, as well as in the question on sales where increased statistics were being collected at more frequent intervals.

The full census for 1963 included detailed questions about the different classes of goods bought and sold, and also questions about certain business expenses. Full particulars were required only from larger firms employing 25 or more persons. Information copies of the forms to be completed by larger firms were issued in 1962 to give firms the opportunity of adapting their records where necessary so that the figures required by the census authorities could be extracted more easily.

In the 1963 census information was obtained about certain business expenses, in particular, transport expenses. Hitherto, there has been little systematic information about the importance of transport costs to different industries, and this is a subject on which the government wants more facts. Moreover, these figures and those of other expenses will improve the accuracy of the indices of production and the national income statistics.

The scope of the census will be similar to that covered by the census for 1958. The census forms for 1963 included questions on: (1) working proprietors; (2) employment; (3) wages and salaries; (4) stocks; (5) capital expenditure; (6) work done by other firms, transport costs and certain other expenses; (7) materials, goods and fuel purchases; and (8) sales work done.

In the purchase section (7), firms were asked for details of the items purchased as well as their total costs, information which has not been obtained since the census for 1954; but the number of items for which particulars of the quantity purchased were required as well as the value was kept to a minimum for 1963. Questions were included about the number of persons engaged on transport work, and about the costs of operating road goods vehicles. Other business costs on which questions were asked include employers' National Insurance contributions and payments to superannuation and other pension funds; payments for rates, postage, telephone, etc.; and the cost of plant and machinery hired.

In this census small firms with fewer than 25 employees were required to state only the nature of their business and the number of persons employed, except in a few industries in which they contribute a substantial proportion of total output. In such industries a sample of firms were asked to complete a simplified version of the standard form.

The data collected in the census of production do not always relate to the calendar year, although we write about the '1948' census. To facilitate the completion of the schedules, the 'establishment' or firm may give figures relating to its *financial* year and not the calendar year. The effect of this concession is that '1957' for example, can mean any twelve-month period ending between 6th April 1957 and 5th April 1958. According to Mr H. Leak, a former director of the census, the mean year-end of the reporting firms is mid-December.

The Board of Trade publishes reports on each industry covered in the census. In addition, for each census at the commencement of the publication of these industry reports, a document entitled 'Introductory Notes' is produced. This contains a detailed account of the scope and scale of the census together with the definitions employed as well as an explanation of the tables published in the individual industry reports. Whenever data are to be extracted from the industry reports, although the latter contain notes relating to the tables, it is advisable to turn to the fuller 'Introductory Notes' to avoid errors in extraction.

Table 64 is taken from the summary industry report on those establishments wholly or mainly engaged in brewing and malting. The table gives an estimate for all firms in the industry. Satisfactory detailed returns were received from firms covering 98% of the industry. The balance was estimated to complete this table. The important definitions to note are as follows:

TABLE 64

CENSUS OF PRODUCTION: BREWING AND MALTING INDUSTRY
Estimates for all firms

Item		Unit	1954	1958
1	Number of enterprises	No.	—	311
2	Number of establishments ..	,,	—	674
3	Sales ⎰ goods produced and work done	£'000	415,675	453,317
4	⎱ merchanted goods and canteen takings ..	,,	41,328	158,132
5	Purchases of materials and fuels ..	,,	47,673	195,455
6	Products on hand ⎰ change during year	,,	—282	+725
7	for sale ⎱ at end of year ..	,,	19,541	23,619
8	Work in progress ⎰ change during year	,,	+38	—83
9	⎱ at end of year ..	,,	2,531	5,516
10	Stocks of materials ⎰ change during year	,,	—3,198	+99
11	and fuel ⎱ at end of year ..	,,	23,921	27,459
12	Payments for work done on materials given out ..	,,	427	473
13	Payments for transport	,,	5,475	6,972
14	Customs and ⎰on beer brewed (net)	,,	242,274	245,720
15	Excise duties ⎨on deliveries for home consumption ⎩of wines and spirits	,,	5,959	6,851
16	Net output	,,	101,754	156,718
17	⎰operatives	Thsds.	53·4	60·3
18	Average number ⎨other employees	,,	15·2	19·0
19	employed ⎬total including working pro- ⎩ prietors		68·7	79·3
20	Wages and salaries ⎰ of operatives	£'000	21,277	30,870
21	⎱ of other employees ..	,,	10,284	15,224
	Capital expenditure:			
22	New building work	,,	2,341	5,121
23	Plant and equipment ⎰ acquisitions ..	,,	4,593	7,894
24	⎱ disposals ..	,,	389	143
25	Vehicles ⎰ acquisitions	,,	1,193	2,073
26	⎱ disposals	,,	231	474

Source: Census of Production 1958. Part 18.

N.B. The 1954 figure for Gross Output in this table differs from that given in Table 63. This is due to the changed basis of the 1958 census which, to permit a comparison of the results with 1954, requires the adjustment of the earlier figure.

(1) Item 1. *Enterprise* consists of one or more firms under common ownership or control as defined in the Companies Act, 1948.

(2) Item 2. *Establishment* comprises the whole of the premises under the same ownership or management at a particular address.

(3) Item 16. *Net output* represents the value added to materials by the process of production. It is arrived at by adding the increase in stock of products (Item 6) and the increase in stock of materials and fuels (Item 10) to the value of sales (Items 3 and 4) and taking from the resulting total the purchases of materials and fuels (Item 5), the decrease in work in progress (Item 8), payments for work done on materials given out (Item 12) payments for transport (Item 13) and Custom and Excise duties (Items 14 and 15).

Table 65 is another standard table reproduced in all the individual industry reports. It provides an analysis by size, *i.e.* labour force, of the larger establishments. The student will note that the class interval in the first column headed 'average number employed' is not constant. It is clearly unnecessary that it should be so, since the information provided in the Table is sufficiently detailed for all practical purposes. The information relating to remuneration of operatives and other employees is a feature of the post-war censuses. Pre-war, the size of the firms' labour force alone had to be returned. The classification then employed consisted of two groups: operatives, covering all manual workers; and administrative, technical and clerical staff. Both of these two groups were further sub-divided as either 'over' or 'below' 18 years of age. It was only with the passage of the Statistics of Trade Act 1947 that firms were compelled to make a return of salaries and wages. This information was first available in the 1948 census.

From the information relating to wages and salaries of persons employed as well as the number of employees a figure described as 'net output per person employed' is derived and given in the final column.

This particular figure, *i.e.* net output per person employed, must be interpreted with great care. It is at best a poor indicator of the relative efficiency of labour in different industries. An obvious and important factor affecting output as between different industries is the degree of mechanisation in them. There is, too, the often overlooked fact that the final products, especially when the comparisons are between different countries, are seldom identical. Contrast, for example, the British and American 'family' car. The justification for this particular figure is that it serves primarily to indicate the changing productivity of labour within the industry. If we assume that the working week remains unchanged then any increased productivity per worker is presumably accounted for either by a greater degree of mechanisation or a more intensive and efficient utilisation of resources.

TABLE 65

CENSUS OF PRODUCTION 1958; BREWING AND MALTING INDUSTRY

Larger enterprises analysed by size of labour force employed

Average number employed by the enterprise in this industry	Enterprises	Establishments	Total Sales	Net Output	Employees		Wages and salaries		Capital expenditure	Net output per person employed
					Operatives	Others	Operatives	Others		
	Number	Number	£'000	£'000	Number	Number	£'000	£'000	£'000	£
25 – 49	36	45	8,755	2,078	1,055	313	488	244	293	1,515
50 – 99	58	81	28,250	6,695	3,168	927	1,511	709	542	1,633
100 – 199	51	94	52,505	14,733	5,639	1,837	2,689	1,520	1,028	1,971
200 – 299	35	65	67,225	17,453	6,292	2,059	3,071	1,510	1,229	2,090
300 – 399	8	13	18,035	5,261	1,947	742	946	543	297	1,957
400 – 499	7	19	23,159	5,465	2,151	854	1,063	554	462	1,819
500 – 749	12	30	65,551	16,831	5,666	1,618	2,783	1,231	2,028	2,310
750 – 999	8	27	47,178	12,860	5,147	1,522	2,464	1,205	968	1,928
1,000 – 1,499	10	93	122,348	27,507	9,287	3,216	5,018	2,716	3,239	2,200
1,500 – 2,999	5	41	94,254	21,236	9,244	2,491	4,874	2,067	1,904	1,810
3,000 – 4,999	3	35	71,325	23,302	9,437	2,998	5,331	2,615	2,767	1,874
TOTAL	233	543	598,586	153,421	59,033	18,577	30,240	14,913	14,758	1,977

Source: Census of Production 1958, Part 18.

The Census of Production, however, suffers from the inevitable disadvantage of any attempt to provide comprehensive and accurate data. It is comparatively expensive to collect, takes some little time to process and publish, and can only be undertaken at intervals of some years. To fill the need for up-to-date indications of trends between censuses the Board of Trade publish in their *Journal* series of sample statistics relating to various aspects of industry. These statistics include information on:

(i) capital expenditure by types of industry;

(ii) industrial building;

(iii) orders, deliveries, production and exports in mechanical and electrical engineering;

(iv) production of man-made fibres;

(v) changes in the volume of industry's stocks at 1958 prices, seasonally adjusted;

(vi) textiles including indices of production of made up clothing and orders and deliveries of textiles, and volume of production of cotton and wool;

(vii) deliveries of scientific and industrial instruments and apparatus;

(viii) sales and stocks of plastic material;

(ix) production and exports of cars and commercial vehicles; and

(x) registrations of business names.

This list is not exhaustive, but it gives an indication of the steps that have been taken to provide comprehensive and up-to-date information on various sectors of the national economy.

The statistics are usually published quarterly but there are some variations, *e.g.*, car production is reported monthly. In addition some annual reviews are carried out covering particular industries.

It is essential that the difference in nature between the census figures and these supplementary statistics is clearly understood. The former are complete enumerations; the latter are more in the nature of estimates and are more significant in the changes they show from period to period than in their absolute totals.

For the benefit of business men the Board of Trade also provide a statistical survey for individual industries on a subscription basis. Details of the scheme were published in the *Journal* of 9th March, 1962.

REFERENCES

Report of the Committee on the Censuses of Production and Distribution, Cmd. 9276, H.M.S.O. 1954.

Guides to Official Sources No.6, Census of Production Reports.

Distribution

Almost a century and a half elapsed after Napoleon described this country as a 'nation of shopkeepers' before official action was taken to ascertain the truth of this comment! In 1963 nearly 14 per cent of the working population was engaged in the distributive trades, a total of 3·3 million persons. A further 5·3 million workers were engaged in service trades and occupations, an increase of 1·2 million, or over 25 per cent, within a decade. With rising living standards and new technological processes in production, the proportion of the working population engaged in distribution and services will continue to increase. The term 'distribution' covers all the various channels through which goods pass from the manufacturer or grower (in the case of food) to the final consumer, *i.e.* all wholesalers and retailers. Included with these for the purpose of the Census are the service trades such as hairdressing, shoe repairing and garages. The first census of distribution ever was taken in Great Britain by the Board of Trade in 1951. Before this, information on the extent of the distributive trades, the number of shops, scale of their activities as reflected in the number of employees, their wage bill and annual turnover were unknown. Even the number of distribution outlets could only be estimated at something over three-quarters of a million as compared with an actual figure in 1951 of about 700,000.

A government committee set up in June 1945 recommended the taking of a census and the government acquired powers to conduct such an enquiry under the 1947 Act. The census was conducted by post during 1951 and the respondents were asked to provide information relating to their activities in 1950. The same concession regarding use of the firm's financial year in place of the calendar year was made as is given in the Census of Production. Traders were given three months to complete the forms but it may be noted in passing that not merely was the response very slow but that a number of prosecutions arose from failure to make the statutory return. Despite the great efforts made to ensure the co-operation of the traders, many were suspicious of the authorities' intentions and unwilling to co-operate. In consequence, the accuracy of some of the returns must inevitably be a matter for speculation.

Before the forms could be distributed it was necessary to carry out a census of the distributive and related service trade establishments in the country. This was done during May to October 1950; enumerators all over the country listed the names and addresses of traders apparently falling within the scope of the census – note – 'as far as could be judged from the outside of trading premises'. This is an interesting example of the difficulties encountered when a sampling frame is either non-existent as in this case and has to be built up, or is seriously defective. In this particular example the funds were made available for a complete enumeration before the census. The enumerators distinguished between shops, stalls, yards, depots and other types of premises. It should be noted that the enumeration staff had instructions not to enter the premises or question traders or their employees; the basis of their description as indicated above, was visual.

The purpose of the census – as distinct from the enumeration which preceded it – was to provide:

(1) information about the number and size of wholesale and retail outlets and other establishments providing consumer services.

(2) information regarding the value of the services rendered to enable more accurate estimates of the national product to be made.

(3) a measure of the relative efficiency of the distributive system as between different regions in the country, *e.g.* which areas have the most shops of certain kinds per head of the population and what is their turnover.

Apart from the above information which was primarily of interest to the government in respect of its economic policies, the census was to provide further information which would be of interest to traders and their trade associations. Quite apart from their natural concern with the distribution of various types of shops throughout the country, they would have an interest in the turnover, wage bills, level of stocks maintained and methods of delivery employed.

The data assembled as a result of the pre-census enumeration are reproduced in a publication entitled *Britain's Shops* and a detailed account is given of the enumeration, its difficulties and methods in the introduction to that report, which is better described by its sub-title, *A Statistical Summary of Shops and Service Establishments*. This particular report is quite distinct from the 1951 Census of Distribution reports themselves. There are, of course, differences in classification and of coverage. For example, the census proper obtained a 91 per cent response (of all the outlets enumerated in the above enumeration) which was estimated to cover 95 per cent of the total trade of the retail establishments enumerated.

The Verdon Smith Committee on the Censuses of Distribution and Production had recommended in 1954 (Cmd. 9276) among other things, that full censuses of distribution should be taken every 10 years and that sample surveys should be taken from time to time in the interval between censuses. The first of these sample surveys was taken for 1957 and the preliminary results published early in 1959.[1]

One of the most interesting features of this survey was the manner in which the sample was selected. The aim was to include all traders with an annual turnover of over £100,000. These included all multiple

[1] Special Supplement to *Board of Trade Journal*, 2nd January, 1959.

retailers, all except the smallest cooperative societies, department stores, etc., but while it is fairly easy to enumerate all these larger traders with reasonable accuracy, there was no way of obtaining an up-to-date list of all other smaller independent traders. This object could only be achieved as in 1950 by enumerating them in a special census for that purpose. The remainder of the sample covering these independents was therefore taken on a geographical basis as follows:

  (i) New Towns, Central London and a few special areas where great changes were thought to have occurred since 1950; all retail trades were enumerated and a sample of one in five taken.

 (ii) Greater London: a sample of electoral wards stratified by size was taken, distinguishing between shopping and mainly residential areas, and all shops in the selected wards were included in the survey.

(iii) Large towns (population 100,000 or more): a sample of streets stratified by size, *i.e.* number of shops in 1950, was taken.

(iv) Other towns were sampled by taking a cross-section of the local authority areas and sometimes stratifying the areas by size, by sales in 1950, and by population change since 1950.

 (v) Rural Districts were also sampled by region after stratifying by size, either by population density or by population change since 1950.

The census was carried out by post with a very energetic follow-up which gave an 89 per cent return from independent retailers, compared with a 96 per cent response from the larger traders, *i.e.* multiples, etc. So thorough was the follow-up that returns were obtained from 75 per cent of street traders, pedlars, hawkers and itinerant market traders! The 1957 totals were estimated by compiling 1950 as well as 1957 figures for the sample and calculating the ratio of 1957 to 1950 figures. These ratios were applied to the known 1950 totals to give the 1957 estimates. The information obtained in the census included the number of establishments, turnover and number of people engaged in the establishments. All this information was analysed both by the form of the organisation and the kind of business.

Since the 1957 survey results were based upon sample data, the published statistics are subject to sampling errors. These have been calculated for each figure under the general classification of 'independent traders', *i.e.* Turnover, Number of establishments and persons engaged, for each category of business or trade. Allowance for the effect of the error on the total results, *i.e.* independent plus large-scale retailers, is also made in the corresponding figures for 'all traders'. However, for all the industry an error of less than half of one per cent is given. The error is larger than this for individual items but considering that the sample was a 12 per cent one, *i.e.* covering about 57,000 establishments out of a total of 480,000, the accuracy seems reasonable and quite adequate for administrative and statistical needs.

The 1957 sample survey covered the retail trade only and, to bring the picture of the distributive field up-to-date, in 1959 an enquiry was carried out into the wholesale trades. The main purpose of this enquiry was to obtain substantially complete figures of stocks and capital expenditure in the wholesale trades as new starting points for subsequent annual and quarterly estimates, rather than to provide a detailed analysis of wholesale trading. To avoid the considerable cost of listing every wholesale business, the register for this enquiry was compiled from the register of companies, known undertakings employing more than five people and other sources such as marketing boards, cooperative societies, etc. It is estimated that companies amounting to only 5·4 per cent of the total activity were excluded. Estimates for non-response and unsatisfactory returns accounted for a further 1·7 per cent. The information collected was limited to figures of receipts and stocks and capital expenditure, together with the particulars necessary for classifying returns by kind of business.

A full Census of Distribution for 1961 has since been taken and the results published over a period of twelve months commencing February 1963. Preparatory work for the census started in August 1959 and an important practical preliminary was combined with the taking of the Census of Population in April 1961, when enumerators were asked to compile lists of shops, both operating and empty premises. To minimise the cost of the Census of Distribution only a 5 per cent sample of trades were asked the full range of questions; the other 95 per cent were asked only for figures of employment and turnover together with descriptive information. The census provides a fresh starting point for the monthly statistics of retail sales, estimates of hire purchase debt, stocks and capital expenditure. These statistics are of the greatest value both to the Government and to the business community.

It may be thought that a gap of nearly three years between the transactions recorded and the publication of results of a census is an unnecessary length of time, and it is worthwhile explaining this apparent delay. The first part of the explanation is that with an operation of this magnitude, half-million retailers are involved, it takes many months for the returns to come in. The returns in the census for 1961 were not all in twenty months after the end of 1961. In all, returns were received from 87 per cent of the establishments accounting for 93 per cent of the total turnover.

Secondly, it cannot be assumed that the information given on every return received is complete and correct. Where errors are found in returns, and the causes thereof are not obvious, it is necessary to write and ask for corrections to be made. This absorbs a good deal of time. A still greater volume of correspondence arises from letters written to the Board of Trade by traders, asking questions about the census forms and their completion.

Apart from dealing with correspondence, there are a number of processes to be carried out on all Census of Distribution returns before any results can be compiled. The first of these consists of a quick examination to see that the returns are complete, and made on the appropriate form; it is at this stage that queries are raised if, for example, turnover figures are missing. The next stage is to assign numerical codes for certain items of information as a preparation for mechanical analysis; in addition to the area codes incorporated in the reference numbers, codes are assigned, for example, for the year of return, legal status, membership of buying groups, and special methods of trading such as by self-service. Additional coding is required in respect of branch shops of multiple organisations, each of which must be given the appropriate area code among others. Classification to one of sixty-nine different kinds of business is carried out as a separate operation, and though the classification rules have been somewhat simplified, they are still quite complex.

Once the data are classified, the information is punched on to Hollerith cards, and are then ready for processing. Preliminary to the final processing, once a sufficiently large batch of returns is available in each kind of business, an analysis is made of certain key ratios, e.g. turnover per person employed. The object of this analysis is to determine the distribution of these ratios and to determine the limits outside which queries will be raised regarding the validity of individual returns. In other words, the statisticians provide themselves with basic standards against which they can judge the validity of their informants' data. Table 66 illustrates some of the information collected in the three enquiries into retail distribution since 1950.

The report on the 1961 Census of Distribution consists of fourteen parts, together with a supplement bringing together the material published in an important series of articles in the *Board of Trade Journal* during 1963 and 1964. Part I of the Census report comprises what are known as the establishment tables, which classify retail and wholesale trade outlets by reference to the type of trade carried on, the form of the undertaking, in addition to classifying them by turnover and the number of persons engaged therein. The remaining parts of the report, except for the last, give a geographical analysis of the Census data, each report covering one of the eleven Standard Regions in Great Britain, plus one containing summarised figures for all regions and

TABLE 66
RETAIL ESTABLISHMENTS
NUMBER AND TURNOVER 1950, 1957, AND 1961

Kind of business	All retail establishments					
	1950		1957		1961	
	Number	Turnover	Number	Turnover	Number	Turnover
		£'000		£'000		£'000
TOTAL RETAIL TRADE ..	583,132	5,000,130	577,405	7,587,154	580,151	8,948,720
GROCERS AND PROVISION DEALERS	143,692	1,222,717	150,552	2,031,251	150,098	2,360,952
OTHER FOOD RETAILERS ..	139,884	997,411	124,602	1,551,825	129,642	1,793,653
Dairymen	10,231	195,382	7,534	322,921	6,710	366,374
Butchers	41,799	287,693	41,698	549,893	44,425	634,219
Fishmongers, poulterers	9,511	61,244	8,108	73,142	7,892	80,919
Greengrocers, fruiterers (including those selling fish)	43,948	193,133	39,515	270,451	42,315	321,942
Bread and flour confectioners	24,181	178,476	17,644	215,889	17,645	234,932
Off-licences	8,197	75,403	8,796	112,408	9,034	141,134
Other food shops ..	2,017	6,080	1,307	7,122	1,621	14,134
CONFECTIONERS, TOBACCONISTS, NEWSAGENTS ..	74,606	502,661	77,437	702,996	70,802	801,130
CLOTHING AND FOOTWEAR	97,162	929,921	94,448	1,146,057	93,068	1,350,291
Boot and shoe shops ..	14,870	138,283	14,451	178,785	14,641	221,656
Men's and boys' wear shops	15,581	199,903	14,892	244,209	14,095	271,782
Women's, girls' and infants' wear, drapery and general clothing shops	66,711	591,736	65,105	723,063	64,332	856,853
HOUSEHOLD GOODS ..	65,795	536,892	65,323	829,721	73,689	1,039,962
Furniture and furnishings shops	18,953	268,888	19,486	328,523	21,308	421,796
Radio and/or electrical goods shops ..	11,929	94,681	13,850	250,577	19,017	337,620
Cycle and perambulator shops (including cycle and radio shops) ..	8,865	46,178	7,497	61,344	5,676	38,612
Ironmongers, hardware shops	26,048	127,145	24,490	189,277	27,688	241,933
OTHER NON-FOOD RETAILERS	60,352	374,413	61,360	563,050	59,125	680,762
Booksellers, stationers ..	10,388	67,359	6,818	81,577	6,303	84,503
Chemists, photographic dealers	18,205	167,037	18,129	269,690	18,481	349,894
Jewellery, leather and sports goods shops ..	18,896	87,914	20,380	137,532	19,441	166,314
Other non-food shops ..	12,863	52,103	16,033	74,251	14,900	80,050
GENERAL STORES ..	1,641	436,115	3,683	762,253	3,727	921,971
Department stores ..	529	308,339	718	453,912	776	539,569
Variety and other general stores	1,112	127,776	2,965	308,342	2,951	382,402

Source: Board of Trade Journal, 8th February, 1963.

towns. The final part of the report, known as the organisation tables, will provide information relating to gross margins, stocks, capital expenditure classified by type of organisation and the kind of business.

An interesting innovation in the 1961 Census is the publication of the proportion of trade of large towns which is carried on in the main shopping centre. The term 'large towns' is generally restricted to towns with 50,000 or more population, but one or two smaller towns which are important market centres are also included. The shopping areas were delimited by the Board of Trade Census office, which identifies the shopping centres on the spot, usually with the assistance of local officials. Such areas are characterised by the presence of department stores or other large shops which attract shoppers from a wide area. There is no precise method for defining such shopping areas, although in practice, states the report, there was generally little difficulty in deciding the limits of such areas. For the most part it was found that each town of any size had one major shopping centre, although in certain large towns, more than one such area was identified. The relevance of such data for what is nowadays termed urban renewal, *i.e.* the re-planning of city centres, is obvious.

With the increasing proportion of the labour force engaged in the distributive and service trades, the importance of the Census of Distribution can hardly be underestimated. The reader wishing to ascertain the nature of the information available should consult Part I of the 1961 Report, together with the report for any one of the Standard Regions. Much of the information to be given in the final part of the report has been published in the *Board of Trade Journal*, starting with the initial article which appeared in the issue of 8th February, 1963, and further articles which have appeared at intervals in the following twelve months. As with all statistical data collected in a census, the data are not always reliable. The references above to the efforts of the census organisers to achieve maximum reliability demonstrate this point. However, there is no reason for believing that the inaccuracies affect the relative order of magnitudes shown in the figures.

REFERENCES

Report of the 1961 Census of Distribution. Part 1. Establishment Tables.

Report of the 1961 Census of Distribution. Parts 2-12 covering the Standard Regions.

Board of Trade Journal, 8th February, 1963 giving first results of the 1961 Census.

National Income

The national income may be defined as the money value of the nation's output of all goods and services in a given period, usually a year. This aggregate is also referred to as the national output, since these incomes represent the cost of producing the output of goods and services. It follows, therefore, that there are two ways of measuring the national income; either all the incomes of the factors of production, or the values of each industry's output, may be aggregated. Before discussing the various problems that arise in measuring these aggregates, it is important to understand their purpose. It is clearly desirable to know what the nation's economy is producing in any year, as well as comparing one year with another to ascertain the rate of economic progress. All production is intended ultimately to satisfy consumer needs. The more that is produced, the more the community may consume, *i.e.* the higher its standard of living. Aggregate consumption is not the only measure of living standards. How hard and in what conditions does the population work in order to achieve a given output of goods? Consumption can be increased in the short run merely by living off capital. Is the community 'better off' if instead of 10,000 new houses it produces 50,000 new cars? In the sum of the national product these will represent about the same money value. It is also important to note the definition of national income used by the statisticians in different countries. For example, in an under-developed rural economy food grown and consumed on peasant holdings would represent a significant part of the national product. In the United Kingdom economy vegetables grown in gardens and allotments are not counted in National Income estimates. Another important omission is the money value of housewives' services in the home which are unpaid. If they were valued in money terms, the national income in money terms would rise by at least 20 per cent. Colin Clark has attempted a more detailed estimate of the value of housewives' services. A full account of this estimate is published in the Bulletin of the Oxford Institute of Statistics.[1] Alternatively, if the housewives went out to work and paid some of the present factory or office workers to do their housework, although the real national income would not have changed, it would have increased considerably in money terms. Much the same conceptual problem arises with work for which no payment is received. Thus, a man working an allotment and consuming the produce at home adds noting to the 'national income'. If, however, he and his neighbour sold each other their respective produce, the value of the produce added should, in theory, be added into the national income. Other problems arise in connection with the services of government, *e.g.* should the salaries of civil servants be included since they add nothing tangible to the national product? The same argument can perhaps be applied with more justice to the payments to members of the armed forces in peace-time.

These conceptual differences as to what should be included in the 'national income' make international comparisons extremely difficult, *e.g.* valuation of home-produced food in an agrarian economy. There

[1] Vol. 20. No. 2, May 1958.

have been several conferences under the auspices of U.N.O., with the purpose of standardising practice. Apart from these problems the differences in the reliability of various statistical data in different countries pose a problem which will probably not be solved for many years yet. Any published international comparisons should be scrutinised with these considerations in mind, particularly estimates made of the annual rate of economic 'growth'.

The national income estimates are so prepared and presented, that they offer a comprehensive picture of the operations of the economy and the inter-relationships between various sectors. To the extent that the statistical data assembled in the annual Blue Book on the National Income and Expenditure are complete and accurate, overall economic planning is greatly facilitated. National income estimates for this purpose were first prepared officially by the Treasury in 1941, and used by the Chancellor as a background to the Budget statement in that year. In each of the next ten years a White Paper on these estimates was published a short while before the Budget statement. Since 1952 a Blue Book, which appears in autumn, has appeared annually. The pre-Budget document is a short White Paper containing preliminary estimates of the main aggregates. Before 1941 there had been several private estimates of the National Income or Output. Various methods were used to arrive at these estimates. Unfortunately the accuracy of the figures was greatly impaired by the shortcomings of the data then available. Despite great improvements, even now the limitations of the data are such that in each year the successive Blue Books contain amended figures for previous years. The volume of data on which these estimates may be based has been considerably expanded and their quality improved, but even twenty-three years after the first White Paper appeared, several of the more important aggregates remain little more than approximations.

The published data are based upon material derived from three main sources, although these must be supplemented by information culled from a wide range of other sources. Even so the coverage is in many cases incomplete, while further difficulties arise from the fact that much of the published data used has in fact been compiled for purposes other than national income estimates.

The three main sources of data are the statistics assembled by the Inland Revenue, the censuses of production and distribution and lastly the accounts of the central government. The significance of the last mentioned source may be better appreciated when it is remembered that the government is responsible for the expenditure of over two-fifths of the national income. Of these data those derived from the Inland Revenue are the most complete and accurate; those compiled from the Census of Production the least reliable.

The national income can be visualised in three ways:
(i) as a sum of incomes derived from economic activity, *i.e.*, from employment and profits;
(ii) as a sum of expenditure, *i.e.* consumption and investment;
(iii) as a sum of the net products of the various industries of the nation.

These three views of the national income tend to explain the ways in which the statistics are presented and the estimates compiled. Some of the more frequently employed aggregates are given in Table 67 below, which illustrates the income approach in practice, *i.e.* those countries with a well-developed fiscal system. The various types of income are given in the upper part of the table and they are largely self-explanatory. The *residual error* is the balancing figure between the

TABLE 67

GROSS NATIONAL INCOME ANALYSED BY FACTORS

Factor incomes	All figures £'s millions				
	1938	1947	1952	1957	1962
Income from employment ..	3,022	6,217	9,107	12,926	17,074
Income from self-employment[1] ..	647	1,232	1,514	1,800	2,148
Gross trading profits of companies[1]..	690	1,694	2,180	3,120	3,531
Gross trading surpluses of public corporations[1]	10	36	277	323	740
Gross profits of other public enterprises[1]	64	119	40	127	74
Rent[1]	470	416	517	823	1,289
Total domestic income before providing for depreciation and stock appreciation	4,903	9,714	13,635	19,119	24,856
Less Stock appreciation ..	80	—450	50	—185	—140
Residual error	—	—	18	207	—136
Gross domestic product at factor cost	4,983	9,264	13,703	19,141	24,580
Net property income from abroad ..	192	150	258	239	325
Gross national product	5,175	9,414	13,961	19,380	24,905
Less Capital consumption ..	359	—	1,275	1,724	2,274
National income	4,816	—	12,686	17,656	22,631

[1] Before providing for depreciation and stock appreciation.
Source: National Income and Expenditure. Table 1.

two separate estimates of the gross national product, the one based on incomes and the other on expenditure. The sub-aggregate is described as the total domestic income before depreciation and stock appreciation. These items inflate all the above incomes except rent and income from employment. An adjustment is made to eliminate the element of stock appreciation which may be defined as the increase in money terms in the value of stock distinct from a change in its physical quantity. The figures given for this item are little better than guesses, 'hazardous approximations' is the official description. Nevertheless, as will be seen from Table 67 it is an extremely important item, more especially in periods of rapidly changing prices.

The figure described as the *gross national product* at factor cost should be distinguished from the total defined as the net *national income*. The difference between the net and gross figures is accounted for by the depreciation of capital equipment in the country. Unfortunately the data relating to depreciation or 'capital consumption' are unreliable and incomplete. Rather than guess at this figure it has been considered better to omit it from the published estimates in some years, *e.g.* in 1947.

However, commencing in 1956 estimates of capital expenditure have been published in the Blue Book. The methods adopted in making the present estimates of capital consumption were first applied to United Kingdom data by Redfern. A paper describing these methods 'Net Investment in Fixed Assets in the United Kingdom 1938-1953' was published in the *Journal of the Royal Statistical Society* in 1955. The methods currently in use are a development of Redfern's ideas but do not depart significantly from his original conceptions. A summary of current methods is given in the 1962 Blue Book.

Very briefly, there are three stages in arriving at a figure for capital consumption:
   (i) estimate gross fixed capital formation in each past year for each class of asset separately distinguished;
  (ii) a length of life of each class of asset is assumed and its capital value written down on a straight line basis (*i.e.* a constant amount is written off each year);
 (iii) indices of building prices for each class of asset are applied to the estimates of gross fixed capital formation (in (i)) thus converting them from a current to a constant (1958) price basis.

Data for these estimates are derived from the following principal sources:
(*a*) census of production for investment by private industry in plant, machinery and buildings.;
(*b*) statistics of road vehicle registration and of houses built for private owners.;
(*c*) statistics of gross capital formation, but these are limited to a few large industries.

It will be appreciated that estimates for long-lived assets, *e.g.* houses, may have to be made for many years back. Hence, these·are of very little validity.

Estimates of capital formation in stocks and work in progress were originally on an 'establishment' basis and on the 1948 Standard

Industrial Classification. They are now on a 'business unit' basis and the 1958 Standard Industrial Classification.[1] Details of this change were given in the *Board of Trade Journal* for 17th March and 21st April 1961, and comparison between figures on the two bases is not possible.

The second method of estimating the gross national income is given in Table 68. The correspondence between the value of the national product and national income was mentioned earlier. In this table the gross products of various industries and sectors of the economy are given for three years. Similar adjustments in respect of stock appreciation and the residual error are made in this table. Most of the data upon which these figures are based are derived from the censuses of distribution and production, but since these are held only at intervals, the reliability of the final results cannot be as great as could be wished. By carrying out sample surveys in the fields of both industry and distribution in the years between the full-scale censuses, the Blue Book estimates for these years are much improved.

TABLE 68

GROSS NATIONAL PRODUCT BY INDUSTRY*

£ million

	1952	1957	1962
Agriculture, forestry and fishing ..	770	862	958
Mining and quarrying	505	707	724
Manufacturing	4,738	6,869	8,651
Construction	752	1,130	1,631
Gas, electricity and water	307	473	728
Transport and communication	1,203	1,619	1,987
Distributive trades	1,737	2,424	3,032
Insurance, banking and finance (inc. real estate)	373	541	788
Public administration and defence ..	917	1,162	1,478
Public Health and educational services ..	462	700	1,071
Other services	1,517	2,073	2,927
Ownership of dwellings	354	559	881
Less Stock appreciation	50	— 185	— 140
Residual error	18	207	— 136
Gross domestic product at factor cost ..	13,703	19,141	24,580
Net property income from abroad ..	258	239	325
Gross national product	13,961	19,380	24,905

* The contribution of each industry to the gross national product includes provision for depreciation and stock appreciation.

Source: National Income and Expenditure 1963, Table 10.

[1] In short, a business unit may comprise several establishments; *e.g.* a limited liability company operates several factories, *e.g.* establishments.

Table 69 shows the gross national product by categories of expenditure. This particular method can be the least satisfactory of the three, particularly in a country where data relating to consumer outlays on various commodities and services are extremely unreliable and subject to a considerable margin of error. The composition of the aggregate in this table requires no explanation, except to point out that taxes on expenditure are the outlay taxes which inflate the prices of goods which are purchased by consumers and public authorities, hence the sub-total of £21,609m. defined as 'expenditure at market prices'. To change 'market prices' to 'factor prices' outlay taxes must be deducted and subsidies added back. In connection with these data, it should be noted that the Blue Books contain detailed analyses of consumer expenditure over a period of years. To bring out more clearly the shifts in consumer outlays between different categories of goods and services, the annual money outlays are corrected for price changes.

The figure of stock appreciation which appears in these tables came into prominence owing to the very sharp rise in commodity prices during 1951 as a result of the Korean War. Note that in Table 69 there is no such figure; only the *real* not money change in stocks is included. The profits arising from the revised valuation of stocks held by companies and trading concerns inflate the annual trading profits for the relevant years, and tax was assessed on these profits. It is a moot point whether such 'income' should be included, but in view of other arbitrary decisions that have to be made in computing the national income, its inclusion in most years makes no significant difference.

TABLE 69

GROSS NATIONAL PRODUCT BY CATEGORIES OF EXPENDITURE

(£ millions)

Industry	1938	1947	1952	1957	1962
Consumers expenditure	4,394	7,987	10,707	14,444	18,452
Public authorities current expenditure	772	1,745	2,894	3,585	4,856
Grossed fixed capital formations at home	656	1,199	2,106	3,390	4,608
Value of physical increase in stocks and work in progress	—	254	50	230	86
Total domestic expenditure at market prices	5,822	11,185	15,757	21,649	28,002
Exports and income received from abroad	976	2,074	4,610	5,884	6,732
Less Imports and incomes paid abroad	−1,038	−2,500	−4,533	−5,598	−6,550
Less Taxes on expenditure ..	− 622	−1,817	−2,293	−2,966	−3,902
Subsidies	37	472	420	411	623
Gross national product at factor cost	5,175	9,414	13,961	19,380	24,905

Source: National Income and Expenditure 1963, Table 1.

The advantages of computing the national income totals by various methods are obvious. The numerous cross-checks which are thereby made possible, especially in the various sub-totals, are invaluable. The modern method of constructing the National Income accounts, known as 'Social Accounting', is simply the adaptation of double-entry book-keeping principles.[1] Its value lies in the fact that every sub-total appears twice in the accounts and if it does not 'fit' in with the expected value as indicated by the size of the other totals in that particular section of the accounts, there is presumably some error. The difficulty is that a system of statistics of national income and expenditure must be comprehensive to be of use and an estimate must be included for each item that appears in a balancing account. It is not possible to base all the estimates on accurately recorded facts nor is it possible to calculate statistical 'margins of error' of the kind derived from random samples. What is done, however, is to form very rough judgements of the range of reasonable doubt attaching to the estimates. Some standardisation is obtained by grading each of the major components as having a margin of error of:

A ± less than 3 per cent,

B ± 3 per cent to 10 per cent,

or C ± more than 10 per cent.

As far as the various methods of deriving the various aggregates are concerned, the correspondence of the three aggregates of the national income, product and outlay, while not necessarily proving the accuracy of any one of them, would suggest that the errors are not such as to invalidate the overall results. It is certain, however, that the major aggregates are much more reliable than the numerous sub-totals in the analyses. There still exist several important gaps in the requisite information for any one of the three approaches. For example, wages in the income approach, distribution in the output data and items such as motoring and holidays in the analysis of current personal expenditure, must be interpreted with caution. The outstanding weaknesses in the over-all aggregates remain the deficiencies in the data from which estimates of the level of savings and net investment in this country can be made. Until more detailed information regarding not merely the volume of savings and investment, but in particular their distribution within the economy is known, economic planning must remain a highly speculative exercise.

Nevertheless, despite all the criticisms of the data as published, it is no exaggeration to state that the Blue Book on the National Income and Expenditure is the most important economic document of the

[1] For an account of the construction of such accounts see 'National Income and Social Accounting' by Edey and Peacock, Hutchinson.

year. The plans of the National Economic Development Council, like the Economic Survey, are based on that information and estimates. Without them fiscal and budgetary policy would be mere guess work. As the volume of statistical data assembled by the government increases, as it undoubtedly will, so the accuracy of these estimates will be improved.

The Central Statistical Office published in 1956 a very full account of the sources and construction of the statistics of National Income entitled 'National Income Statistics: Sources and Methods'. The first three chapters are probably the most useful for students as the remainder of the book goes into great detail. It is, however, an excellent source of reference. The notes which accompany the Blue Book on National Income consists:

(1) of definitions of items in the summary tables in the Blue Book;

(2) of revisions made in the previous year's estimates; and

(3) changes in treatment and definition made since the publication of the Central Statistical Office study mentioned above.

An especially important development in the field of short-term forecasting has been the development of quarterly estimates of the national income. These started in 1957 with the publication of quarterly estimates of consumer expenditure and since then similar estimates have been made of factor incomes, *i.e.* wages, salaries, profits and rent. These have been published since 1958 and in the following year quarterly estimates of all forms of final expenditure and the G.D.P. were started to be followed by seasonally adjusted series of both factor incomes and consumer expenditure. In 1963 fuller information on the invisible items on the balance of payments made it possible to give quarterly estimates of the gross national product (G.N.P.) as well as the gross domestic product (G.D.P.). All these quarterly series are published in the Monthly Digest of Statistics and with appropriate commentary in *Economic Trends*. Bearing in mind that the annual series are themselves liable to revision after publication, it must be recognised that some of these quarterly series are at best tentative estimates. However, given the need for economic planning, even such data are worthwhile.

REFERENCES

Annual White Papers and Blue Books on National Income and Expenditure.

'Use and Development of National Income and Expenditure Estimates' by R. Stone in *Lessons of the British War Economy*, Ed. D. N. Chester.

British Economic Statistics by C. F. Carter and A. D. Roy, Chapter IX.

'National Income Statistics: Are they accurate or useful?' C. T. Saunders in *The Incorporated Statistician*, March 1954.

'National Income and Related Statistics.' J. E. G. Utting. *J.R.S.S.* 1955, Pt. IV.

'Input – Output Tables for the U.K. 1954'. H.M.S.O. (These show the extent to which each industry was dependent on others for the sale of its output and for purchases of inputs in 1954). Summaries of these tables appear in the 1958 National Income Blue Book.

Quarterly Estimates of Personal Income and Expenditure – New Series. In *New Contributions to Economic Statistics. First Series.*

Overseas Trade

Statistics of overseas trade are among the oldest to be prepared in the United Kingdom. They date from the establishment in 1696 of the office of 'The Inspector-General of the Imports and Exports'. Their origin was to be found in the need to collect revenue and even today, despite the considerable changes and improvements, the classification still bears traces of this purpose. The statistics as compiled at the present time effectively start in 1871 when the statistics were based upon importer's and exporter's declarations of value (as well as of quantity) collected by Customs Officers at the ports and transmitted to the Customs Statistical Office for compilation. It is of some interest to note that since 1871 these data have been affected by two major changes only; the inclusion of exports of ships and boats as from 1899 which then represented about $3\frac{1}{2}$ per cent of the total value of exports and again in 1923 when the Irish Free State was created and trade with that country then became part of the United Kingdom external trade. The only other changes concerned classification, in particular, of countries due to changes in their frontiers.

The statistics of United Kingdom overseas trade are based upon the official certificates or declarations which must be made by both importers and exporters. These certificates give details of the nature of the merchandise together with figures of quantity and value. Imports are valued c.i.f. and exports and re-exports are valued f.o.b. The first abbreviation means 'carriage, insurance and freight'; the practice of including these items with the cost of the commodity imported follows logically from the definition of the value of imports required by the Customs; *i.e.* the 'open-market' value or price inclusive of all costs of importation, which the merchandise would fetch if sold on the open market at time of entry into this country. The valuation placed upon exported merchandise represents the cost of the goods packed and delivered to the ship, *i.e.* f.o.b. The c.i.f. basis of valuing imports is something of an anachronism; it started as a result of the 1932 Import Duties Act which created a general *ad valorem* tariff, *i.e.* a tax on imports based upon the value thereof and clearly some standard method of valuation was needed. Actually this method did not differ greatly from the mode of valuation employed up till 1932, which was the cost to the importer including freight, insurance, etc. Nowadays about 85 per cent of the imports are duty free and the need for precise valuation in accordance with the formula is weakened. In practice, it is the cost price, *i.e.* the price actually paid for the goods to the port of entry, which the importer records on his certificate. This may differ substantially from the 'open-value' price if for example, as occurred in 1949, a devaluation takes place and goods are imported at their cost before the devaluation occurred. For the dutiable goods, the authorities are satisfied that the 'total values reflect fairly accurately the actual c.i.f. cost of imports'.[1]

The statistics are first published in the monthly Trade and Navigation Accounts which appear about the twentieth day following the end of the month covered by the Accounts. The monthly Accounts cover imports, exports and re-exports; the latter comprising that merchandise which is exported in virtually the same form as it was imported. The classification is extremely detailed, with about 800 headings for imports and some 1,200 for exports. Details of the country of origin or destination for imports and exports respectively are given for the main headings. In addition to the monthly figures, the Accounts contain cumulative figures for the expired part of the year, *e.g.* the June accounts give the first half-year's aggregate, and these data for the current year are repeated for the preceding year for comparative purposes. The tables are preceded by a short introductory Note; any changes in classification are always announced in these Notes. These monthly tables are extremely detailed and not suitable for all requirements. Since March 1950 the Board of Trade has produced a condensed monthly Report on Overseas Trade based upon the monthly Accounts which re-classifies and condenses the original data into more useful form for general purposes, *e.g.* commodity analysis by different trading areas. See Table 71 below. These are supplemented at quarterly intervals by statements and tables on the nation's trade, *e.g.* value and volume in the main categories of merchandise, which appear in the *Board of Trade Journal*. These are accompanied by the indices of unit values, prices, etc., and volume for both imports and exports discussed below.

The companion volumes of the monthly Accounts are the four massive volumes which comprise the Annual Statement of Trade which appears usually more than a year after the end of the relevant year. The amount of information included in these volumes is such that 'the difficulty in use may be because of an excess of detail rather than a lack of it'.[2] Each volume contains comparable data for the preceding four years. The first volume provides an analysis of the value and volume of goods traded for each commodity or product. The second volume is concerned only with imports and re-exports,

[1] International Trade Statistics bv R.G.D. Allen and J. E. Eley. Section on U.K. statistics prepared by J. Stafford, J. M. Maton and M. Venning, p. 302.
[2] J. Stafford and others *op. cit.* p. 306.

classifying them by country of origin and for each commodity giving value and quantity imported. Volume 3 is identical with the second, except that it analyses exports by country of destination. The last of the four volumes summarises the details of goods traded and gives a detailed country analysis of United Kingdom trade; *e.g.* imports and exports by value and volume for Sweden, Switzerland, etc.

The classification of both imports and exports is regularly brought up to date, the list of headings being revised annually; a balance being kept as far as possible between the need for continuity and for an up-to-date system, and between the need for detail and what it is practical to require from traders. On the import side, foodstuffs and raw materials form the main bulk of the trade, but on the export side, a list of 2,500 headings of which over 2,000 relate to manufactured goods, offers scope for inaccurate classification by the trader.

The main commodity headings under which United Kingdom imports and exports are classified were revised as from 1954 to correspond with the Standard International Trade Classification. As from 1st January 1963 changes were made in the commodity classification used for the United Kingdom overseas trade statistics. These changes were made to correspond with the revisions made in the Standard International Trade Classification. The reasons and extent for these revisions were detailed in the *Board of Trade Journal* of 27th April 1962, but the main consequence of the introduction of these revisions were to increase the number of commodity headings by 30–35 per cent.

TABLE 70

VALUE OF U.K. IMPORTS AND EXPORTS 1957 AND 1ST QUARTER 1958

Class and Division	Exports		Imports	
	1957	1958 1st Qtr.	1957	1958 1st Qtr.
	£000	£000	£000	£000
A. Food, beverages and tobacco	206,196	44,354	1,496,441	353,235
B. Basic materials	122,986	28,292	1,169,361	241,403
C. Mineral fuels and lubricants	152,704	38,213	466,302	104,469
D. Manufactured goods ..	2,754,375	690,703	928,315	230,767
E. Postal packages	82,716	18,073	7,838	2,543
Live animals not normally used for food	6,003	1,380	7,331	1,260
Total U.K. exports and imports	3,324,981	821,015	4,075,588	933,677

Source: Report on Overseas Trade, Vol. IX, No.7.

In the case of both imports and exports the bulk of the trade is clearly concentrated into the first four classes, *i.e.* food, basic materials, fuels, and manufactures, Tables 70 and 71 provide a summary of the main groupings. For exports the fourth class is by far and away the largest, for imports the first and second predominate. The classes are further sub-divided: Class A into 10 divisions, B into 12 and C and E 2 each, while D contains 23 divisions. Thus division D 16 is 'electric machinery, apparatus and appliances. The description 'postal packages' which is one of the categories used for analysing both exports and imports is misleading. All the contents of parcels liable to duty are actually classified under the appropriate commodity heading and the figure entered as Parcel Post is simply an estimate based upon the product of the number of parcels containing non-dutiable goods and an average parcel value. The information upon which this estimate is based is derived from the Customs declaration which accompanies every parcel imported or sent overseas.

In passing, it should be noted that details of volume and quantity are available for about 98 per cent of all imports (by value) and 90 per cent of exports. This is important since such data enable index numbers to be calculated which permit the money aggregates over a period of years to be adjusted for changes in both price and quantity. The difficulties arising from changes in quality and type of product cannot be completely overcome by an index, hence it is sometimes difficult to be sure that one is comparing like with like. For example, the value of machinery which is adjusted by reference to its weight, will be affected by the growing use in recent years of lighter alloys for its construction.

So brief an account of the statistics of overseas trade can do little more than warn the reader who anticipates consulting any of the references given that the utmost care in extraction of figures is necessary. Comparability over a period of years is often more apparent than real and the notes to the various Tables and Accounts must be examined for changes, especially in classification. The difficulties of the student are intensified by the importance of the balance of payments problem; it is tempting to regard the statistics of overseas trade as a means of interpreting the balance of payments. This is far from being the case; the quarterly and annual figures on the balance of payments are very different from the publications discussed above. In fact, not even the expert can reconcile the documents since they are compiled on different bases. A major difficulty to which reference is sometimes made in the press discussions of monthly trade accounts, is the problem of adjusting the cost of imports from the c.i.f. valuation to that for the goods themselves, *i.e.* ex insurance and transport costs. These costs are estimated to represent between 10 and 13 per cent of the total c.i.f. value, but the percentage varies as between the various commodities and as between different countries. At best the correction can only be approximate.

TABLE 71
UNITED KINGDOM VISIBLE TRADE
(Monthly averages)
EXPORTS BY AREA

£ million

	1959	1960	1961	1962	1963	1962	1963				
	Year	Year	Year	Year	Year	4th qtr.	1st qtr.	2nd qtr.	3rd qtr.	4th qtr.
Sterling Area ..	112	119	117	112	121	114	119	119	123	122
Australia.. ..	19	22	17	19	20	20	21	20	19	19
New Zealand ..	8	10	10	9	10	8	9	10	10	9
India	14	13	13	10	11	11	11	11	11	11
Western Europe ..	76	86	99	113	127	116	121	126	130	129
Economic Community.. ..	39	43	51	60	69	62	65	69	71	70
Free Trade Assocn. (inc. Finland) ..	32	36	40	43	46	44	44	45	47	49
North America ..	48	45	42	43	43	41	39	41	45	47
U.S.A. and Dependencies ..	30	27	24	28	29	28	25	28	30	31
Canada	17	18	18	16	14	14	14	14	15	16
Latin America ..	13	14	14	13	12	10	11	12	12	12
Soviet Union and Eastern Europe ..	5	6	9	9	10	9	9	12	10	10
Rest of World ..	24	25	26	25	27	26	26	26	27	30
Total	278	296	307	316	340	317	327	335	346	348

EXPORTS BY COMMODITY

£ million

	1959	1960	1961	1962	1963	1962 4th qtr.	1963 1st qtr.	2nd qtr.	3rd qtr.	4th qtr.
Food, beverages and tobacco	16	17	18	18	21	19	18	21	24	22
Basic materials ..	11	11	11	12	13	12	13	14	14	14
Fuels	10	11	10	12	14	12	13	15	14	13
Manufactures ..	232	247	257	263	281	264	272	276	285	290
Machinery and transport equipment ..	120	128	135	138	151	140	150	146	151	156
Machinery ..	70	77	88	91	98	92	97	98	97	100
Road motor vehicles	34	37	32	38	41	37	39	40	41	45
Other transport equipment ..	16	13	15	10	11	10	14	8	13	11
Chemicals ..	25	27	27	29	31	29	29	31	31	32
Metals and misc. metal mfrs. ..	37	38	39	39	38	39	37	38	39	39
Textiles	20	21	21	20	21	20	20	21	22	22
Other mfrs. ..	30	33	35	37	40	36	37	40	41	41
Miscellaneous ..	8	11	10	10	11	11	11	10	11	11
Total	278	296	307	316	340	317	327	335	346	348

IMPORTS BY COMMODITY

£ million

	1959	1960	1961	1962	1963	1962 4th qtr.	1963 1st qtr.	2nd qtr.	3rd qtr.	4th qtr.
Food, beverages and tobacco	127	128	124	131	140	123	126	141	148	143
Fuels	39	40	40	44	47	46	48	49	46	46
Basic materials ..	79	90	84	77	83	79	77	76	85	91
Chemicals	12	15	14	14	17	15	15	17	18	19
Other semi-manufactures	44	61	56	56	59	57	55	58	60	64
Finished manufactures	31	43	46	50	54	52	52	54	55	57
Miscellaneous ..	1	1	1	1	2	2	1	2	2	2
Total	332	378	366	374	402	374	375	395	412	422

Source: Board of Trade Journal, 31st January, 1964,

The Balance of Payments

The Balance of Payments accounts provide a summary financial record of the overseas trading activities during the past year. Until 1939 it was based upon the data provided by the Customs Statistical Office, *i.e.* the Trade Accounts described above. With the introduction in 1939 of Exchange Control, a new basis for these statistics became available. An importer requiring foreign exchange to pay for goods had to make a detailed application to the authorities for the currency. Similarly, the exporter had to account for the proceeds in foreign currency of any exports. The authorities found themselves in possession of far more detailed and accurate data about the nation's overseas financial transactions than ever before. Since actual payment for goods usually takes place after receipt or despatch of the goods concerned, the Balance of Payments account before 1939 was in the nature of a revenue and expenditure account with debits accrued and credits outstanding. When the Exchange Control data became available the account became in effect a cash account, reflecting the timing of the payments rather than the actual movement of the goods giving rise to payments and receipts. When in October 1950 the new series of half-yearly White Papers on the Balance of Payments was introduced the basis was changed. The current accounts record transactions in goods when a change of ownership takes place. In the case of certain imports such a change of ownership takes place in the country of origin, with exports the change is assumed to be effected on or after the arrival of the goods in the port of destination. The balance of payments accounts will differ from the Trade accounts not only in respect of valuation of imports, which for the latter are c.i.f. and for the former f.o.b., but also in respect of timing. There are also differences in the goods covered by the two sets of accounts, *e.g.* precious stones and gold are excluded in the Trade Accounts but included in the White Paper. The latter document contains a table showing the adjustment between the two sets of accounts, but it is clearly of little help in trying to determine the trend of payments when only the monthly figures of visible trade are available.

A series of changes in the sources and methods of compiling the balance of payments figures have been made in recent years. A full description thereof was contained in the publication *United Kingdom Balance of Payments 1946-57*, which was published in March 1959. In August 1963 there appeared the first issue of a new publication, this time called the *U.K. Balance of Payments, 1963*, prepared by the Central Statistical Office. It is intended that this shall appear annually, to be supplemented by quarterly figures which will appear in the May, June, September, and December issues of the journal *Economic Trends*.

TABLE 72
GENERAL BALANCE OF PAYMENTS OF U.K.
£ million

	1952	1954	1957	1960	1962
Current account:					
VISIBLE TRADE					
Imports (f.o.b.)	3,048	2,989	3,538	4,106	4,059
Exports and re-exports (f.o.b.)	2,769	2,785	3,509	3,728	3,991
Total	− 279	− 204	− 29	− 378	− 68
INVISIBLES					
Government:					
Debits	219	231	253	336	404
Credits	165	105	106	49	37
Transport-Shipping:					
Debits	425	484	652	689	690
Credits	559	520	659	639	645
Transport-Civil Aviation:					
Debits	38	38	52	79	66
Credits	38	38	49	95	88
Travel:					
Debits	83	101	146	188	214
Credits	80	95	129	171	194
Other Services:					
Debits	190	189	221	264	275
Credits	304	349	407	465	490
Interest, profits and dividends:					
Debits	255	311	361	452	455
Credits	513	566	600	688	780
Private transfers:					
Debits	65	66	110	96	99
Credits	63	76	90	103	111
Total invisibles:					
Debits	1,275	1,420	1,795	2,104	2,203
Credits	1,722	1,749	2,040	2,210	2,345
Net	+ 447	+ 329	+ 245	+ 106	+ 142
CURRENT BALANCE	+ 168	+ 125	+ 216	− 272	+ 74
Long term capital account[1]					
Inter-Government loans (net):					
By United Kingdom government	+ 16	+ 34	+ 16	− 20	− 47
To United Kingdom government	− 16	− 54	+ 59	− 72	− 44
United Kingdom subscriptions to I.M.F., I.F.C., I.D.A., and European Fund	—	—	—	− 10	− 9
Other United Kingdom official long-term capital (net)	− 20	− 8	− 9		− 5
Private investment:					
Abroad	− 127	− 238	− 298	− 313	− 259
In the United Kingdom	+ 13	+ 75	+ 126	+ 228	+ 274
BALANCE OF LONG-TERM CAPITAL	− 134	− 191	− 106	− 187	− 90
Balance of current and long-term capital transactions	+ 34	− 66	+ 110	− 459	− 16
Balancing item	+ 61	+ 49	+ 97	+ 269	+ 115

TABLE 72 (continued)

	1952	1954	1957	1960	1962
Monetary Movements:[1]					
Miscellaneous capital –					
Change in acceptances outstanding	+ 42	— 33	— 21	+ 26	— 4
Other	+ 18	+ 43	+ 11	+ 119	+ 119
Change in overseas sterling holdings:					
Countries	— 358	+ 210	— 149	+ 376	— 45
Non-territorial organisations	+ 1	— 35	— 24	— 156	— 353
United Kingdom balance in E.P.V.	+ 53	— 78	+ 11		
United Kingdom official holdings of non-convertible currencies	— 26	— 3	— 22	+ 2	+ 1
Gold and convertible currency reserves	+ 175	— 87	— 13	— 177	+ 183
BALANCE OF MONETARY MOVEMENTS	— 95	+ 17	— 207	+ 190	— 99

[1] Assets: increase — / decrease +.   Liabilities: increase + / decrease —.
Source: United Kingdom Balance of Payments 1963 (H.M.S.O., August 1963).

These quarterly estimates will replace the series of half-yearly White Papers which appeared until 1963.

The main feature of the new publication, as with the earlier White Papers, is a series of accounts which summarise the transactions between the United Kingdom and other countries. The balance of payments account summarises the more detailed accounts of different aspects of the United Kingdom's international trade and financial activities. Among these individual accounts there is that dealing with visible trade, which covers the imports and exports of goods but excludes freight. The second major account is that which deals with the so-called invisible items. The invisible account incorporates all payments and receipts in respect of services as distinct from goods, such as shipping and air freights, insurance and tourism. They include all interest and profits received from United Kingdom investment overseas, as well as that paid to overseas owners of investments in the United Kingdom, together with expenditure on the armed forces overseas, and aid to underdeveloped nations in the form of grants (but not loans). The two accounts described above, *i.e.* visible and invisible transactions, when amalgamated produce what is known as the balance on current account.

This 'current' account is supplemented by tables which estimate the balance of long-term capital transactions. The long-term capital account includes payments and receipts of a capital nature, such as investment by United Kingdom firms overseas, or by overseas firms in United Kingdom industry; government loans and repayments, and borrowing by overseas governments on the London market. As will be seen from Table 72, which incorporates the figures for most of these items, the accounts are extremely complex. The major complicating factor in the compilation of these accounts is the inadequacy of some of the basic data. Thus, even in the current account, information relating to the invisible items is incomplete and in some cases inaccurate. It is the official view that these inadequacies lead to an understatement of the true position, *i.e.* in all probability the country has done better than the recorded figures suggest. The data in the capital account have been greatly improved in recent years, primarily as the result of the Board of Trade's special enquiries into capital movements, prompted by the recommendations of the Radcliffe Committee. Nevertheless, as will be seen from Table 72, there is a substantial 'balancing item' for each year. In fact, the balancing item is often larger than the surplus or deficit on the overall account. This balancing item reflects the net total of errors and omissions in the various figures which make up the balance of payments account. In the view of the official statisticians, the item is normally positive, *i.e.* indicating a net unrecorded inflow. In other words, the position of the United Kingdom overseas payments account is probably better in any year than the recorded figures would suggest.

The current and capital accounts are supplemented by a final account which is referred to as 'monetary movements'. Most transactions, whether current or capital, give rise to corresponding movements of funds affecting, in particular, the United Kingdom gold and foreign exchange reserves, or the overseas residents' holdings of sterling. The balance of monetary movements reflects the changes in these funds; a minus sign in front of the figure indicates a favourable change in the United Kingdom's position, and *vice versa* for a plus sign. The official view is that the balance of monetary movements has been a better measure of the United Kingdom's overseas trading and financial activities than the balance of the current and capital accounts. Unfortunately, in some years there have been substantial unrecorded short-term flows, so that if the balancing item has been exceptionally large and positive, the balance of monetary movements has given too favourable an impression. On the other hand, if the balancing item has been unusually small, and positive, or even negative, the balance of monetary movements gives too unfavourable an impression.

The foregoing cursory description of what are probably the most important economic data produced by the Central Statistical Office makes two points clear. First, these accounts are extremely complex and subject to a margin of error which may at times prove seriously misleading to anyone not familiar with the statistics. In particular, any assessment of the United Kingdom's overseas payments position should not be made without studying all aspects of these accounts. The second point to note is that these figures have been subject over the past two decades to extensive revisions and amendments as new data have become available. It is essential to consult the explanatory memoranda which accompany the published tables whenever these figures are quoted.

REFERENCES

Economic Trends, March 1961.
Balance of Payments White Paper, Cmnd. 1329.
U.K. Balance of Payments, 1963.

Hire Purchase Statistics

Since 1945 hire purchase has come to play an increasingly important part in the national economy. By 1963, credit outstanding amounted to over £1,000m. In consequence of this rapid expansion and its impact on the economy, it has become necessary to find some measure of the changes in both the volume and character of hire purchase debt. The *Board of Trade Journal* is publishing statistics monthly which cover the period since October 1955. These relate to the hire purchase trade of retailers in kinds of businesses where substantial sales of goods on hire purchase terms are made (*e.g.* furniture, radio and electrical shops) as well as to the hire purchase business of finance houses.

The information is collected on an extensive sample basis similar to that used for obtaining the figures of retail sales. In January 1958 about 665 independent retailers were making returns although it is hoped to raise the number to 800. They are divided into some 284 furniture and furnishing shops and 381 radio, electrical and hardware shops. The department store sample alone accounts for about a third of the total turnover of goods usually bought on hire purchase, while the multiple organisation return covers about half the turnover in furniture. The Cooperative Union returns details of over half the hire purchase business done by cooperative societies. Information is also collected from the area boards of the nationalised gas and electricity undertakings. Over 500 finance houses contribute to the Board of Trade's information on direct credit retailing.

Data are collected on a voluntary basis and are subject to possible errors of bias, *e.g.* the sample is not random; some shops with a lot of hire purchase business may not bother to make a return. Consequently too much significance must not be attached to any particular figure but generally the trends in the volume of business are fairly indicated.

The information is presented in four tables which show:
(1) estimated total outstanding hire purchase debt;
(2) hire purchase and other instalment credit extended and repaid;
(3) index numbers of value of goods sold on hire purchase by household goods shops;
(4) index of new hire purchase extended direct to hirers by finance houses;
Explanatory notes are appended to the published tables.

TABLE 73

INDEX NUMBERS OF VALUE OF GOODS SOLD ON HIRE PURCHASE

Average of 1957 = 100

	Year	Jan.	Oct.	Nov.	Dec. *
FURNITURE AND FURNISHING SHOPS:					
Total (multiple and independent retailers, cooperative societies)	1962	98	124	131	140
	1963	93			
of which Multiple retailers† ..	1962	103	131	139	151
	1963	101			
Independent retailers ..	1962	89	108	113	122
	1963	81			
HARDWARE, RADIO, ELECTRICAL GOODS, CYCLE AND PERAMBULATOR SHOPS:					
Total (multiple and independent retailers, cooperative societies, gas and electricity showrooms)	1962	89	120	127	136
	1963	97			
of which (multiple and independent) radio and electrical goods, cycle and perambulator shops ..	1962	68	82	86	106
	1963	71			
DEPARTMENT STORES					
Household goods departments ..	1962	142	153	160	155
	1963	138			
TOTAL HOUSEHOLD GOODS SHOPS ..	1962	97	124	131	139
	1963	98			

* Average of five weeks.
† Multiple retailers are defined as those having ten or more branches.
Source: Board of Trade Journal.

The two *indices* measure changes in value of new hire purchase business, the first at household goods shops broken down by type of shop, and the second by finance houses broken down by type of goods. They are now based upon July 1957 = 100, the choice of base year being determined by the fact that a Census of Distribution was taken for that year. Extracts from tables of both indices are given in Tables 73 and 74. For the first two main categories of shop in Table

73 there is given not only a total index, but separate indices for sectors of each main category of retail outlet, *e.g.* multiple and independent radio and electrical goods shops. The content of Table 74 requires no comment.

The original base of both indices was December 1955 = 100. The changes introduced when the base was revised to July 1957 = 100 made comparison between the new and old based figures difficult, although the changes were not large enough to alter the character of the statistics entirely. In consequence, general trends since 1955 can still be traced with some confidence.

TABLE 74

INDEX NUMBERS OF NEW HIRE PURCHASE EXTENDED DIRECT TO
HIRERS BY FINANCE HOUSES

Monthly aver. of 1957=100

	Year	Jan.	Oct.	Nov.	Dec.
Private cars – new	1962	124	158	148	102
	1963	159			
Private cars – used	1962	96	133	111	78
	1963	81			
Commercial motor vehicles – new and used	1962	155	143	136	94
	1963	122			
Motor cycles, side-cars, power assisted cycles – new and used	1962	43	66	51	33
	1963	30			
Caravans – new and used	1962	95	153	115	76
	1963	74			
Farm equipment and tractors	1962	93	153	126	75
	1963	90			
Industrial and building plant and equipment	1962	254	235	209	181
	1963	233			
Household goods*	1962	118	120	117	110
	1963	128			
All goods (including goods not shown above)	1962	121	140	126	94
	1963	115			

* Includes pedal cycles and some mopeds.
Source: Board of Trade Journal.

It is important to note the extent to which other forms of credit are not covered by these statistics. In fact, the published statistics do not cover instalment credit in durable goods, or credit trading in trading and weekly and monthly accounts owed to retailers. An estimate of the coverage of the published statistics based on information collected in the Census of Distribution is shown in Table 75 relating to the end of 1957 taken from *Economic Trends*, September 1961.

TABLE 75
AMOUNTS OUTSTANDING AT END OF 1957

	£ million
Covered by monthly statistics:	
Instalment credit business of finance houses ..	253
Instalment credit business of household goods shops	195
Not covered by monthly statistics:	
Loans by finance houses	5
Credit traders calling on customers	42
Mail order houses	26
Check traders	17
Other retailers other than dealers in household goods	46
Other weekly and monthly accounts owed to retailers	156

Following the recommendations of the Radcliffe Committee on the Working of the Monetary System attempts are being made to improve the range of hire purchase statistics. As part of these efforts a quarterly survey is now being made by the Board of Trade of selected assets and liabilities of hire purchase finance companies. The results of the first survey were published in the *Board of Trade Journal* of 19th October 1962. Returns were received from 570 finance houses out of the original 800 asked in the first survey. The results from those replies were inflated to represent the assets and liabilities of all

TABLE 76
SELECTED ASSETS AND LIABILITIES OF FINANCE COMPANIES
£ million

	Dec. 31 1961	Mar. 31 1962	June 30 1962
Selected Assets			
Hire purchase, credit sale, etc.	674	652	652
Assets with U.K. financial institutions, not banks	10	5	7
Trade investments and investments in unconsolidated subsidiaries	36	37	37
Other Securities	5	4	8
Other advances and loans	108	113	114
Total selected assets	833	811	818
Selected liabilities			
Deposits	337	351	377
Bills discounted with banks and Discount Houses	80	72	61
Other borrowing	179	165	152
Issued capital and reserves	148	148	148
Total selected liabilities	744	736	737

N.B. Owing to rounding totals are not always the exact sum of components.

finance companies, statistics of whose hire purchase transactions are already published monthly. In fact only 440 of the 570 replies were used. The remaining 130 replies were not used for a variety of reasons, *e.g.* the company was a subsidiary of a manufacturing company and could not separate its assets from those of the parent company. The type of information published from this survey is shown in Table 76.

The rapid growth of hire purchase transactions and the recommendations of the Radcliffe Committee have given the collection of statistical information on this type of trade a tremendous fillip. A detailed account of developments up to mid-1961 in this field is given in an article in *Economic Trends*, September 1961.

Statistics of Profits

Quite apart from the importance of measuring the level of profits in industry at the present time, when there is so much discussion of an incomes policy, such statistics are of especial importance to the business man and the economist. Fluctuations in the level of profits provide a useful indicator of changes in the level of production and productivity. Although gross trading profits of companies are only about one-sixth of total domestic income, compared with two-thirds taken by income from employment, they are much the more sensitive indicator of the level of economic activity. There are various definitions and measures of profit which can be used for different purposes.

The simplest series is one based on the figures of gross and net profits of all trading bodies, incorporated and otherwise, in the U.K. The nearest approximation to these figures is provided by the quarterly series prepared by *The Economist* and the *Financial Times*. These are formed by aggregating the accounts of all companies which report, *i.e.* publish their balance sheets, in successive quarters of the year. Since such series relate only to public companies, they are inadequate as a measure of the actual share of profits in the gross domestic product. For this purpose, the Inland Revenue compiles data based upon the Schedule D returns of companies and individuals, which give a more reliable guide to the total of profits in the economy.

Another way of studying profits is to examine what are termed 'profit margins', *i.e.* the rate of profit on turnover. Successive issues of the recent annual reports of H.M. Commissioners of Inland Revenue classify such data by reference to industrial groups. Lastly, it is possible to relate profits to the capital employed. Such data have been prepared by the Board of Trade and published in *Economic Trends* (December 1961). In this analysis net income is expressed as a percentage of net assets and gross income of gross assets. In addition, dividend and interest payments are also expressed as a rate percent of net assets.

Increasing interest by the government in the trend of profits, as reflected by a speech from Mr. Macmillan in August 1956, when he was Chancellor of the Exchequer, has stimulated a considerable volume of statistical activity. After the publication of *Company Income and Finance 1949-53*, a study sponsored by the National Institute of Economic and Social Research, the Board of Trade undertook the continuation of these analyses of the accounts of public companies. These regular analyses by the Board of Trade's Statistics Division are based upon the accounts of some 3,000 quoted companies, *i.e.* those companies the shares of which are quoted on the Stock Exchange, with assets of £0·5 million or more, or with an income of £50,000 or more. The smaller companies, which are excluded from these quarterly analyses, actually account for only about one-tenth of total corporate income and assets. It should, perhaps, be noted that not all the quarterly 'samples' are representative in the sense that they are random samples. Their inclusion is determined by the date of publication and within each quarter's sample the accounts are classified by reference to the accounting year-end. In some quarters results of the numerous smaller companies may be swamped by the results of a handful of industrial 'giants'. Each year's analysis covers the accounting years finishing between the 6th April of the year shown and the 5th April of the following year. Although the accounting years of companies end on different dates throughout the year, about 40 per cent of quoted companies have accounting periods ending in the 4th quarter of the calendar year and about 30 per cent end in the first quarter.

At quarterly intervals, the *Board of Trade Journal* publishes analyses of the accounts which have appeared in the preceding quarter, showing the appropriation of income, balance sheet summaries, and the sources and uses of capital funds during the accounting year.[1] The considerable volume of detail derived from these analyses is reclassified annually by industrial groups, and published in *Statistics of Incomes, Prices, Employment and Production*. The relevant annual figures relating both to income and assets are given back to and including 1950. The same material, but in somewhat different form, is published annually in *Economic Trends*, in which the accounts of about 3,000 companies engaged mainly in the U.K. in manufacturing, distribution, construction, transport, and certain

---

[1] Note that these analyses include only the large companies while those prepared by *The Economist* and the *Financial Times* include all accounts received. The various series should, however, reflect a common trend.

other services are analysed and published with comparable figures for the previous year. Where possible, the accounts used in these analyses are the consolidated accounts of groups of companies. Thus, in the case of companies with overseas interests, the profits do not reflect only domestic economic activity.

Because these quarterly and annual analyses of profits and assets are based on aggregation of individual financial accounts, it does not follow that the aggregation of accounts drawn up on widely divergent basis produce perfect statistics. There is no standard form of presenting accounts of public companies, apart from the basic legal requirements, hence the figure of profits shown in one company's accounts may not be comparable with that in another company's accounts. It is impossible to judge from published accounts the extent to which hidden reserves have been created, for example, by excessive writing down of certain fixed assets. This has the effect of undervaluing the assets in the balance sheets, so that when the net assets figures are aggregated, they are too low and in future years the rate of earnings on those assets will tend to be unduly inflated. In other words, the same warning applies to a profits series as to most other time series of economic data. It is the major fluctuations and the general trend of the series to which one should attach importance. Minor movements from quarter to quarter or year to year should be taken with the proverbial pinch of salt.

In brief, therefore, it is evident that recent years have witnessed a very marked improvement in both the quantity and quality of published data on profits, mainly in respect of quoted companies. As in many other fields of published statistics, considerable care is required in extracting and using published data. Attention must be paid to the definitions used, for example, of gross or net income and net assets. In particular, the reader should be careful not to confuse the national income estimates with the analysis by the Board of Trade, the latter are usually referred to as *Income and finance of quoted companies, New series*, as distinct from the national income figures defined as *gross trading profits*.

REFERENCES

Company Income and Finance 1949/1953, B. Tew and R. F. Henderson, National Institute of Economic and Social Research.

Economic Trends, February 1958, February 1959, December 1959, December 1960.

Relevant Tables in *Statistics of Incomes etc.* (Ministry of Labour).

Annual Reports of H.M. Commissioners of Inland Revenue.

Board of Trade Journal Quarterly analyses.

Statistics of Transport

Since 1957 the Ministry of Transport has carried out a series of surveys of inland goods transport and passenger transport. The purpose of the enquiries is partly to fill in the gaps that existed in the available transport statistics before that date and also to meet the growing demands of the government for more information in order to devise what has been defined as a coordinated transport policy.

Goods transport

Prior to 1957, when the above surveys were started, there was no comprehensive series of statistics measuring the volume of goods carried by inland transport. The only available data related to rail transport. Virtually the only information collected in respect of goods transport by road consisted of fairly detailed analyses of the number of vehicles on the road classified by reference to their unladen weight. There was no information however about the kind of work which the vehicles were performing.

A sample survey carried out by the Ministry in 1958 required operators of selected vehicles to give particulars of the work performed during one week during 1952 and in 1958. These two years were chosen as being as free as any year could be from special influences. The sample of vehicles selected was stratified according to the unladen weight and the type of carrier's licence. To encourage a good response rate, it was decided to limit the enquiry to the one week of the year and also to ask only for information that the operators could reasonably be expected to supply quite easily from their own records.[1] Such enquiries, however, are relatively expensive, and hence they cannot be undertaken very often. Nevertheless, a further sample survey similar to that of 1958 was carried out in 1962, but the full results thereof were not available by the end of 1963.

In the intervals between such surveys, a system of traffic counts is employed. Traffic counting was inaugurated in January 1958, on both a manual and automatic basis. In other words, in some cases an automatic counting device which recorded the passage of a vehicle was used, and elsewhere enumerators kept records of the vehicles which passed them. The information derived from these counts is used to measure the developments since the 1958 survey. One of the limitations of this use of traffic counts as an estimate of the volume of goods transported by road, is that it necessarily assumes that ton mileages change between one point and another in the same proportion as does the vehicle mileage.[2] However, by the use of such

---

[1] The results of this sample survey were published in *The Transport of Goods by Road*, H.M.S.O., July, 1959.

[2] This point and similar technical matters are discussed in 'Statistics of the Transport of Goods by Road', *J.R,S.S.*, Pt. I, 1960.

traffic counting methods, it is now possible for the Ministry of Transport to prepare monthly estimates of the volume of inland road transport. Furthermore, a monthly index of inland goods transport is prepared for both road and rail, the rail figures being prepared with the aid of information produced by British Railways. The index is published monthly in the Monthly Digest of Statistics in the form shown in Table 77 on p. 432.

Passenger transport

In 1962 the Ministry of Transport produced a comprehensive series of statistics on passenger transport and travel over the last ten years.[1] The publication contained detailed information for each mode of transport, rail, public service road vehicles, private road vehicles, and domestic air transport. In these statistics, passenger movement is measured in terms of the number of journeys undertaken and the passenger miles performed by each mode of transport.

Public service transport

The collection of information on the number of passengers carried by public service transport is relatively simple, since an estimate of passenger movement is given by the sales of tickets. Special provision is made for assessing the number of journeys made under weekly, monthly, etc., season tickets. By classifying the statistics of passenger journeys for the country as a whole into two groups, the one to cover the main urban areas and the other covering the rest of the country, an estimate of total passenger mileage in Britain can be derived. In the case of the urban areas, the total number of passengers carried is multiplied by the estimated average length of journey as shown by the London Transport statistics, and in the case of the provincial journeys the total of passengers is multiplied by an estimate of average mileage for each journey.

Private road transport

With the growth of a car-owning democracy, any estimate of road traffic would be incomplete without data relating to the private motorist. The existing information is based on information derived from the periodic traffic counts at selected points, and on an assumed average occupancy per car or motor cycle. The 1961 survey of motoring carried out by the Ministry of Transport revealed that the average car occupancy over the selected week in October that year was 1·93 persons, although the day by day occupancy varied markedly between 1·54 in respect of journeys to work, and 2·63 on Sundays. The traffic counts, both automatic and manual, at selected points on the road system, enable estimates of vehicle mileages by the day of the

[1] Passenger Transport in G.B., 1962.

TABLE 77

INDEX NUMBERS OF ROAD TRAFFIC AND INLAND GOODS TRANSPORT

Monthly average 1958 = 100

	All road traffic	Pedal cycles	Index of vehicle miles travelled on roads in Great Britain								Index of ton-miles of inland goods transport[1]		
			Motor traffic										
			All motor traffic	Cars	Mopeds	Motor scooters	Other motor cycles	Buses and coaches	Light vans[2]	Other goods vehicles	Total	Rail	Road
1960 :: :: ::	115	85	120	122	147	183	103	100	125	113	110	103	115
1961 :: :: ::	123	76	130	136	140	194	95	102	135	117	110	96	120
1962 :: :: ::	126	65	135	146	135	189	81	101	136	120	108	89	123
1962 October	129	64	139	149	133	200	81	101	138	132	113	92	129
November	111	52	120	126	114	157	61	86	125	120	110	94	122
December	104	42	113	124	65	99	48	91	114	104	99	87	108
1963 October	139	59	151	166	137	177	74	104	148	140	120	98	137
November	123	45	134	147	127	140	58	97	135	129	121	102	134
December													

Source: Ministry of Transport and Road Research Laboratory.

[1] The railway figures up to December 1962 are derived from statistics compiled by the British Transport Commission and from January 1963 from statistics compiled by the British Railways Board. The figures for road transport are estimated from a sample enquiry and traffic counts. This index is adjusted so as to give the average four-weekly rate in each month.

[2] Not over 30 cwt. unladen weight.

week, month and year from 1958 onwards to be made. By combining the occupancy rates of such vehicles with the vehicle mileages an estimate can be made of the total passenger mileage by this mode of transport. The following table, taken from *Economic Trends*, shows the type of information available on the basis of such statistics and the relative importance of each main type of transport for passenger movement over the decade 1952-62.

TABLE 78

ESTIMATED TOTAL PASSENGER TRANSPORT IN GREAT BRITAIN 1952-1962
Thousand million passenger miles

	Air (including N. Ireland and Channel Islands)	Rail	Road		Total
			Public service vehicles	Private transport[1]	
1952 ..	0·1	24·1	50·1	37·9	112·2
1953 ..	0·2	24·1	50·7	42·1	117·1
1954 ..	0·2	24·2	50·0	47·2	121·6
1955 ..	0·2	23·8	49·8	54·3	128·1
1956 ..	0·3	24·5	48·6	59·5	132·9
1957 ..	0·3	25·9	45·9	59·9	132·0
1958 ..	0·3	25·5	43·4	72·9	142·1
1959 ..	0·4	25·5	44·1	82·1	152·1
1960 ..	0·5	24·8	43·9	88·9	158·1
1961 ..	0·6	24·1	43·1	97·7	165·5
1962 ..	0·7	22·8	42·4	103·7	169·6

Source: Economic Trends, November 1963.
[1] Including taxis and private hire cars.

Index of Industrial Production

The purpose of the Index of Industrial Production is to provide a general measure of monthly changes in the level of industrial production in the United Kingdom. The index is prepared by the Central Statistical Office in collaboration with the various statistical divisions of certain Ministries, in particular the Board of Trade and the Ministry of Public Buildings and Works. The index is published monthly in the *Monthly Digest of Statistics* and the *Board of Trade Journal*. An official account of the construction of the index is published by H.M.S.O. in Studies in Official Statistics No.7: The Index of Industrial Production.

The index of industrial production covers mining and quarrying, manufacturing, building and the public utilities, gas, electricity and water, but excludes trade, agriculture, transport and all other public and private transport. The precise coverage of the index is brought

15

out in the following Table 79, while the composition of these main groups in the index is indicated in more detail in Table 80.

TABLE 79

PERCENTAGE CONTRIBUTIONS TO GROSS DOMESTIC PRODUCT, 1961

Industries included in the index of industrial production:

	%
Manufacturing	34·7
Mining and quarrying	2·9
Construction	6·4
Gas, electricity and water	2·8
Total	46·8
Industries excluded from the index:	
Agriculture, forestry and fishing	4·0
Distributive trades	12·7
Transport and communication	8·2
Other services	28·3
Total G.D.P. at factor cost	100·0

While the index is designed to reflect changes in the level of industrial production from month to month, the individual series or indicators used in the construction of the index are often weekly or quarterly, and sometimes annual, rates of production. In fact, about half the series upon which the various indicators are based are for calendar months; the remainder are for weekly or quarterly periods. Since it is the purpose of the index to compare the level of production in different months, corrections have to be made for the fact that calendar months do not all contain the same number of working days. Furthermore, some contain four and others five Saturdays, a day upon which production is likely to be considerably lower than on other days of the week. Such vagaries of the calendar have as far as possible to be eliminated.

The index has been calculated with the average monthly production for 1958 as the base period, for each month from January 1958. This base will continue to be used until the results of the full Census of Production held in 1963 are available. Previous bases of this index have been 1954 and 1948 and on the occasion of each change of base the Central Statistical Office has calculated a number of back years on the new base to provide a sufficient overlap on the old and new bases to effect some continuity. This is reasonably effective for the 'all manufacturing industries' index, but much care is needed in attempting long term comparisons of movements in the indices of individual industries.

The index is a weighted arithmetic mean combining individual production series weighted in proportion to the net output of each industry covered by the index. The latter figures are derived from the 1958 Census of Production and are adjusted by deducting the estimated amounts paid for services rendered to the industries by firms outside the field covered by the index, *e.g.* insurance and advertising. About 880 production series are incorporated in the index, most of which represent physical quantities produced. Where output figures are not available alternatives such as the consumption of raw materials, or the numbers of persons employed, have been used. For some industries, however, value series are used, adjusted to eliminate changes in price by using the index of Wholesale Prices. If the individual product of an industry takes a long time to produce, as in shipbuilding and construction of buildings, the amount of work in progress is taken into account. Large engineering contracts are treated in the same way.

Not all the series on which the indicators are based are available when the index for each month is first prepared, since a large number of sources are used and some of them are available only quarterly. Thus the advantage of an up-to-date index of production has to be weighed against the dangers of early estimates based on insufficient data, which may prove wrong. At present the index is published six to seven weeks after the end of the month to which it relates. By then about 40 per cent of the data used are in final form and there are provisional figures for another 40 per cent of the indicators.

A complete list of the series and weights used in the index on the 1958 base was published in the March 1962 issue of *Economic Trends*. Table 80 shows the industries and industrial groups for which separate indices are published each month in the *Board of Trade Journal*, together with the weights given to each index. The indices are published in two series; the second of which is merely the first adjusted for holidays and other seasonal causes of variation in production. The second series is designed to eliminate normal month-to-month fluctuations and thus to show the trend more clearly. Nevertheless, the seasonally corrected series should not be regarded as in any way more reliable than the uncorrected series, nor are they intended to replace them.

The correction for seasonal variation is based on the assumption that the seasonal pattern for recent years will recur in the current year. Nevertheless, the seasonal pattern may be changing and hence it must be kept constantly under review and reassessed each year.

The adjustments for seasonal variation are made in two stages. The first step is to adjust the indices for those movements which arise from regular public holidays and annual holidays. When the effect of holidays has been removed the annual average is adjusted so that it equals the average of the unadjusted indices. The second step is to calculate a seasonal adjustment factor for each month using the moving average method (see Chap. XI). The vast amount of calcu-

TABLE 80

INDEX OF INDUSTRIAL PRODUCTION: DISTRIBUTION OF WEIGHTS

Industries or Industrial Groups for which separate Indices are published each month	Weight		
TOTAL All industries	1,000		
Mining and quarrying		72	
TOTAL All manufacturing industries ..		748	
TOTAL Food, drink and tobacco ..			86
Food				55
Drink and tobacco ..				31
TOTAL Chemicals and allied industries			68
Coke ovens, oil refineries, etc.				9
General chemicals, etc. ..				59
TOTAL Engineering and allied industries			310
Engineering and electrical goods				167
Shipbuilding and marine engineering				22
Vehicles				79
Metal goods n.e.s... ..				42
TOTAL Metal manufacture ..			68
Ferrous				55
Non-ferrous				13
TOTAL Textiles, leather and clothing			92
Textiles				58
Leather, leather goods and fur				4
Clothing and footwear ..				30
TOTAL Bricks, pottery, glass, etc. ..			28
Bricks, cement, etc. ..				17
Pottery and glass				11
Timber, furniture, etc.			20
Paper, printing and publishing ..			55
Other manufacturing industries ..			21
Construction		126	
Gas, electricity and water		54	

Source: 'Economic Trends', March 1962.

lation involved in the second step has been greatly facilitated by performing the work on an electronic computer.

While the substantial improvement in this index since its introduction is freely acknowledged, it would be a mistake to place too much emphasis on month to month changes. For both the ordinary monthly, as well as the 'seasonally adjusted', indices it is the general persistent movement in one direction or another which is the most satisfactory and reliable guide to production levels. Thus, an official comment on the index advises that a better indication of the underlying trend of industrial production is probably obtained by comparing a run of months, taking say the average of the last three or four.

REFERENCES

Studies in Official Statistics, No. 7.
Economic Trends, March 1962.

Retail Sales Indices

For more than thirty years there have appeared monthly in the *Board of Trade Journal* indices reflecting changes in the level of retail trade. The primary purpose of these indices is to provide up-to-date information on the short-term trends in the trade of retail establishments. Inevitably over this long period these indices have been subject to several revisions; the more recent of these were made in 1952, in 1959 and in 1963. On the occasion of the last change, the base year of the index was altered to 1961.[1] One other important change which should be noted if longer period comparisons are to be made was the change made in 1955. This involved a change in the basis of presentation of the indices from a *commodity* basis, *i.e.* sales of furniture, groceries, etc., to the present *kind of business basis*, *i.e.* the sales by furniture shops, grocers, etc.

With only a few exceptions, *e.g.* coal merchants and florists, the whole field of retail trade in Great Britain is covered. Retail establishments are divided into four groups: independent retailers, multiple retailers, *i.e.* those with chains of ten or more branches, cooperative societies, and general department stores. These traders contribute a voluntary monthly return of their sales[2] inclusive of purchase tax, to the Board of Trade. To this extent, *i.e.* that the returns are made voluntarily, it is arguable that the sample is not completely random. In fact, the coverage of the index reflects the bulk, particularly that of large-scale retailers, of retail trade in Great Britain. The panel of independent retailers numbers some 7,500 shops and the large scale retailers account for about 60 per cent of the total trade of their class. Continuing attempts are being made to improve the representativeness of the panel of retailers which constitutes the sample.

The indices are compiled using a ratio method, *i.e.* linking the monthly sales of the current year with the sales of the corresponding month in the previous year. The calculation of the index consists of dividing the total sales in any one category of shops in the current period by their total sales in 1961, the base year, and multiplying the result by the total sales of all shops in that category in 1961. The latter data are derived from the Census of Distribution. The following formula illustrates the basis of construction of the index for all classes of retailer, the retailers being first classified by size of turnover. This revised series of indices, introduced in May 1963, was not com-

$$\begin{array}{l}\text{Index of the} \\ \text{sales in} \\ \text{January 1964}\end{array} = \frac{\text{Sales of the sample in January 1964}}{\text{Sales of the sample in 1961}} \times \begin{array}{l}\text{Sales of} \\ \text{all shops} \\ \text{in 1961}\end{array}$$

---

[1] Details of the change and the design of the new index were published in the *Board of Trade Journal*, 10th May, 1963.

[2] Inclusive of service charges, rentals of goods, e.g. TV. Goods sold on H.P. are included at their cash value plus any credit charge.

pleted by early 1964. Deficiencies in the information available and some changes in the character of the trade of some of the contributory shops have affected the representativeness of the panel. Bearing in mind the limitations of these data, considerable caution is required in interpreting month by month movements in the index. It is probably safest to take note only of marked longer run trends shown by the index until the new index is completed.

REFERENCES

Board of Trade Journal, May 10, 1963, contains an account of the construction and nature of the current index.

A fuller description of these statistics is given in *Economic Trends*, May, 1962.

Board of Trade Wholesale Price Indices

Between 1951-5 the Board of Trade was producing two indices of wholesale prices. The first, which was the revised version of its original 1921 index, was finally discontinued in 1955. Table 81 shows the annual averages for the constituent groups of that index for the concluding years of its life. The index incorporated some 258 quotations classified into 200 commodities, *i.e.* some commodities being quoted several times to obtain an average. Fully finished articles were not included but were indirectly represented by weighting those semi-finished articles and raw materials which entered primarily into manufactured articles. The weighting in the index was effected by including in each of the groups of commodities several quotations for particularly significant commodities. The object was that each commodity should be weighted in proportion to its significance in the overall net value of all manufactured goods produced in the United Kingdom as given by the 1930 Census of Production. Each month's average of prices was compared not with that of the preceding month, but with the average of the same month in the preceding year. This ensured that the changes in prices shown by the relatives each month were between the same goods, since many products, for example fruit, are seasonal. The index for the year was the geometric mean of the twelve monthly indices. In other words, the index was of the chain base variety and this fact enabled considerable variations in the constituent items to be made, whilst ensuring that in the short period at least the changes indicated by the index were between comparable sets of prices.

A detailed account of the construction of this index was given in the supplement to the *Board of Trade Journal* of 24th January 1935. This index was replaced by the new index of wholesale prices which appeared initially in the *Board of Trade Journal* of 19th May, 1951. The replacement of the old index was urgently required, since both the

TABLE 81

WHOLESALE PRICE INDEX NUMBERS (OLD SERIES)

(Average 1930 = 100)

Annual average	Total all articles	Inter- mediate products	Iron and steel	Total food and tobacco	Cereals	Meat, fish and eggs	Other food and tobacco
1938	101·4	104·5	139·1	97·3	109·9	85·9	97·5
1949	230·0	260·3	252·9	196·7	196·7	156·1	230·6
1950	262·4	294·5	260·8	221·1	235·3	173·6	251·3
1951	319·5	371·8	292·4	246·9	287·0	179·3	278·6
1952	327·6	364·3	353·5	284·1	313·6	222·1	316·4
1953	327·9	355·9	362·1	307·4	341·3	247·0	334·6
1954	329·9	385·9	366·2	307·5	320·1	237·4	358·9

Source: 'Board of Trade Journal.' This series has been discontinued since 1955.

composition of the old index and the weighting system, which was based on 1930, were unsatisfactory and out of date. The only reason for including the above account of the old index is if long period comparisons of price trends are required. The new index of wholesale prices, which is explained below, does not go back earlier than 1950, although some of the indices have been calculated for the years 1946-50 where the data permit.

The old index was to be a means of answering the question 'what is the average change in the value of money relative to other things?'; a reflection of the acceptance of then current quantity theory of money. The new index is based on an entirely different conception of the functions which the index should perform. The new indices reflected the view that there is no such thing as *the* price level. At best the majority of prices move in the same direction, but always in varying degrees. The indices were to be related to major economic groupings, for example industries, and constructed 'as far as possible so that they may be of direct help to the government, to industry and to economists in studying the effects of price changes'.[1]

A new index was introduced in 1951 and the base date was subsequently revised to 1954. The current index based on 1954 was introduced and described in some detail in the *Board of Trade Journal* of 21st March 1958 and is in fact a collection of index numbers which have been classified into three main groups. First, there are price indices of commodities and materials which are important in the production processes of certain industries. These commodity price indices relate to materials such as aluminium, brass and copper among metals. Among the staple fibres there is an index for raw cotton which is supplemented by separate indices for the main types of raw cotton just as the index for raw wool is supplemented by three separate

[1] *Board of Trade Journal*, 19th May, 1951.

indices for the different varieties. The second group of index numbers are to a certain extent based upon the first group; they are termed indices of basic material prices. Among the first of these indices to be produced were those based on the prices of materials used in the mechanical engineering industry and building and civil engineering respectively. It was intended that these particular indices would be sufficiently reliable to permit price revision clauses to be inserted into contracts for public works, whereby an agreed basis for adjustment of prices would be available to contractor and the authority placing the order. The last of the three groups of indices are designed to reflect the price movements of the total output of certain important industries. For example, there is an index for the china and earthenware industry, for iron and steel (tubes) and for tinplate. Examples of the indices taken from each of the three groups are given in Table 82.

These index numbers are calculated from the price movements of some 7,000 closely defined materials and products representative of goods purchased and manufactured by United Kingdom industry. The index for an individual commodity is its current home market price expressed as a percentage of its annual average price in the home market in 1954. In compiling group indices the percentage changes are combined in proportion to the value of purchases or sales of the individual commodities in 1954. The data for weighting the constituents of the group indices are culled from the census of production, other short-term production indices, the Trade and Navigation accounts, together with information supplied by trade and industry. The base date for all the indices is 1954 = 100. The indices are arithmetic means of the percentage changes in the prices that have taken place since the base date.

The prices used in the calculation of these indices are the 'ex-works' prices of the commodities. If it is the practice for the industry to quote the price for the commodity 'delivered', then that quotation is used. The weighting is determined by the information derived from the Census of Production, although supplementary information (the source of which cannot apparently always be disclosed) has also been utilised to obtain correct weighting. In the case of the commodity price indices (group 3) which are compiled upon the basis of a number of types, e.g. the raw cotton or wool index, the weighting is determined by reference to the value of the sales of each constituent commodity in 1954. For the basic materials price index (group 1) the weighting is determined by the value of the relevant materials actually consumed in the appropriate sector of industry in 1954. For example, in the case of the house building materials index, bricks form 12·8 per cent of the total weighting, softwood 0·9 per cent and sand and gravel 8·2 per cent.

TABLE 82
BOARD OF TRADE WHOLESALE PRICE INDEX
1954 (Annual Average) = 100

Indices relating to:	Annual Averages					Monthly Averages			
						1963		1964	
	1959	1960	1961	1962	1963	3rd Qtr.	4th Qtr.†	Jan.†	Feb.†
1. Materials Purchased by Broad Sectors of Industry:									
Basic materials and fuel used in manufacturing industry	103·4	104·5	104·4	103·0	104·5	101·5	106·3	107·4	107·3
Materials and fuel used in the electrical machinery industry	115·6	116·9	118·3	120·0	120·9	120·9	121·5	122·3	122·6
House building materials	111·0	114·0	118·0	120·2	121·7	122·2	123·0	124·0	124·1
2. Output of Broad Sectors of Industry:									
All manufactured products: Total sales	111·4	113·1	115·7	118·0	119·8	119·9	120·8	121·3	121·4
Iron and steel: Total sales	125·4	125·6	126·1	128·7	128·8	128·7	128·8	129·0	129·1
Paper industries: Home sales	107·6	107·6	109·8	111·0	111·6	112·3	113·3	114·3	115·3
3. (a) Commodities produced in the United Kingdom:						Aug.	Dec.		
Coal	135·4	137·4	144·9	151·8	153·7	151·4	156·2	156·2	156·3
Soap	127·9	128·9	129·9	134·1	135·3	135·3	136·2	136·8	137·0
Beer	95·5	92·1	96·5	102·2	106·3	107·1	107·6	107·6	107·6
3. (b) Commodities wholly or partly imported into the United Kingdom:									
Cotton, raw	68·6	75·0	74·8	73·2	71·4	70·9	72·4	72·6	72·8
Wool, raw	76·6	76·3	76·2	76·5	89·0	87·8	98·2	99·7	102·5
Wood pulp, imported	97·0	98·3	100·3	94·5	95·5	96·4	98·7	100·3	100·3

Source: Board of Trade Journal, 21st February and 13th March 1964.
† Provisional figures.
N.B. Although the monthly indices are given to one decimal place to facilitate further calculation, small month to month movements have little significance.

15*

The price indices of output of broad sectors of industry (group 2) derive their weights from data relating to the sales of the product by the corresponding sector of industry in 1954. For example, the price index of the output of the iron and steel industry is based upon the combined prices of the commodities contained in the list of commodity price indices, *i.e.* iron castings, sheets, tinplate and tubes each. It should be noted that the prices and weights of the materials used in this particular index, *i.e.* product of broad sectors of industry (group 2), relate only to the output sold outside the industry and not to that sold between firms within the same industry.

The monthly indices are published in the *Board of Trade Journal*. In the mid-July issue each year there is a detailed review of the movements in the indices during the past eighteen months. In the mid-February issue, complete tables for each group of indices showing the annual averages for the latest year together with comparative figures for earlier years are given. Table 82 is a combination of the information given in these issues; it shows the annual averages for selected groups and commodities as well as the quarterly and monthly indices for selected quarters and months in 1963 and 1964. In the same issue of the *Journal* any major changes or additions to the current indices are discussed. A selection of the more important indices in each of the three groups is published in the *Monthly Digest of Statistics*. As with so many other economic series these indices have been subject to periodic revision since they were first introduced in 1955. The number of prices included has been increased from 5,000 to 7,000, while the revision of the Standard Industrial Classification has meant that pre- and post-1958 indices for certain industries and trades may not be comparable. Before using these data the notes in the *Journal*, which accompany the annual analyses, should be studied.

Import and Export Unit Value Indices

Since 1946 the *Board of Trade Journal* has published a series of monthly indices which are used to measure the short-period changes in the United Kingdom's terms of trade. The 'terms of trade' is simply the ratio of import to export prices; if the former are rising more rapidly than the latter, then the terms of trade are said to be moving against the United Kingdom. In other words, the United Kingdom is receiving a smaller quantity of imports for a given volume of exports. In view of the nation's balance of payments problem this particular index is quoted regularly in discussions of the overseas trade statistics. The indices are designed to measure the monthly changes in the aggregate value of a *fixed* but representative selection of imports and exports. The basic data used in the construction of the indices is published in the Trade and Navigation Accounts (Trade Accounts). One of the weaknesses of the data, however, is that some of the commodity headings used relate to products which are not closely homogeneous. This results in certain fluctuations over a period of time which are not true price movements and the accuracy of the indices is consequently reduced. It should be noted that the indices are called 'unit value' indices to distinguish them from price indices which are normally constructed from data on suppliers' price quotations for closely defined products.

The commodity headings in the Trade Accounts are selected for use if their behaviour over a period of time leads one to suppose that they are reasonably homogeneous. The import index is based on 352 headings covering 78 per cent by value of the imports in 1961 while the export index is based on 538 headings having a coverage of 62 per cent of total exports by value. Table 83 gives details of the coverage by value for the principal categories for which sub-indices are published.

TABLE 83

COVERAGE OF THE HEADINGS DIRECTLY USED IN U.K. TRADE INDICES

	Number of commodity headings	Coverage by value in 1961	
	Number	%	
Imports			
Food, beverages and tobacco (Sections 0 and 1)..	92	90	
Basic materials (Sections 2 and 4)	90	85	
Fuels (Section 3)	9	98	
Manufactured goods (Sections 5 to 8) ..	161	53	
TOTAL (Sections 0 to 9)	352	78	
Exports			
Non-manufactured goods (Sections 0 to 4) ..	70	76	
Manufactured goods (Sections 5 to 8) ..	468	60	
of which			
Chemicals (Section 5)			42
		41	
Textiles (Division 65)			76
		41	
Metals (Divisions 67 to 69)			74
		90	
Machinery and transport equipment (Section 7)			61
		225	
TOTAL (Sections 0 to 9)	538	62	

Source: Economic Trends, September 1963.

Although the basis of selection of the headings used in the indices is that they are reasonably comparable over a period of time, there arises occasionally sizeable fluctuations that would not cancel out in aggregate. Such fluctuations are usually smoothed by discounting half or more of the price change if it represents a change of more than

TABLE 84

IMPORT AND EXPORT UNIT VALUE INDEX NUMBERS[1]

Selected Years and Months 1954 – 1963

(1961 = 100)

	Imports					Exports									Terms of trade[2]
								Manufactured goods (5 to 8)							
	Total	Food beverages and tobacco (0 and 1)	Basic materials (2 and 4)	Fuels (3)	Manufactured goods (5 to 8)	Total	Non-manufactures	Total	Chemicals (5)	Textiles (Div. 65)	Metals (Divs. 67–69)	Machinery and transport equipment (7)	Other	
Weights ..	1,000	338	230	110	318	1,000	129	838	89	67	126	442	114	
1954 : :	104	108	104	110	98	91	105	89	114	95	90	83	87	87
1956 : :	110	109	111	119	107	95	112	93	113	94	99	88	92	87
1958 : :	103	104	100	120	96	99	108	97	109	96	103	95	94	96
1960 : :	102	104	101	104	101	100	102	99	104	96	101	99	97	97
1961 : :	100	100	100	100	100	100	100	100	100	100	100	100	100	100
1962 : :	99	102	96	98	99	101	99	102	98	100	101	103	102	102
1963 : :	103	111	97	97	101	104	103	104	100	102	102	106	104	101
1963 Oct	105	116	97	96	101	105	104	105	101	103	102	106	105	100
Nov	106	118	99	96	102	105	105	105	101	103	103	106	106	99
Dec	106	116	100	95	102	105	106	105	102	104	103	106	105	100

Source: Board of Trade Journal, 13th March, 1964.

[1] The figures in parentheses relate to Section or Division code numbers in the Standard International Trade Classification Revised.

[2] Export unit value index as a percentage of the import unit value index.

0·1 per cent in the total unit value index. Similarly if a product of a marked seasonal trade is used where for some months no quantities enter into trade, the usual practice is to repeat the unit value of a month in which trade of reasonable proportions did take place.

TABLE 85

U.K. TRADE INDICES CHANGES BETWEEN 1954 AND 1961
IN UNIT VALUE AND VOLUME

(1954 = 100)

	Unit value		Volume	
	1954 based index	1961 based index	1954 based index	1961 based index
Imports				
Food, beverages and tobacco (Sections 0-1)	94	92	121	121
Basic materials (Sections 2 and 4)	95	96	102	104
Fuels (Section 3)	96	91	164	160
Manufactured goods (Sections 5 to 8) ..	107	102	195	201
TOTAL (Sections 0 to 9) ..	97	96	135	137
Exports				
Non-manufactured goods (Sections 0 to 4) ..	97	95	118	121
Manufactured goods (Sections 5 to 8) ..	115	113	129	126
of which				
Chemicals (Section 5) ..	95	88	198	177
Textiles (Division 65) ..	106	105	73	71
Metal (Divisions 67 to 69)	117	111	139	133
Machinery and transport equipment (Section 7)	122	121	133	133
TOTAL (Sections 0 to 9) ..	112	110	125	123

Source: Economic Trends, September 1963.

The selection of the unit values under each heading is so devised that a collection of commodities for both indices is derived which is representative of the current pattern of trade. The weights employed are 'fixed' base-year weights, determined by the pattern of trade in 1961. Thus the weighting employed for the indices in 1962 and later is given by the pattern of goods trade in 1961. The index itself is derived by calculating the geometric mean of the products of the unit values and their respective weights. In other words, the resultant indices measure the change from period to period in the value of a fixed selection of commodities, regardless of the fact that the composition of the goods traded in any period differs from that in others. This weighting system is therefore adequate only for as long as the pattern of commodity trade remains constant from year to year. If in any particular year there are marked changes, the use of the weights based upon the 1961 pattern will distort the indices in the later period.

Although the price indices are published monthly, separate indices – based on the monthly data – are published for successive quarters and for each year. A selection is given in Table 84 for both imports and exports. The annual indices – those for some years back are also reproduced in Table 84 – are used for measuring long period changes in the terms of trade of the United Kingdom.

The 1961 based indices were introduced and described in the *Board of Trade Journal* of September 13th, 1963. Their construction was discussed in an article in *Economic Trends* for September 1963. The 1961 based indices replaced those based upon 1954 and to provide continuity were calculated back to 1954. Table 85 shows the effect in the overlap period of 1954-1961 of the change in base period.

Prior to 1955, when the 1954 based indices were introduced, a system of average value indices was used. The average value index, now discontinued in the published tables, was a form of price index derived by dividing the declared value of a given quarters' imports by their value at 1950 prices. Comparison with indices relating to periods before 1954 is therefore very difficult.

The differences in both unit value and volume over the period are comparatively small and are largely explainable by the change in the pattern of trade.

Import and Export Volume Indices

Index numbers which measure the changes in the *volume* of both imports and exports are also prepared by the Board of Trade. They are published monthly in the *Board of Trade Journal*. The indices are designed to show the variations in imports and exports after eliminating price variations, *i.e.* volume changes only. This is done by recalculating the value of the imports (or exports) for any quarter at the average prices of the year 1961. By expressing the corrected value of imports (or exports) as a percentage of the 1961 value an index of *volume* change is derived.

The monthly figures used in calculating the index are derived from the Trade and Navigation Accounts. As with the import and export unit value indices, adjustment of the contents of various headings is necessary. For those items for which only value and not volume is given in the Accounts, estimates of the probable changes are made by assuming that they move in the same manner as do related items for which both value and volume figures are available. This procedure is

TABLE 86
IMPORT AND EXPORT VOLUME INDEX NUMBERS[1]
1961 = 100

	Imports					Exports		Manufactured goods (5 to 8)					
	Total	Food, beverages and tobacco (0 and 1)	Basic materials (2 and 4)	Fuels (3)	Manufactured goods (5 to 8)	Total	Non-manufactures	Total	Chemicals (5)	Textiles (Div.65)	Metals (Divs. 67-69)	Machinery and transport equipment (7)	Other
1954	73	82	96	63	50	82	83	79	57	141	75	75	82
1956	81	90	99	71	63	92	89	91	69	129	97	89	90
1958	84	98	92	76	67	90	88	90	71	109	93	90	89
1960	102	100	107	96	103	98	92	98	93	109	98	97	96
1961	100	100	100	100	100	100	100	100	100	100	100	100	100
1962	103	104	96	112	104	102	110	101	107	99	100	100	103
1963	107	102	101	120	113	108	120	106	113	101	100	107	110
1963 October	117	112	109	113	130	112	128	110	116	105	105	109	117
November	111	105	105	116	119	111	126	109	119	108	109	105	120
December	111	100	117	122	113	115	126	110	117	104	99	111	117

Source: Board of Trade Journal, 13th March, 1964.

[1] The figures in parentheses relate to Section and Division code numbers in the Standard International Trade Classification Revised.

adopted to a rather greater extent for the export index since a larger proportion of total exports are given in value terms only. Both volume and value figures are available for all but 3 per cent of imports (by declared value).

These indices have been published for many years, but since in the view of the Board of Trade a change in both the base year and structure of the index should be made at least every five years (owing to the changes in the pattern of overseas trade) the index is not comparable as between different base years, except where the change of base has been accompanied by a revision of indices for the earlier years. The latest base year is 1961 to which the index was revised from 1954 in September 1963.[1] At the same time the weights were revised on the basis of the average prices ruling in 1961 instead of 1954. In the current index the weights are based on the relative values of the individual categories of goods comprising the total value of trade in the year 1961.

The volume indices as may be seen in Table 86, are given for the same sub-headings, *i.e.* raw materials, manufactures, etc., as the import and export *price* indices, which were discussed earlier. Like the monthly price indices the volume indices are separately calculated for each month, quarter and for each year.

Security prices

Fluctuations in share prices and the level of stock market activity may be used as indicators of changes in the level of economic activity and business confidence generally. At the present time, there are no adequate published data relating to the volume of business undertaken on the stock exchanges, either London or provincial. The published figure of daily bargains in London is based on voluntary returns made by members and, quite apart from the fact that not all bargains are recorded, no indication is given of the size of the bargain struck. Movements in share prices are faithfully recorded in the daily Stock Exchange Official Lists and the columns of the financial press. These prices are used for purposes of compiling indices of share movements.

The two best known indices at the present time are *The Times* Index of Industrial Ordinary Shares, and the *F.T.-Actuaries* Share Indices. Detailed accounts of the construction of both these indices are available.[2] *The Times* index number is a weighted arithmetic average of the prices of all shares included in the index. The index covers 150 shares of all classes, 50 large companies, large being defined as those

---

[1] *Board of Trade Journal,* 13th September, 1963.

[2] *The Times* daily index numbers of Stock Exchange security prices (The Times Publishing Co. Ltd.) and Guide to the *F.T.-Actuaries* Share Indices.

with over £30 million of capital at market prices, and 100 smaller companies. There are additional indices for capital goods and consumer goods industries, based on 43 companies apiece which produce wholly or mainly capital goods or consumer goods respectively. The weight given to each share in each index is proportionate to the average market value of the issue on two dates, on 1st July 1958 and 27th October, 1959. The base date for each index number is 2nd June 1959. It should be noted that whereas before March 1960 the indices were based on quotations on alternate Wednesdays, since that date the monthly indices are the average of working dates.

Since November 1962, a new index has been prepared by the Institute of Actuaries, in conjunction with the *Financial Times*. The latter newspaper still publishes its old Index of Industrial Ordinary Shares based on the geometric mean of 30 blue-chip equities. As a measure of share price changes this has, however, been superseded by the new index.[1]

The new *F.T. – Actuaries* Index is strictly speaking a series of 50 price indices based upon about 690 securities quoted on the London Stock Exchange. The two main equity price indices are '500-industrial' and an 'All-Share' index of 594 equities. The 500-industrial and the index of the financial group, when combined form the above 'All-Shares' index. Other than these indices there are commodity share groups and fixed interest indices. The object of the price indices is mainly to reflect the performance of the ordinary share market of the London Stock Exchange as a whole. According to the official account, changes in the indices record how much better or worse off investors are as a whole. The choice of the shares comprising the main indices was determined by the fact that out of the 1900 quoted companies with equity capitalisation greater than £1 million apiece, some 650 with a market capitalisation greater than £4 million had a total value approaching 90 per cent of the 1900 companies. The choice of shares to be included in the two main indices was therefore made from this group of 650 large companies. At 10th April 1962, which is the base date, the market valuation of the 594 shares included in the 'All-Share' index was £18,170 million. This was approximately equal to 60 per cent of the value of all quoted equities in the sections covered by the index. The calculation of the index is basically simple, since it is merely the total current market valuation of the shares included in each index related to the corresponding aggregate valuation of the same shares at the base date.

The '500-industrial' share index comprises three main groups of companies: those producing capital goods, consumer durable goods;

and consumer non-durable goods, together with chemical, oil, shipping, and miscellaneous groups. Separate indices are calculated for each group. The remaining 94 quoted shares, which are added for purposes of the 'All-Share' index, relate to financial and property companies. In view of the size of this index, the daily computations are performed on a computer. The results are published daily in the *Financial Times* and the Institute of Actuaries supply subscribers with further details thereof.

Statistical Sources

The largest body of published statistical data is assembled as a by-product of the government's daily administration of the economic life of the nation. Some of the data are derived from the routine adminis-tration of government departments, *e.g.* the overseas trade statistics from the Customs office, unemployment figures from the Ministry of Labour Employment Exchanges. Other data are derived from specific enquiries conducted by government departments and are provided by members of the public, as on the occasion of the population census, or by the business community as with the census of production.

A particularly useful publication for the student contemplating a study of official statistics is the pamphlet prepared by the Treasury entitled 'Government Statistical Services'.[1] This document provides a succinct description of the statistical work in government depart-ments. It explains the origins of many series of data, *e.g.* day to day administration and special enquiries either by census or sample. It also contains some useful comments on the problems which arise when data are to be collected. Apart from an explanation of the legal powers under which information is obtained from industry and the public, there is a short account of the organisation of statistical work in government departments and an account of the origins and functions of the Central Statistical Office. It includes two very useful appendices which alone would justify its publication. The first appendix outlines the various statistical data collected by each government department. The second gives a classification of all pub-lished statistics under subject headings, *e.g.* agriculture, education, etc. and for each subject shows the principal publications containing the statistics together with a note of their frequency of appearance and the department responsible for producing the data.

The need to control the economy during the last war led to a very considerable improvement in the flow and quality of statistical in-formation relevant to the war effort. The body responsible was the then statistical unit in the Cabinet Office which after the war became the Central Statistical Office. This body has over-riding responsibility

[1] Revised edition 1962 published by H.M. Stationery Office, 3s.

for the production of statistical information to the Cabinet, or any government committee which may be formed. It also co-ordinates and assists the various statistical departments in the various Ministries, and when several such departments are involved in statistical enquiries. While it is probably true to say that Mr. Macmillan's famous comment on the state of U.K. national statistics in 1957, *i.e.* that using them for formulating budgetary policy was like looking up trains in last year's Bradshaw, gave the production of economic statistics a major fillip,[1] without the co-ordinating and research activities of the C.S.O. the results of the efforts of recent years to improve the data would not have been as impressive as they have been.

Similarly, the creation of the National Economic Development Council in 1962 stimulated economic research which drew attention to the deficiencies of the available data relating to many spheres of the economy. The Radcliffe Committee on the Working of the Monetary System had, some years earlier, been highly critical of the shortcomings of financial statistics.[2] The government responded to the criticisms and since Spring 1961 there is a monthly publication, entitled *Financial Statistics*, produced by the C.S.O. and other departments as well as the Bank of England statistical department. The latter body now makes a significant contribution to a new enlarged *Quarterly Bulletin* from the Bank which first appeared in December 1960. Quite apart from up-to-date statistical data tabulated in the statistical annex to the *Bulletin* it often contains very useful articles on the sources and new developments of financial data.

The most useful single source of published statistical information compiled by the government is the annual *Abstract of Statistics*. This is a joint publication by the Central Statistical Office and the Statistical Divisions of the various government departments. Although most of the data are collected by government agencies, some information is provided by private organisations. The 'Abstract' is published annually and for many series the data for a ten-year period are brought together. As a general rule, when any information is required, the annual Abstract should be the first source to be consulted. Since 1946 the Central Statistical Office has produced the *Monthly Digest of Statistics*, which, like the *Annual Abstract*, has a wide coverage. It excludes, however, the data relating to social conditions and deals primarily with economic data. Monthly figures of supplies of fuel, raw materials and finished products together with indices of output in selected manufacturing industries are provided. These are supplemented by data covering labour, wages, transport, and foreign trade, as well as important financial statistics. For most of

[1] For an account of the work then in progress and projected see article in *Economic Trends*, May 1957.
[2] *Cmnd. 827*. H.M.S.O. August 1959, Chapter X.

these items the *Monthly Digest* gives monthly data for the last one or two years, together with comparable figures for the earlier years. To ensure that the data given in the *Digest* are correctly interpreted, a supplement is issued in January each year. This provides detailed definitions of units and items given in the *Monthly Digest*.

Another useful monthly publication entitled *Economic Trends* is produced by the Central Statistical Office in collaboration with the Statistical Divisions of government departments. It provides both charts and statistics illustrating current trends in the United Kingdom economy. Each issue contains at least one article commenting on features of current economic statistics or introducing a new series, or describing methods used by a Statistics Division of a government department in the preparation of their particular data.

More detailed information on a very large variety of economic subjects is published at intervals in the weekly *Board of Trade Journal*. For current data this is more up-to-date and useful than either the *Annual Abstract* or the *Monthly Digest*. Much of the in-dustrial and commercial information contained in the *Monthly Digest* appears in rather greater detail and earlier in the *Board of Trade Journal*. Important in its specialist field is the *Ministry of Labour Gazette* published monthly which, with the new Statistics on Incomes, Prices, Employment and Production, constitute the source of all labour statistics.

As may be judged by the length of this chapter alone, the sheer volume of economic statistics contained in official publications is at best impressive, at worst it is utterly confusing. It is perhaps pertinent to note that the volume is rather more impressive than the reliability of some of the series. The impression of limitless data is not lessened by the fact that several of the statistical publications overlap. For example, quite detailed financial series are reproduced in the *Monthly Digest*, *Economic Trends*, and *Financial Statistics*. In the case of labour statistics, the *Gazette*, the new *Bulletin*, and the *Monthly Digest*, as well as *Economic Trends*, all contain some common data. The average statistician, having for so many years pleaded for better official statistics, should hardly complain now that the government is starting to meet his wishes. Nevertheless, the only solution is for the reader to decide which of the several publications in the field of statistics in which he is especially interested is most suited for his particular purposes.

It is, of course, quite impossible for anyone to become familiar with all economic series. Generally speaking, statisticians tend to work in particular fields, for example, on overseas trade and balance of payments statistics, internal trade, production, or on financial data with special reference to the banking system and stock market. Even when the field of study is restricted, keeping abreast of the continuing development in most of these fields can be almost a full time job. In practice, what happens is that by a concentration on particular series the statistician becomes familiar with the various snags and limitations of his data. When his work takes him into another field of statistics, which may possibly be virgin territory to him, he is by virtue of his training well aware of the limitations of published data, and thus before long he can find his way through the mass of information available in that field.

It should be taken as a guiding principle that eternal vigilance, coupled with an eye for detail, will alone ensure accurate extraction of published data in the face of footnotes denoting changes in definition, a broken series, incomplete series or change in coverage, which characterise most published data.

Numerous references have been given to a wide variety of publications and articles which will enable the student of economic statistics to familiarise himself with their background and limitations. It is helpful at any time, however, to ensure easy access to the supplement entitled *Notes and Definitions* of the units and data which appear in the *Monthly Digest of Statistics*. The supplement appears in revised form in January each year. For keeping abreast of new developments and new series, the most useful single publication is undoubtedly *Economic Trends*. A number of useful articles on developments in economic statistics from *Economic Trends* have been brought together in two publications prepared by the C.S.O. and published by H.M. Stationery Office. These are entitled *New Contributions to Economic Statistics:* First and Second Series, being numbers 5 and 9 in the Studies of Official Statistics. Other issues in this series are available on labour statistics, population census, local government, agricultural statistics, social security statistics, and the indices of Retail Prices and Industrial Production. Others will no doubt appear in due course.

Conclusion

In recent years especial attention has been paid not merely to producing the existing series of economic data more quickly, but to preparing new short-term indicators, *e.g.* hire-purchase statistics and quarterly estimates of the gross national product. Not merely are these series prepared in crude form, *i.e.* the actual figures for the period in question, but they are also corrected for seasonal variations and changes in prices. Such refinements have been made possible by the use of electronic computers. More attention too, is being paid to the production of what are often called 'forward-looking' statistics. Most economic series merely reflect the events which are past.

Such 'forward-looking' statistics are collected in order to make some assessment of future developments. For example, business men are asked about the state of their order books for the home and export markets, and their intentions regarding investment in new plant, factories and vehicles. While there are obvious limits to the value of such forecasts, experience in their interpretation suggests that in due course they can serve a very useful purpose. Likewise, more detailed and continuous surveys of consumer expenditure enable estimates to be made of changes in the pattern of consumer spending habits and their implications for production and the balance of payments.

While the main concern of the government statisticians must be to ensure a flow of reliable data published without undue delay, new fields are being continuously explored in response to new problems. For example, the rising level of government expenditure is causing some concern, not least to the taxpaying public. As tax rates tend to rise there is greater interest in where the burden actually lies. There have been studies in *Economic Trends* of the incidence of taxation by reference to the type of household linked with an assessment of the extent to which government social and welfare services compensate for such taxes.[1] If and when a new government introduce a new tax on wealth, the recent studies of personal wealth in the U.K. published in recent reports of the Commissioners of Inland Revenue should be especially valuable in estimating the prospective yield from such a tax.[2]

In brief, the quantity of economic statistics is steadily growing and, at the same time, their quality is being improved. There is an ever-increasing reliance upon sample rather than census data.[3] It would be silly to suggest that given adequate statistics the government's economic problems will be at an end. Ministers, no less than other human beings, are quite capable of ignoring the statistics. What better statistical information does ensure, however, is that decisions which affect the lives of the entire population can be made in the light of knowledge rather than on the basis of a few facts mixed with personal beliefs. They also provide the basis for informed public discussion of issues which are the publics' concern. Each successive government enquiry into different aspects of the nation's economic and social life reveals ever more clearly the quite appalling lack of information that existed when policies were first formulated years ago and which, it must be conceded, still exists today in many fields, *e.g.* crime, education, rates and taxes, road accidents, overseas capital transactions, private and personal savings. If the list is not endless, at

[1] *Economic Trends*, November 1962 and February 1964 relating to 1957 and 1959 respectively.
[2] 104th, 105th and 106th Reports for the years 1961-1963 inclusive.
[3] Sampling for Current Economic Statistics, R. G. D. Allen, *J. R. S. S.* 1964, Part I.

least it is true to assert that it is still far too long. As long as the electorate insist on unemployment being kept at a level below 2 per cent of the insured population, regular intervention in the economy is unavoidable. The scale and timing of such intervention, *e.g.* change in Bank Rate, increases in outlay taxes, etc., needed, if worse confusion is to be avoided, can only be ascertained in the light of detailed and up-to-date information. If the achievements of recent government economic policies in stabilising the economy are any guide, the extent to which such policies are formulated in faith and statistical ignorance is all too evident. Likewise, in the field of social policy, there is ample evidence of wasted effort and money arising from policies often based largely on misguided sympathy and half-truths rather than on the facts of the situation.[1]

[1] For example, how much do we really know about the extent of poverty in the so-called affluent society?

CHAPTER XIX

STATISTICS IN BUSINESS AND INDUSTRY

The efficiency of any industrial or commercial undertaking depends ultimately on the quality and efficiency of its management. To function properly, management must have available to it, as and when required, all information which is relevant to the conduct of the affairs of the undertaking. These range from the state of the labour force or the cash requirements of the firm to the results of the latest market research survey for a new product. If the firm depends upon raw materials from overseas, then it will have an interest in the state of world trade and its effects upon commodity prices and, if it operates overseas, it will need to know something about economic conditions in the countries where its subsidiaries and associated firms are situated. With the tendency for industrial and business organisation to expand, not least through the medium of subsidiary and ancillary undertakings, the time has long since passed when the manager not only knew the name and face of every workman on his staff, but his every customer as well. In the modern large scale industrial or commercial undertaking, the statistician can serve in two ways. He can feed management with data and information relating to its commercial activities, *i.e.* its selling costs and the distribution of its markets, and in industry he can help achieve the maximum efficiency on the production side. These various aspects of the statistician's work can now be discussed separately.

Desk Research

It is a commonplace that many firms maintain only the minimum of records sufficient to satisfy their accountants and auditors. Relatively few use such material to break down their turnover according to the size of individual orders, periodicity of re-ordering by the larger customers, the distribution of orders in the home and/or overseas markets, etc. On the labour side few records are kept regarding absenteeism and sickness. Admittedly, the maintenance of such records entails expenditure, but if the expenditure of a few thousand pounds per annum can achieve either savings in costs of production or increase efficiency amounting to many more thousands of pounds, then clearly such expenditure is fully justified. Most industrial consultants called in to examine the affairs of an undertaking invariably start by learning all there is to learn about its affairs from its internal records, *i.e.* costs, sales, etc. Their recommendations are often based on the facts to be found in the data already available to management!

The Z Chart

Graphical methods are especially valuable in business as a means of conveying information to management.

A type of graph which enjoys a considerable vogue in business rather than in statistical circles, is that known as the Z chart. It derives its name from the form made by the lines on the graph, as will be seen from a scrutiny of Figure 22. The Z chart is merely a method of graphing a time series in such a way that the totals for successive periods are plotted and in addition the cumulative total and a moving annual total.

TABLE 87
ABC COMPANY LTD. SALES RECORD, 1963

Month	Monthly Sales	Cumulative Monthly Total	Moving Annual Total
January..	9,378	9,378	138,680
February	7,624	17,002	138,827
March ..	9,310	26,312	138,965
April ..	12,851	39,163	139,633
May	14,394	53,557	140,172
June	17,839	71,396	142,619
July ..	15,674	87,070	142,206
August ..	15,301	102,371	141,977
September	12,219	114,590	143,869
October	10,046	124,636	144,705
November	8,917	133,553	144,147
December	11,463	145,016	145,016

The data on which the graph is based are given above in the table, showing the sales of ABC Co., Ltd. The first column gives the monthly turnover, and the second the cumulative total as from January. The final column provides the annual total of sales for the twelve months ended in any month of the current year. Thus against June the figure in the third column is £142,619, *i.e.* the total sales for the twelve months ended 30th June 1963. The total for the period ended 31st July is £142,206, which is smaller than the preceding figure by £413. Since the sales for the current August were £15,301, the sales for the August in the previous year were £413 greater, *i.e.* £15,714. The figures for all the months for the preceding year can be so computed if necessary from the table, except for January. The real value of the moving annual total is that it indicates the trend of sales relatively to the preceding year's experience. If the moving annual total line is rising, it indicates that each month this year is an improvement on the same month of the preceding year. If required, a series of such charts for successive years can be set side by side for comparative purposes.

Figure 22

Z CHART

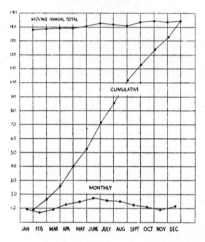

Sales record of ABC Company Limited 1963

External Statistics

Quite apart from the information relating to the operation of the business, every business undertaking has some interest in the overall state of the economy and, in all probability, in particular parts thereof. For example, an engineering firm would be interested in the trends in prices of particular metals, of machine tools and small parts. Likewise, the level of unemployment in the country has a direct effect upon the availability of labour, and not least upon its cost. While the value of an exchange of views on the state of the economy and its immediate prospects with other business men may be interesting and even informative, there is really no adequate substitute for a careful appraisal of all the relevant facts, as presented in documents such as the *Economic Survey*, the Board of Trade industrial returns on capital expenditure, hire purchase, etc. Likewise, for firms interested in overseas markets, the study of market reports such as appear regularly in the *Board of Trade Journal* is essential for the efficient conduct of business.

Closely related to the collection of information on the state of the economy, many firms also conduct either through the medium of their own research department, or by the services of a market research agency, market research surveys. Such research is not restricted to consumer goods. In recent years the many producers of capital goods such as large machines and equipment, have been using market research to ascertain the views of users, both actual and potential, of their products on the desirability of changes, adaptation to different requirements, etc. Such surveys are not only informative but, if they are undertaken at fairly regular intervals, changes over time may be more indicative of the future of the undertaking than a mere snapshot of its market at a particular point of time. Indeed, most market research consultants complain that some firms regard their function in much the same light as people regard their doctors; *i.e.* to be called in only when there is trouble. As long as all seems well, the firms are happy to drift along. Yet only by periodic surveys is it possible to assess how well a firm is doing. For example, in a growing and expanding market, while a single firm's sales may be increasing very satisfactorily, it may be steadily losing ground to a competitor whose share of the market is growing even more rapidly. Without continuous or at least periodic surveys such facts may be overlooked, with dire consequences in the long run.

Some market research agencies maintain consumer panels, which are used to test their client firm's new products. Such panels are, at least in theory, random samples of households or consumers who at intervals receive samples of a product and return a pre-paid business card indicating their views on the product. The services of such a consumer panel can be obtained more cheaply than a full scale survey, since once established, the survey organisation can keep the panel busy on behalf of a number of client firms. The main problem of such consumer panels – and it is for the survey organisation to meet this problem – is that first, members become conditioned to acting as guinea pigs and that second, the membership tends to fluctuate. Thus the members no longer tend to behave as would the average consumer while to the extent that the composition of the panel changes, the substitute may not exactly match the unit it replaces, and thus the representativeness of the panel may be adversely affected. The results of such enquiries are a valuable counter to the director who has a penchant for arguing dogmatically from a sample of one. We have all heard the individual who knows a friend who knows someone else whose experience is then used to reflect the state of current opinion on a particular topic.

As has been shown in the preceding chapter and elsewhere in this book, there is no real shortage of information relating to the state of the economy, trade internal and external, industry or distribution in this country. Admittedly the data required for the use of a particular firm or undertaking may not be available in just the form it would like, but substitute data or series can usually be found. The real problem confronting the economic statistician is not usually, however, a shortage of data, so much as the plethora of published statistics from which he must choose those most suited to his needs. The statistician soon learns in industry and business that the fewer figures he can present to his board the better. He learns too that the more conviction with which he can support his conclusions, the shorter will be the discussion thereof. As every statistician in business knows, his biggest problem is the man who wants 'facts', and not intelligent deductions from the available evidence. There are times when the conclusions from the available data must be qualified, and great care is required on the part of the statistician in the manner he qualifies his evidence. On the other hand, he should not allow himself to be bulldozed into statements which he knows cannot be substantiated. For example, in making estimates of the future trend of sales, he is at best making an informed guess, and it is important that this point be got across to the board without at the same time, however, implying that any damn fool's guess is as good as his own.

In short, the problem of the statistician in industry or commerce is not so much statistics, as the need to be able to present to his colleagues and superiors the information he has collected in such a way that it is clear and capable of only one interpretation. In this respect, diagrams are an especially valuable adjunct. Overloaded tables must be avoided.

One last point is worth making since it tends so often to be over-looked. Generally speaking, the value that a firm derives from its statistician, particularly if he is the first appointment, is dependent on the statistician's own efforts. Usually he can expect little guidance as to what is required from him. It is up to him to decide what information is most likely to help in the formulation of policy, and as already stated, what manner of presentation will find readiest acceptance.[1]

Business Forecasting

One of the most interesting applications of statistical techniques in business is known as forecasting. It is a commonplace that anyone who can predict the future accurately will soon make a fortune for himself and those associated with him. The mere fact that there are so few millionaires in this world makes it clear that there are few such people. However, some estimates of future trends of sales, consumer wants, etc., have to be made. There are various types of forecast,

---

[1] For a rather different situation in a major undertaking, see 'Statistics in Imperial Chemical Industries Ltd.', *The Statistician*, No.1, 1962.

ranging from the hunch of the business man that prices are going to rise, to the rather more sophisticated efforts of the statisticians at the National Institute of Economic and Social Research and the Central Statistical Office, whose labours are directed to the determination of short-run forecasts. These techniques are directly related to economics and are described as econometrics. Between these two extremes, forecasts can be made on the basis of extrapolating trends either by sketching in a freehand line following the path of the trend on a graph or by more complicated mathematical methods on the lines of those discussed in Chapter XI.

Nevertheless, for all the energy and thought expended in this particular field of study, it can hardly be asserted that the achievements of statistical forecasting have added any laurels to the subject of statistics. While the hunch is a plain undiluted guess, the complex econometric model is also ultimately dependent upon numerous hunches, insofar as one must make guesses or estimates about the relationship between and the magnitude of certain key variables in the economy.

Modern economic planning or forecasting depends largely on the development of an economic 'model' which enables the workings of the economic system to be simulated in such a way that the consequences of given policies, e.g. a Budget surplus, or import surplus, can be predicted. The accuracy of the prediction depends, of course, on the extent to which the model can reproduce the manifold complexities of the economic system. How far is this a practicable proposition?

The basis of any economic model is known as a 'social accounting matrix' and at the present time a group of economists under Professor J. R. N. Stone are working at Cambridge on what is the most complicated and largest model of the economic system ever devised in this country, graphically referred to as SAM. This is being used by the National Economic Development Council and takes the form of drawing up a set of accounts on double entry principles for each main sector of the economy, e.g. agriculture, constructional trades, public authorities, exports, etc. The object of this set of accounts, i.e. SAM, is to reproduce the inter-relationship of each sector, e.g. the output of the mining and quarrying industry is the input of the building and constructional industry among others.

Once such a social accounting matrix has been devised, it is possible to follow through the effects of changes introduced in one sector of the economy and their consequences on other sectors. This is done by setting up a system of simultaneous equations which reflect these inter-sector relations. The electronic computer into which these data are fed enables the economists to simulate a whole range of situations which might emerge given various conditions and responses. Unfortunately, the reliability of the predictions, *i.e.* the behaviour pattern of the economy reproduced in the model, depends on the quality of the information put into the matrix and the computer.[1] While a computer can perform miracles of high speed sorting of facts or calculations, it cannot 'think'. It can only work on the information fed into it by its operators.

With economic model building there are two basic problems. The first is to know exactly how the economy works and the precise nature of the inter-relationship between different sectors of the economy. For example, if the UK had a Budget deficit of £200m. in 1964–65 what would the results be for prices, capital investment, saving, on imports, on exports, etc. ? While it is true to state that the modern economist knows a great deal about the working of the economic system, and his knowledge is continually expanding, it has not yet reached the stage at which he can talk of economic relationships which approximate to the laws of physics or chemistry. There are so many influences in the economy, from the state of business confidence, the state of the world markets, to the volume of bank credit and even political considerations, that it is quite impossible to predict with any degree of confidence their consequences in a given framework of economic facts.

Apart from deficiencies in our knowledge of the precise workings of economic system, our statistical information is, as was indicated in the previous chapter, still limited. When Mr Macmillan presented his 1957 Budget he complained that formulating his budget policy with the then available statistical indicators was akin to looking up this year's trains in last year's Bradshaw. Since then, there has been a very real improvement both in the quality and coverage of official statistics. But there are still gaps and, more to the point, the statistics often appear long after the events they describe.

'Forward-looking' Statistics

Because of the difficulties of making reliable forecasts, usually on the basis of extrapolating, *i.e.* projecting, the experience of the past, recent years have witnessed an attempt to collect what are sometimes called 'forward-looking statistics'. These data are based generally on surveys of consumers' and industrialists' expectations and intentions regarding the future. In the United States such surveys are frequent, but they are relatively new in the UK. One of the best known is the Board of Trade's sample survey among large industrial

---

[1] For a clear and non-technical description of such econometric models see 'Econometric and Sample Survey Methods of Forecasting' by L. Klein in *Business Forecasting*, published by the Market Research Society 1958. Also R. J. Ball's article 'The Construction and Uses of Statistical Economic Models', *The Statistician*, 1963, No.4.

TABLE 88
INDUSTRY'S FIXED CAPITAL EXPENDITURE*
£ million

Period	Expenditure at 1958 prices seasonally adjusted						Expenditure at current prices		
	Analysis by Industrial Sectors			Analysis by Type of Asset					
	Manu-facturing Industry (b)	Distributive and Service Industries (c)	Total	New Building Work	Vehicles	Plant and Machinery	Manu-facturing Industry (b)	Distributive and Service Industries (c)	Total
1958	922	699	1,621	449	378	794	922	699	1,621
1959	871	799	1,670	457	422	791	867	789	1,656
1960	1,016	873	1,889	530	469	890	1,021	862	1,883
1961	1,195	927	2,122	634	445	1,043	1,239	931	2,170
1962	1,098	884	1,982	629	372	981	1,168	914	2,082
1963									
1st quarter	241	216	457	138	98	221	241	221	462
2nd quarter	253	226	479	148	104	227	265	229	494
3rd quarter	246	227	473	143	105	225	267	226	493
4th quarter									

The total of manufacturing industry and of the distributive and service industries covers a wide sector of the economy, accounting–after exclusion of dwellings – for 56 per cent of total capital expenditure in 1960. The total comprises most industries and services outside agriculture, forestry and fishing, coal mining, public utilities, public transport corporations and public administration. The excluded parts are, except for agriculture, very largely in the public sector of the economy and though some public expenditure is included, it is small; the figures are therefore very largely representative of the trend of expenditure of private industry and business.

* Expenditure on vehicles (including ships) and plant and machinery less receipts from sales of these assets and on new building work.

undertakings to assess their intentions in the field of capital expenditure. The results of a recent survey, as published quarterly in the *Board of Trade Journal*, are shown in Table 88. In the past year these estimates of capital expenditure in the manufacturing, distributive and service industries have been improved in a number of ways. First, new industrial groups have been included, and industrial groups already covered by the survey have been extended to bring in unincorporated businesses. Second, the panel of firms supplying the data making up each industry's figures has been increased, so that it is estimated that the panel of informants now covers about two-thirds of the total fixed investment. Finally, whereas initially information on projected capital expenditure was given 6–9 months ahead, the Board of Trade is now asking for intentions relating to the period 12–15 months ahead. The aggregate figures are adjusted to 1958 prices, and classified by industrial sectors and by the type of asset.

Directly related to the Board of Trade survey is the Federation of British Industries' 4-monthly survey, which seeks to evaluate industrialists' views regarding their short-term prospects, regarding output, orders and exports. (A copy of the Schedule used in this survey is reproduced on pp. 27-8.) A recent review in the National Institute's *Economic Review* (Nov. 1963) of the usefulness of the F.B.I. Survey suggests that it has been helpful in assessing national economic trends. In contrast, the authors of a paper in the Market Research Society's collection of essays entitled *Business Forecasting* comments that 'We found by past experience that we are not able to base future projections of our requirements upon consolidated manufacturers' plans. Manufacturers tend to be incurably optimistic, and tend to talk in terms of maximum capacity only'.[1]

A recent official description of the methods of short-term economic forecasting makes the point that 'while the relationship between forecast and out-turn has varied very much from company to company it has been found that, in the aggregate, firms' forecasts are in excess of actual expenditures in a fairly systematic way',[2] Thus it is possible to make reasonably reliable assessments of future trends even if the basic information is biased but, of course, the bias must be consistent in direction!

Linked to this field of enquiry is the recent and considerable improvement in the quality and coverage of statistics with special reference to industry. An important point here is the need for early publication. Thus, the index of industrial production is published monthly, with a time lag of only six weeks, and this is increasingly used as an indicator of trends. The index numbers of the volume of

---

[1] C. P. Davidson and J. J. Joyce, p. 19, *op cit.* See, however, the concluding paragraph of the essay by C. T. Saunders in the same collection.

[2] Short-term economic forecasting in the U.K., *Economic Trends*, August 1964.

TABLE 89

Index Numbers of the Volume of Orders and Deliveries in the Engineering and Electrical Goods Industries

	Total			Export			Home		
	Orders-on-hand (end of period)	Net new orders*	Deliveries*	Orders-on-hand (end of period)	Net new orders*	Deliveries*	Orders-on-hand (end of period)	Net new orders*	Deliveries*
	December 1958 = 100	Average 1958 deliveries = 100		December 1958 = 100	Average 1958 deliveries = 100		December 1958 = 100	Average 1958 deliveries = 100	
1958	100	92	100	100	88	100	100	93	100
1959	101	105	104	102	103	102	101	106	105
1960	117	123	114	119	122	111	117	123	115
1961	120	123	122	126	122	118	117	123	123
1962	115	120	123	129	128	126	109	117	122
1963 1st quarter	117	132	126	123	121	139	115	135	122
2nd quarter	123	138	123	128	146	133	121	135	120
3rd quarter†	129	136	119	134	137	122	128	136	118
1963 October ..	118	115	125	129	127	129	114	111	123
November	116	115	126	130	142	130	111	105	125
December									

* Adjusted to allow for differences in the length of calendar months, but not for holidays or for other seasonal variations.

† Provisional.

16

'Orders and Deliveries' (by adjusting money values to 1958 prices) from a wide range of industries where the production process takes a certain time, *e.g.* engineering, textiles, clothing industries, are valuable short-run indicators of activity. The *Board of Trade Journal* publishes such indices as shown in Table 89 relating to the engineering industries for both home and export markets. The Ministry of Works collects statistics of new orders obtained by the constructional industry. Formerly such data were collected quarterly, but since January 1964 the series has been collected monthly. Likewise, the Ministry of Works carries out a half-yearly enquiry into builders and property developers' intentions concerning private house building. A tremendous amount of working capital is tied up in stocks held by British industry and firms engaged in wholesale and retail trade. Important data are collected thereon by the Census of Production and the Census of Distribution. In addition to these, however, figures are collected each quarter from a sample of manufacturers, wholesalers and retailers whose total stock represent about two-thirds of the national stocks, and the results are published periodically in the *Board of Trade Journal*. Table 90 illustrates the type of information collected and its mode of publication. These figures do not show the absolute value of the stocks held, but only changes in volume measured in

TABLE 90

CHANGES IN THE VOLUME OF INDUSTRY'S STOCKS AT 1958 PRICES
SEASONALLY ADJUSTED*

(£ million)

Period	Manu-facturing	Wholesale distribution	Retail distribution	Total
Level of stocks at 31st December 1959	4,952	810	946	6,708
1958	+ 37	+ 14	+ 16	+ 67
1959	+ 73	+ 15	+ 49	+ 137
1960	+ 520	+ 48	+ 37	+ 605
1961	+ 230	+ 36	+ 4	+ 270
1962	+ 33	+ 22	+ 40	+ 95
1963	+ 89	+ 15	+ 27	+ 131
1963 1st quarter	+ 56	− 16	− 3	+ 37
2nd quarter	− 9	+ 22	− 3	+ 10
3rd quarter	− 33	− 19	− 8	− 60
4th quarter	+ 75	+ 28	+ 41	+ 144

Source: Board of Trade Journal, 10th April, 1964.

* Estimates are shown to the nearest £ million but should not be regarded as accurate to this degree.

terms of 1958 prices. A description of the sources of information on manufacturers' and distributors' stocks was given in *Economic Trends* (March 1959).

In themselves, such series and data may not provide a complete answer to the problem of economic forecasting. However, insofar as they are published relatively quickly after the events they describe, and to the extent that experience shows some inter-relationship between changes in these forward-looking data, there is every possibility that a valuable tool may be devised which will improve the accuracy of short-run predictions. At least, if the interpretations of these data and the findings of the more sophisticated techniques such as the econometric model of the U.K. and its quarterly forecasts, coincide, then it is a reasonable assumption that the findings are reliable. Always assuming, of course, that the model is not based on the same data as are the other estimates.[1]

Government economic policy is based upon short-term forecasts which are themselves formulated on the basis of two main sets of economic data. The first stage in the formulation of a forecast is the diagnosis of recent trends. The statistical evidence for this is derived from a variety of economic series, in particular those relating to the labour situation such as the number of registered vacancies and unemployed, together with figures of wage rates. In addition, retail sales, which account for some fifty per cent of personal consumption, new car registrations, hire purchase transactions, as well as export and import data, are all examined to assess the volume of demand for the economy's resources. The second branch of the exercise is based upon the predictive data available. Such information is derived from two main sources, surveys of intentions such as those described above and the programmes of planned and scheduled expenditure developed by government departments and public boards, *e.g.* the nationalised industries. These predictive data are then linked with other estimates, *e.g.* the quarterly figures of the Gross Domestic Product[2] which then enable preliminary forecasts of all expenditure other than personal expenditure to be estimated.

Estimates of prospective personal consumption are built up by making individual forecasts of all the main components of personal income, direct taxes, consumer prices and personal savings. Such estimates are then reconciled with estimates of the aggregate expenditure and income derived by estimating total personal income from the prospective total of national output. This is possible because the level of factor incomes is obviously directly linked with the volume of output. A key figure in the estimate of prospective

---

[1] The cynic would, however, argue that the mere fact of several economists agreeing on the significance of a set of data is in itself conclusive!

[2] See p. 434.

income is that for wages and salaries which together account for almost two-thirds of total personal incomes. This is based on estimates of rates of change in wage rates which are themselves estimated in the light of recent trends. This is not as haphazard or arbitrary as it may appear since wage bargaining processes are considerably influenced by habit and convention.[1] Hence past experience is a reliable guide, although special attention must be paid to the trend of prices which can affect the pattern of wage bargaining. These estimates of wage rates and earnings are then linked up with the estimates of prospective demand for labour.

The pattern of price movements is a fundamental factor in any economic forecast. The trend of consumer prices is evaluated from a break-down of the goods and services which are included in the retail price index. Thus the future movement of food prices which are sensitive to changes in domestic supplies are evaluated independently of changes in rents and rates, as well as of other goods and services. Of these, the official report notes that U.K. businessmen generally fix their prices on the basis of a standard, or normal, degree of capacity working. In other words, given the cost of labour, short-term variations in the level of output make little difference to prices charged.[2]

The entire structure of the forecasts relating to the internal domestic economy is dependent upon changes in the balance of payments. For this purpose exports are estimated by breaking down the aggregate for the main markets. For example, U.K. exports to the U.S.A. are correlated with changes in the U.S. gross domestic products. There is a similar breakdown of imports, in this case by class, e.g., foodstuffs, commodities and materials plus manufactures. For each class an estimate is made on the basis of price trends and demand.

For all the apparent comprehensiveness of this approach to economic planning, the fact remains that all the component estimates are no more than informed guesses. In addition, a serious error in any one major component, e.g. the balance of payments, can vitiate the results of the entire exercise. Nevertheless, the techniques and information available are being continuously improved.

Industrial Statistics

So far the discussion has concentrated primarily upon what has been called earlier in this book 'descriptive statistics'. Generally speaking, particularly in commerce and even in industry, a large volume of the statistics collected and prepared are of this type.

---

[1] Op. cit. *Economic Trends*, August 1964.
[2] Op. cit.

Sometimes, where sales forecasting is an important part of the work, more refined statistical techniques will be used. Generally speaking, however, a familiarity with statistical sources, the presentation and interpretation of such data are the prerequisites of a competent statistician. In industry, however, at least on the production side, the main emphasis is laid upon statistical techniques. In recent years the importance of these techniques has been growing apace, and there is, as will be seen, tremendous scope for the mathematical statistician in industry, who is interested in the growing field of operational research.[1]

Quality Control

Probably the most widely known of all statistical techniques employed in industry is statistical *quality control*. This is a simple application of the theory based on the Normal curve (see Chapter XII). Modern mass production techniques involve ensuring that each item of output has standard dimensions or other physical properties within certain limits of a pre-determined ideal standard. In theory, the product can be checked when it comes from the machine. But at this stage the damage may have been done. It is far more economic to try to check any fault in the production process at the earliest possible moment, and statistical quality control, because it draws attention at an early stage to a situation in which the quality of production is beginning to deteriorate, is extensively employed. The basis of the technique is illustrated in Figure 23, which shows two charts, the Mean Chart and Range Chart. Time is measured along the horizontal axis, while the variation in specifications of the product are measured along the vertical axis. The so-called process average or mean range represents the approximation to the standard product which experience with the manufacturing process has shown can be achieved. Furthermore, on the basis of past production experience, the variation about this standard quality has been measured and is indicated by the addition on both charts of two further lines referred to as the inner and outer limits. These limits, usually referred to as control limits, are equal to 1·96 and 3·0 standard errors of the sampling distribution. Once the production process is under way, samples of output are taken at fixed intervals and the average dimension of the sample and the range of dimensions thereof are plotted on the two charts. If either the average or the range begin to drift outside the control limits, this signifies that the production process is tending to go out of control. If left unchecked, therefore, the process will begin to produce defective components. The size of the sample and the periodicity of the sample taken depends on the process, and must

[1] Some of the methods are outlined below.

Figure 23

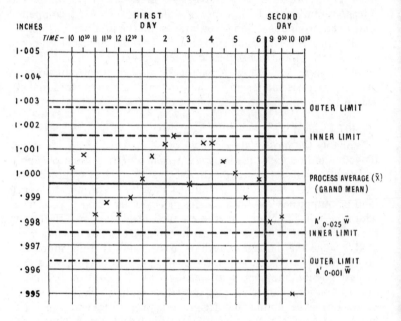

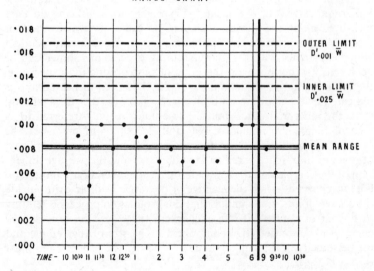

be determined by the statistician in conjunction with the engineer. The reader will doubtless appreciate the reasons for using two charts for this particular technique. Successive samples may in fact produce the same mean size, but as was shown in the discussion of averages, it is possible for two samples to have the same means but very different dispersions. Thus, one could have a situation in which the sample mean was within the inner control limits, and yet every one of the items in the sample was defective, since it fell outside the limits set on the range chart. The great advantage of this particular system is not only that it draws attention almost immediately to a situation in which the production process is getting out of control, but once installed, it can be operated by relatively unskilled workers, provided they adhere strictly to the rules laid down by the statistician.

The foregoing example illustrates the simplest type of quality control. In recent years, however, there have been extensive adaptations of this system to all kinds of industrial processes and the study of industrial quality control forms an important part of the application of statistical theory.

REFERENCES

Quality – Its Creation and Control, Institution of Production Engineers.
Quality Control, G. Herdan. Nelson.
A First Guide to Quality Control for Engineers, E. H. Sealy, H.M.S.O.

The Electronic Computer

Just as it would have been quite impossible for men to have launched satellites into space and sent rockets to the moon without the employment of high speed electronic computers, so some of the new statistical developments in industry would be impossible without them. Before reviewing briefly some of the ideas and methods known generally as 'Operational Research', some description of the modern computer is necessary. It is not suggested that the reader for whom this book is written needs to understand the workings of a computer. It is, however, essential that he has some understanding of what the computer is and is capable of doing. The origins of the electronic computer can be traced back to well over 100 years ago, when Charles Babbage first constructed his calculating machine. It is, however, only in recent years that modern engineering techniques have made the modern computer a practical possibility.

There are, in fact, two main types of computer. The more common is the *digital computer*, which is no more than an extremely fast calculating machine with a built-in memory. This is the type of computer generally used in the commercial and industrial undertaking. The other type of computer is known as an *analogue* computer. The characteristic of this instrument is that it can be connected up to provide an electronic network analogous in its behaviour to a particular physical system. This means, in effect, that one can set up a machine to behave as, for example, an aircraft wing in flight and simulate conditions similar to those which the wing will meet in flight. The importance of such an instrument in engineering research and design is obvious. Since, however, the analogue computer is a highly complex and specialised instrument, no more need be said about it here. More relevant to present requirements is a consideration of the digital computer.

Such machines are often referred to as electronic brains. Nothing could be more mistaken. The computer cannot think constructively or originally, as can the human mind. It merely has the ability to arrange knowledge on pre-determined lines and to derive conclusions from the experience stored in its 'memory'. This is possible because the computer is a high speed calculating machine, using a memory into which data can be fed. It can then classify this information, interrelating different branches of it at extremely high speeds. In other words, this is not thinking, but merely a high speed process of sorting and comparison. Herein lies the main advantage of the machine for clerical work. The work of filing and maintenance of records, for example, hitherto done by hand, can now be achieved by the computer.[1]

The computer itself is not a single machine, but rather a collection of machines, some of which are linked together and others of which operate in isolation. The first unit which goes to make up a computer system is an input device, whereby information and instructions for action are fed into the computer. This may be done on the basis of a punched card, paper tape or magnetic tape. Before information can be fed into the computer, it must be codified, *i.e.* translated into a language which the computer can deal with, which means, of course, some form of figures or symbols. There is now a British Standards specification for a uniform input code. A term often encountered in connection with computers is 'programming'. This merely refers to the preparation of instructions to the machine. This is not a particularly skilled or complicated job, but nevertheless it is a highly specialised activity. Young women with a flair for arithmetic and figures can, with proper training, become excellent programmers for a particular type of machine. The remainder of the computer system proper usually consists of a processing unit which performs the arithmetic and sorting activities, a storage unit which can be termed

[1] As is being done for the National Insurance records and is shortly to be used by the Board of Inland Revenue for its P.A.Y.E. records. Many large firms have installed smaller machines to handle their payroll, invoicing and ledger procedures.

the memory of the computer, and an output unit which produces the results of the work. The latter is really a very high speed printing machine. This is obviously necessary since the calculations performed by the processing unit are carried out at such a rate that without high speed printing the advantages would be lost.

The pace of development in the computer field in recent years has been remarkable. The latest computers are capable of storing very much more information than the earlier models, and above all of working at very much faster rates. Such highly specialised and complex machines are obviously expensive, and the decision as to whether or not a computer should be installed is not always easy. Generally speaking, unless there is sufficient work in an organisation to keep the computer working continuously, then it is more economical to hire the services of a computer. The advantages of hiring computers are very real, and an increasing number of firms in this country are providing such facilities. It obviates the situation which has arisen in some cases where, in order that the computer should not be left idle for long periods, more work is being processed through the computer merely in order to occupy it rather than for any intrinsic value that such additional data may possess for the firm.

Operational Research

The major benefit in the statistical field as distinct from the savings on routine clerical work which industry and business have derived from the introduction of the computer, is the application of operational research techniques to their problems. Evolved during the last war, O.R. as it is often termed, has been adapted to assist management in two ways. First, it can help management in arriving at policy decisions in the light of all the known facts, and with the full knowledge of the consequences of the alternative policies available to the management. Second, it can assist management to secure better methods of carrying out existing operations. The significance of the computer lies in the fact that operational research uses mathematical models in the study of industrial and commercial problems. These models are often difficult to construct and invariably involve innumerable calculations which, without a computer, could not be resolved in any reasonable time. The basic principle underlying O.R. is the mathematical model based upon a set of equations which reflect different aspects of the working of the system. For example, in problems concerning the level of stock of component parts which a manufacturing firm should hold, the size of its stocks at any time determines the amount of working capital tied up in this way. Likewise, the cost of running out of the stock of a given component in terms of production delays can be calculated and expressed in the form of an equation. Thus, these mathematical models comprise a number of such equations, to which an optimum solution may be derived. The optimum solution in question may be in terms of that allocation of resources which yields a manufacturer the highest profit. Or it may, in another set of circumstances, be the solution which reduces to a minimum delays in the production process. The essence of O.R. techniques, the mathematics apart, is precise information and data relating to the problem concerned. Whether the problem be one of stock levels, transport, or production, the first stages must be to define the problem and obtain sufficient information relating to the relevant variables in order to construct the model, *e.g.* cost of resources used. One of the major virtues of O.R. is that before any mathematical work can be started, the team and management are forced to consider all aspects of the problem and what its solution may entail. Such detailed analyses of given situations are in themselves very useful in clearing the mind! Reference is occasionally made in literature on this subject to the 'O.R. team', and it is important that the reader should appreciate that operational research is essentially inter-disciplinary. It uses the skills and techniques of mathematics, statistics, economics, engineering, and other sciences, so that whatever the type of problem encountered, the skills are available to analyse it.

Linear Programming

One of the earliest and the most widely applied O.R. techniques is known as *linear programming*. This is a technique for allocating limited resources among a number of competing demands in such a way as to maximise the output or minimise costs. The complication here lies in the fact that alternative uses exist for the limited resources. Furthermore, there are substantial costs involved in allocating these resources more to one use than to another. The computer is eminently suited for this type of problem involving factors which can vary within specified limits. The machine repeats the calculation over and over again, varying each factor by pre-determined amounts, and an optimum solution is derived. One of the earliest applications of linear programming has been that used by the National Coal Board to determine the best policy of marketing coal. The facts are that a number of different grades of coal can be produced from a number of different coal mines and are required in varying quantities by a number of different customers. Linear programming techniques enabled the decision to be made as to which grade of coal should be produced from which coal mine, and sent to which customer in such a way, either that output from some of the mines is maximised, or that output and transportation costs are minimised.

Transportation problems lend themselves especially well to linear programming. Thus, a large undertaking which maintains depots in different parts of the country in order to meet the needs of its various customers in those areas, can by the use of linear programming determine whether these depots are the best-sited to keep transportation costs to a minimum yet satisfy their customers' requirements with the minimum delay.

Queueing Theory

One of the most interesting applications of O.R. is that known as *queueing theory*. Motorists are accustomed to delays at toll bridges or ferries; many of us are familiar with the delays on a busy telephone switchboard; with waiting in canteen queues for lunch, etc. The general issue dealt with in this particular type of problem is how to reduce waiting time to the point at which the costs thereof are minimised. One of the simplest illustrations concerns a post-war study of ore handling in British ports, where substantial imports of ore were handled by special equipment. The ideal situation would be, of course, that ships would arrive at regular intervals, come straight into port, be unloaded immediately and allowed to depart. Furthermore, the intervals of arrival should be such that at no time was any of the unloading equipment in the docks idle. In practice, the situation is of course rather different. Ships arrive at irregular intervals and often have to wait because there is no berth available. This is an expensive process, since the ship is thereby losing revenue. On the other hand, to expand the port facilities to meet any feasible number of ships arriving more or less at the same time so that none of these ships shall be delayed, would be extremely expensive and would mean that the port facilities would be under-utilised for long periods of time. O.R. techniques enabled a considerable amount of information relating to costs of the ship's time, as well as the operation of unloading equipment, to be calculated. It was possible to estimate the minimum discharging capacity that needed to be made available in order to keep waiting time of ships at a minimum. In other words, this type of queueing problem illustrates the classical problem of conflicting interests between the two parties; the port operator who wishes to see all his berths occupied and busy, and the ship owners or charterers who wish to ensure that their ships are never kept waiting to turn round.

The same techniques can be employed in the designing of landing facilities at airports, of taxi parking areas, the movement of travellers through customs, or out-patients at a hospital; and even the timing of traffic lights to facilitate movement of traffic in such a way that queues do not pile up on either side of the intersection.

Stock Control

Most business organisations, whether they are manufacturers, wholesalers or retailers, carry a stock both of components and raw materials for production purposes, or finished goods for supply on demand from their customers. The published statistics available on stocks by firms in the U.K. suggests that the value of such stocks represents many thousand millions of pounds of working capital. As any accountant knows, a firm which is short of working capital can usually meet part of its needs by running down its stocks. The essence of stock control, with which is usually linked the buying policy for stock, is to evaluate on the one hand the relative costs of maintaining so large a stock that there is no danger whatsoever of being unable to satisfy demand, whether that demand be from the producer or a customer, and on the other hand the cost in terms of tied-up capital, factory space, etc. An optimum solution to this type of problem can be obtained by an analysis of information relating to stocks and withdrawals from them, as well as the annual cost of providing storage space, of interest on capital tied up in stocks, loss arising from obsolescence or deterioration, and the risk of loss due to falling prices. On the other hand, with a large stock advantages may accrue from the elimination of the danger of run-outs; possibly lower prices from suppliers in the case of raw materials where a regular order is placed and, in the case of the manufacturer, the risks and cost of failing to meet his customers' requirements on demand. Linked with the problem of what level of stocks should be held to balance the relative costs of the two extreme policies, is the problem as to when new orders for the replacement of stock should be made. The problem here arises because if there is a sudden demand for a particular component, and stocks are depleted, production may be held up since there is always some delay before new materials can be ordered and delivered. Furthermore, an examination of this particular problem will also bring to light the problems of ensuring that the person in charge of stock is kept fully aware of the position of each component or type of stock held. Where the stock consists of a wide range of heterogeneous products, it is all to easy to overlook that some stocks are near run-out. Thus, by ascertaining the optimum size of stock to be held and the point at which a re-order should be placed for such stock, substantial savings may be made in a firm or organisation where stocks are high. For example, Mr. R. A. Ward reports on the basis of practical studies carried out over the last three years, 'In most of the stores at least 20 per cent of the commodities stocked were found not to have been issued in the preceding 12 months and those in charge of the store had not realised that so high a proportion of the stock had remained static'.[1] He goes on to add that the results so far achieved by such methods indicate that between 2 and 4 per cent of the annual value of the store's turnover can be saved.

Critical Path Analysis

Towards the end of the 1950s, both in Britain and the United States a new management technique was evolved which, unlike most of the O.R. methods, does not necessarily entail the application of mathematical statistics. In practice, however, the more complex problems necessitate such techniques and a computer! This technique, known as *Critical Path Analysis*, has been used with especial success with large contracts where the basic problem comprises the organisation of a large number of interrelated jobs in such a way that one part of the work cannot always be started until another is finished, and where there is a fixed point of time at which the entire project must be completed. An obvious example would be a large construction project, *e.g.* an oil refinery or electricity generating station, and the U.S. Navy used this technique in the construction of the Polaris submarine. The essential principle of critical path analysis is the setting down in simple graphical form – in the form of a network of 'paths' – all the available information relating to the project in such a way that the time required for each single stage of the project is evaluated. The chart shows the interdependence of each stage and, given the precedence of the various sections of the work, an estimate can be derived of the probable minimum time required for the project. The sequence of events required to minimise the time required is known as the critical path. The term critical path analysis (C.P.A.) stems from the fact that by continuously checking the progress of the work, if at any point any of the sections of the work take longer to complete than anticipated, then the original schedule will be thrown out. By use of a computer, which is fed continuously with the latest information on progress achieved, the model can be revised in the light of such unexpected changes and a new plan of operation formulated to minimise the time needed to complete the job. If, however, the time taken on particular sequences of the job is reduced, then it may prove possible to follow a slightly different schedule, dependent on the revised programme put out by the computer, which will have the effect of reducing the overall time for the entire project. This will mean that a new sequence of work becomes relevant and hence critical.

The main advantage of this technique is that it enables the pro-

[1] Operational Research in Local Government, R.I.P.A.

ductive efforts to be concentrated on those points where the maximum benefit in terms of time saving are likely to be achieved. A more sophisticated application of the same technique developed in the United States is known as *project evaluation and review technique* (known as P.E.R.T.). In this case three separate time estimates are submitted for each activity or section of the work, the most optimistic, the most likely, and the most pessimistic. As the project proceeds, information relating to progress is assembled, and the probability of achieving the target dates is then reviewed. The obvious advantage of using three estimates of the time required for each section of the work is that it avoids the situation in which the engineer or technician in charge plays safe if he only has to give one estimate of the time he will require. An offshoot of the same technique is the so-called least-cost programme, wherein both the normal time and cost for a given project are estimated and in addition a 'crash' time and cost are submitted for each activity, so that in the event of any unexpected delay on a particular section of the work, the relative advantages of expending more money on a crash effort to offset the delay and the consequences of deferring the final completion date may be compared.

Apart from the foregoing, O.R. techniques there are many others, such as the Monte Carlo method, replacement theory, and dynamic programming. The theory of games is a relatively new development, which undoubtedly has a future as an aid to business decision-making, but some of the theoretical problems involved are highly complex and as yet commercial application of this technique is limited. However, the essence of all O.R. techniques is that they are primarily mathematical in character. The foregoing sections have sought merely to indicate in very brief outline the nature of the problems which such techniques may help resolve. The management trainee or accountant who has some mathematical ability, *i.e.* at least Additional Maths at 'O' Level, preferably 'A' Level Pure Maths, can consult with profit and interest some of the references given hereunder. For the student, however, whose mathematics is merely sufficient to acquaint him with the elementary statistical methods outlined in the earlier part of this book, the only solution is to read one of the simpler introductions to O.R. in the hope that he will at least understand what his statistical colleagues are talking about.

REFERENCES

One of the least technical introductions to O.R. is *A Manager's Guide to Operational Research* by Rivett and Ackoff (John Wiley and Sons). This is completely non-mathematical and eminently suited to the non-mathematician. A slightly more advanced introduction is *A Guide to Operational Research*, by E. Duckworth (Methuen). This contains a number of simple illustrations of various techniques with no more arithmetic than the average 'O' Level candidate can understand.

At about the same level, *An Introduction to Critical Path Analysis*, by K. G. Lockyer (Pitman) deals with this one particular technique.

For the reader with some mathematics, *Management and Mathematics* by Fletcher and Clarke (Business Publications) is a useful introduction to the various O.R. techniques referred to above. All these works provide references to further and more advanced reading, of which there is a growing quantity, particularly from the United States.

Acceptance Sampling for Accountants

Every accounting student will be familiar with test checks on the books and vouchers of client firms. The principle underlying test checking is the same as that underlying sampling. But, whereas in statistical sampling the sample is selected in such a way as to ensure that the choice of population units is random, with the conventional test checks everything depends on the auditor's own judgement. For example, he may decide to vouch two or three months purchase invoices and a different two months of sales invoices, and perhaps one month of the petty cash vouchers. In other words, a very large element of intuition and subjective judgement determines the actual scope of the audit.

In recent years, statistical sampling techniques have been evolved to assist auditors and accountants who are confronted with large quantities of checking, as for example in the records of a large organisation. One of the most generally used techniques is known as *acceptance sampling*. This is a statistical approach to the problem of controlling the accuracy of clerical work by means of sample inspection of the records. The results of the inspection are then compared with a previously determined objective standard, and in the light of the results of that comparison, the decision is made as to whether or not more checking is necessary.

The basic problem, as with any significance test, is to decide at what level of confidence the auditor may accept the sample and thus run the risk of accepting an unsatisfactory population. Alternatively, if he rejects the sample, he runs the risk of rejecting a satisfactory population.

The basic principle underlying all acceptance sampling methods is to achieve a balance between two extremes, *i.e.* a complete checking of all items, which is largely unnecessary and expensive to the client or the organisation concerned, or a sample enquiry which can in theory be so small that a considerable risk is run of serious errors being overlooked. Since these errors will cost the organisation money, they may considerably outweigh any savings achieved by the reduced audit.

The application of acceptance sampling necessitates certain information before a sampling scheme can be devised. First, the size of the population of vouchers, invoices, etc., must be determined.[1] This may seem to be at cross-purposes with what has been written earlier about sample sizes being independent of the population size. These auditing techniques involve the use of specially prepared tables (see below) based upon a formula which uses the population size to provide the auditor with the sample size appropriate to the expected proportion of errors in the population and the level of confidence at which it will be decided to accept or reject the sample. The actual choice of the sample units, once the size has been determined by reference to the specially constructed tables, is best carried out by normal systematic sampling, *i.e.* a random starting point and every *nth* item selected. In drawing his sample, the auditor will probably stratify his population, deciding to take a small, *e.g.* 2 per cent, proportion of the very small invoices, and a slightly larger proportion of the invoices for larger sums. The acceptance level is simply the highest number of errors which can be accepted without rejecting the sample.

Second, an estimate must be made of the relative quality of the vouchers, *i.e.* what proportion are likely to contain errors. This can normally be estimated on the basis of past experience of that particular audit. Thus, if the client has his own internal audit it is reasonable to assume that the chances of errors are less than in a firm where financial control is less satisfactory. Next, the auditor must determine what is defined as an *acceptable error risk*.[2] This is merely a guide to action in the sense that once this pre-determined number or proportion of errors has been detected in the sample the decision is made to extend the audit. In practice, however, the accountant uses one of the published sets of tables[3] to guide him as to the appropriate sample size, given the population and the acceptance rate for that population that he considers appropriate.

The foregoing technique may be illustrated by a simple example. Let it be assumed that the bought ledger comprises some 4,800 accounts with balances of varying sizes. Let us assume, further, that the population can be divided into two strata, one of 4,000, another of 800, the former containing relatively small accounts, and the other rather larger balances. It may be assumed, further, that experience shows that the expected rate of errors in both strata will be 3 per cent and 2 per cent respectively. The auditor would be prepared to accept a variation about this expected error rate of 2 per cent, and he is prepared to accept or reject the sample subject to this degree of precision at the 95 per cent level of confidence. A reference to the

---

[1] Acceptance sampling can, of course, be applied to documents other than vouchers. It can also be applied to postings, costings, the castings of cash books, ledgers, etc.

[2] Some writers in this field use the tem *acceptance quality level*.

[3] *Sampling Inspection Tables*, by Dodge and Romig (John Wiley & Sons) 1944, or *Sampling Tables for Estimating Error Rates or Other Proportions*, R. G. Brown and L. L. Vance (University of California), 1961.

appropriate tables shows that a sample size in the case of the 4,000 population will be 261, and in the case of the 800 population a sample of 152 will be required. Thus a total sample of 413 out of the 4,800 balances in the population will need to be checked. The samples are then drawn by some systematic method of sampling from the strata, and the balances verified. Provided the errors detected do not exceed 5 per cent in the case of the smaller accounts and 4 per cent in the case of the larger accounts, the auditor will be prepared to pass the bought ledger. If, however, the error rate detected is in excess of the limits set, then further work will be required.

There are three main types of sampling techniques employed in this field. First, the single sample characterised by a fixed sample size and single acceptance number. Second, a double sample where the sample is divided into two parts and the second part is examined only if a clear decision cannot be arrived at upon the results derived from the first sample. Third, sequential sampling where, instead of choosing a sample of fixed size, the population is sampled item by item, the work being terminated as soon as there is sufficient information for a significant conclusion, *i.e.* that the records are so accurate that no more work need be done, or so inaccurate that a detailed investigation is necessary. Sequential sampling may take some considerable time, and consequently modifications of this technique have been evolved. Basically the method consists of taking a sample of a fixed size and then, dependent upon the number of errors found in that sample, the decision is made whether to accept the entire population or whether to take another sample.

Where double sampling is used, the tables again are consulted to determine the appropriate sample size given the acceptable error rate. Double sampling is sometimes preferred to single sampling, because a smaller size of sample can be used since there is the possibility of further check. Since in practice it is often not required, the amount of checking to be done is kept to a minimum. In the case of double sampling, however, it is generally agreed that the quality of the sample must be much higher, *i.e.* the proportion of errors detected in the first sample must be much lower, than with single sampling. What that level should be, if a second sample is not to be used, must be determined in advance.

Acceptance sampling, whether single or double sampling is used, only enables the auditor to decide in the light of pre-determined criteria to accept or reject the population as satisfactory or unsatisfactory. It does not enable him to decide how satisfactory or inaccurate any work is. For this purpose, a technique known as *estimation sampling* is used. This involves the taking of a sample to estimate the average characteristic of the population, which when taken together with the standard deviation will give an estimated value of the whole with a degree of probability of its accuracy. Estimation sampling is especially useful, for example, in stock valuation, particularly if there are numerous items. A relatively small sample thereof will enable a reliable estimate to be made of the whole, the size of the sample selected for purposes of estimation clearly depending on the limits within which any acceptable estimate of the population parameter must fall.

REFERENCES

Statistical Sampling for Auditors and Accountants, L. L. Vance and J. Neter. Introductory Chapters to Sampling Tables, op. cit.

SOME SUGGESTIONS FOR FURTHER READING

As was explained in the Preface, this text is designed primarily for students studying part-time for examinations in which Statistics is an optional or subsidiary subject. The coverage is sufficient to meet the needs of most of the professional associations' examinations, while the student working for a London External Degree in which Statistics is a subsidiary subject may also find some parts of the text useful to him.

Experience shows that few such part-time students find it possible to read extensively on any single subject. All too often, particularly in a subject such as statistics, even if they have the time, the choice of suitable literature, given their limited background knowledge of the subject, is often difficult. In fact, further reading of unsuitable texts may often result in greater confusion. However, since most examination candidates' papers reflect the absence of reading beyond a single textbook and/or tutorial notes, some selective reading on particular topics which recur regularly in most examination papers is strongly advised.

In order to assist the student who wishes to extend his reading and to offer other readers whose interest in this subject may have been sufficiently aroused by this text to read more widely, the following suggestions are made.

For the student who has no mathematics beyond the arithmetic in this text, the best book is *Principles of Medical Statistics* by Professor Bradford Hill (*The Lancet*, 7th edn.). The adjective 'medical' should not be allowed to put off the prospective reader. This book provides a brief but simple exposition of the principles of sampling and the various significance tests, some beyond the scope of the present book. In addition, it introduces the reader to some elementary ideas in the problems of statistical inference. For the reader who wishes to read more about the logic of significance tests and the basis of modern statistical analysis, *An Introduction to Statistical Science in Agriculture* by D. J. Finney (Oliver and Boyd, 2nd edn.) can be recommended. As with the first book mentioned, the title should not deter the reader. For the student interested especially in the application of statistical methods to economic data, a reading of selected chapters of *Applied General Statistics* by Croxton and Cowden (Prentice-Hall), or *Applied Statistics for Economists* by P. H. Karmel (Pitman, 2nd edn.)

is instructive. The reader must be prepared to work through the numerous but brief examples which illustrate points made in these texts, but few of them are difficult.

As indicated at the end of Chapter XV, there are many studies both of the theory and practice of sample surveys. However, the average reader of this book will undoubtedly find all he needs in Professor C. A. Moser's work, *Survey Methods in Social Investigation* (Heinemann). This has a detailed bibliography attached to it with suggestions for further reading. Selective reading on Interviewing, Questionnaire Design, etc., in the book *Research Methods in Social Relations* by Jahoda, Deutsch, Cook and Sellitz (Methuen) is also informative.

Extensive references to the literature on social and economic statistics have already been given in the relevant chapters, and there is no need to add to them here. An excellent book for the student of economic statistics, in which many of the problems of using such data are well explained and illustrated is *The Use of Economic Statistics* by C. A. Blyth (Allen & Unwin). The standard work on Social Statistics is the joint study by Carr Saunders, C. Jones and C. A. Moser referred to in Chapter XVI. Among periodicals and journals, *Economic Trends* is invaluable to the student of economic statistics, as are the two collections of articles from that journal which have been prepared by the Central Statistical Office, entitled *New Contributions to Economic Statistics*, First and Second Series, published by H.M. Stationery Office.

On industrial statistics and computer techniques, there is now a fairly extensive range of publications. As explained in the relevant text, however, these are mainly for readers equipped with a fair degree of mathematics (at least 'Advanced' level) and it is left to the reader so equipped to browse and to decide for himself which of these publications will meet his needs. From the titles cited in the relevant chapter he will be able to obtain further references.

ANSWERS: TABULATION

CHAPTER IV

1.

Period	Fuel used for production (000 tons)		Type of gas produced at gas works (mn. therms)				Purchased from (mn. therms)		Total available	
	Coal	Oil	Coal	Water	Oil	Total	Coke ovens	Other sources	Therms mn.	cu. feet mn.
1955										
1956										
1957										
1958										
1959										
etc.										

2.

Period	Entered			Cleared		
	Number	Net Tonnage		Number	Net Tonnage	
		Com'n-wealth	Foreign		Com'n-wealth	Foreign
		'000	'000		'0000	'000
1959 1st Quarter	11,945	9,926	9,465	9,472	7,635	4,568
2nd ,,	13,206	11,186	10,692	10,248	8,339	5,242
3rd ,,	13,815	11,445	11,270	11,091	8,816	5,757
4th ,,	12,145	10,753	10,043	10,551	8,335	5,262
1960 1st Quarter	12,696	10,456	10,786	10,674	8,151	5,297

3.

Family background	Private School				Council School				Total	
	Only children		More than one child in family		Only children		More than one child in family			
	Unit	%	Unit	%	Unit	%	Unit	%	Unit	%
Children from manual working class homes	7	0·1	33	0·4	567	7·7	5,150	69·8	5 757	78·0
Children from other than manual working class homes	37	0·5	142	2·0	209	2·8	1,236	16·7	1,624	22·0
TOTALS	44	0·6	175	2·4	776	10·5	6,386	86·5	7,381	100·0

Comment: Small percentage of working class children at private schools with advantage enjoyed by only children.

4.

Trading Area	Company					
	A		B		C	
	Amount	% of total	Amount	% of total	Amount	% of total
	£	%	£	%	£	%
United Kingdom	100,000	77	67,000	88	40,000	76
Commonwealth	10,000	8	2,500	4	5,700	11
European Free Trade Area	3,000	2	1,500	2	5,000	9
European Economic Commn.	15,000	12	5,000	6	2,100	4
United States of America	2,000	1	—	—	—	—
	130,000	100	76,000	100	52,800	100

Comment: Note differing area contributions in each company.

5. Production of gramophone records by type 1955-1958 (thousands):

Period	Type of record			
	78 r.p.m.	45 r.p.m.	33⅓ r.p.m.	Total
1955	46,347	4,587	8,989	59,923
1956	47,508	6,903	12,116	66,527
1957	51,359	13,161	13,766	78,286
1957 Jan.-Oct.	41,257	10,273	11,163	62,693
Nov.	5,224	1,466	1,394	8,084
Dec.	4,878	1,422	1,209	7,509
1958 Jan.-Oct.	24,387	20,141	12,488	57,016
Nov.	1,881	3,463	1,644	6,988

Comment: Note fall in 1958 and switch to 45 r.p.m. from 78.

6.

Disease	1935 - 39			1940 - 44		
	No. of Cases	Deaths	% Deaths in total	No. of Cases	Deaths	% Deaths in total
Lead Poisoning	677	74	10·9	326	29	8·9
Other „	111	7	6·3	859	52	6·1
Anthrax	144	18	12·5	109	16	14·7
Gassing	843	82	9·7	3,288	148	4·5
Totals	1,775	181	10·2	4,582	245	5·3

Comment: Note that proportions of deaths virtually halved; sharp increase in gassing cases due probably to wartime production conditions.

7.

Household Classification	Unit	Numbers of Households				Numbers of individuals in occupation				Numbers of Individuals per Household
		With married heads	With children under 5 years	Others	Total	Married heads of households	Children under 5 years	Others	Total	
Primary Family Unit Households	'000	9,848·4	2,516·4	136·4	12,501·2	9,848·4	3,306·6	23,870·0	37,025·0	2·96
	%	78·8	20·1	1·1	100·0	26·6	8·9	64·5	100·0	
Composite Household	'000	1,125·2	162·4	692·7	1,980·3	1,125·2	209·6	7,168·5	8,503·3	4·29
	%	56·8	8·2	35·0	100·0	13·2	2·5	84·3	100·0	
TOTALS	'000	10,973·6	2,678·8	829·1	14,481·5	10,973·6	3,516·2	31,038·5	45,528·3	3·14
	%	75·8	18·5	5·7	100·0	24·1	7·7	68·2	100·0	

Points of comment: (i) Density of population much higher in composite households.
(ii) Lower proportion of children under 5 in composite households.
(iii) Lower proportion of married heads of households in composite households.

8. The procedure for checking should be as follows:

(a) Check the addition of the total acreage per farm. This check reveals Farm C totals 58¼ acres (+2) and Farm F totals 71 acres (—10).

(b) Total acreage per farm should be checked against previous year's returns. If Farms C and F acreage is the same for both years as originally submitted the current year's return must be sent back to the farmers for the detail to be amended so that it sums to the total. If Farms C and F previous acreage is as the corrected figure on the current returns the original figures can be altered so that the columns sum correctly. The total acreage of the remaining farms should be checked and any significant differences queried with the farmers. (A question on variations in total size should be included on the census form.)

(c) A similar process should be followed with the total acreages per crop. This reveals Wheat 145¾ acres (—10); Oats 47½ acres (+2); Potatoes 90½ acres (—3); Sugar Beet 97 acres (+3).

It should be noted, however, that there is a far greater likelihood of a change in the distribution of the acreage between crops in the total than in the total itself. Therefore variations in crops per farm should be regarded as only worthy of query if they involve a considerable change in the nature of the farming. Here again the census return should include information to indicate changes in crop distribution.

ANSWERS: GRAPHS AND DIAGRAMS

CHAPTER V

1. These data are best represented by four line charts, plotting the years along the base axis, with the Y axis marked off with appropriate break at the foot between 50 and the origin as in Fig. 10.

Alternatively, use two vertical axes (as in Fig. 10) to bring out the relative movement in the individual series.

2. Use a bar diagram dimilar to that in Fig. 7. Since the change in the production index, with but one exception, is larger than the employment index, use the same bar, marking the employment index within the length of the bar as in Fig. 6. Alternatively, but since there are so many industries this would take time, repeat Fig. 4, i.e. contiguous bars for the two indices for each industry.

3. Two line graphs with the quarters marked off along the base and the employment figures along the vertical axis, which should be marked off from just under 200,000 to 550,000, with a break of the origin as in Fig. 10. The point to note is that there is an inverse movement between the two series. When unemployment is high, e.g. 1958 fourth quarter, vacancies tend to be low, and vice versa.

4. (i) A bar chart, preferably horizontal, on the lines of Fig. 6. The progress being marked within the length of the bar which marks the time required for the entire plan.

   (ii) A map of England and Wales with the circulation figures under the names of each county.

   (iii) Column diagram if half-yearly figures, line chart if monthly.

   (iv) Similar to Fig. 11, i.e. heavy line for Bank Rate and bars for gold and dollar reserves, since these are measured at half-yearly intervals.

5. The histogram is explained on page 82. The frequency polygon is derived by linking by straight lines the mid-points at the top of each column and the frequency curve is the result of smoothing the frequency polygon into a smooth shaped curve. The two frequency polygons of the two distributions can be superimposed, using a continuous line in the one case and a broken outline for the other. Note, however, in the last three classes the class/interval is twice as wide as the earlier classes, and therefore, since the areas of the bars are proportional to the frequency, the height measured against the Y axis of these bars is halved.

6. See pages 78-81.

7. Since the purpose of these diagrams is to show the proportionate change in the contribution from each source, the best method is to draw two bars of equal length, distributing the relative, i.e. proportionate contribution of each source to the revenue within its length. There will thus be two bars for the 4-week periods and two bars for the thirty-six week periods.

8. The answer to this question is to be found in Fig. 16, page 108.

9. Age distribution, car registrations or population growth over the last 150 years; preferences for holidays as in Fig. 6; distribution of population as in Fig. 5; and drinking habits as in Fig. 3.

10. See Fig. 15, page 85. Note that each axis is marked off in percentage terms so that the values would have to be converted before plotting on the graph.

11. One method is to draw a bar for each age group, corresponding to the total column and each bar being sub-divided by reference to 1sts and 2nds and others. Alternatively, one could draw four bars (one for each class and the total column) and sub-divide these by reference to age groups. The former of these two is preferable. Another method is shown in Fig. 12 with age groups along base.

12. Pie charts, in which the circle for each year would be divided into four segments, is less satisfactory in this case than bar or column diagrams. A bar can be drawn for each year, its length corresponding to the total column and the sub-divisions corresponding to the three main classes. This would facilitate comparision between years.

ANSWERS: TYPES AND FUNCTIONS

CHAPTER VI

1. The answer will be affected slightly by the choice of the mid-point for the first class, since the interval is clearly not ten units. If a mid point of six is taken, the mean is just over £15 and the standard deviation £4·1.

2. The definitions can be derived from the previous text; given the data in the question the mean is 8, the median is 7, after rearrangement in order of magnitude. Since the only item which occurs more than once is six, this is the mode.

3. (a) The mean is thirty-five years, the standard deviation is twenty-three years
   (b) thirty-five and eighteen. Note that the class-interval is 10, not 9.
   All figures given to the nearest year.

4. The average size of dwelling occupied by more than one household is 6·3 rooms, that occupied by one household 4·7 rooms.

5. The mean earnings are £18, and the median £17. The lower quartile £13, and the upper quartile £22. All figures to the nearest £. Note the variation in the class interval and the two open ended classes at the beginning and end of the table. The mean will be affected only slightly by the choice of mid-points made here.

6. The ratio of receipts to stocks of all businesses is derived by aggregating the receipts and stocks and dividing the latter into the former. The ratio is 19·3 : 1. This particular ratio is most likely to occur in businesses with receipts of about £2½ million. For businesses with receipts of only £50,000, the ratio is 13·9:1.

7. The two averages of particular use are the cost per ton produced and the man hours worked per ton produced. The averages show that the variation in cost per ton is due to variations in labour productivity and this should be investigated to explain the differences between the two works.

8. (a) 532s.   (b) 852s.

9. Tabulate the figures for each of the following:
   Industry X; all other industry; total industry; days lost; men involved; days lost per man.
   The total is the sum of the first two except for 'days lost per man'. 'Days lost per man' is derived by dividing the number of men involved into the number of days lost for the industry and the other groups.

10. (i) The histogram should be as in Fig. 14 noting in particular, however, the variable class interval with the resultant effect on the height of the corresponding columns.
   (ii) The mean age at death is approximately thirty years.
   (iii) The mean remains the most suitable measure of central tendency since an estimate of the upper limit can be made since, even if it were incorrect the number of frequencies in that class is not so large that it would greatly affect the result.

11. The mean age for males is approximately $18\frac{3}{4}$ and for females $18\frac{1}{2}$. The assumption is made that the mid-points of these classes reflect the age distribution within the class and that, for the final class, the maximum age is 23. To calculate the overall mean of both sexes, add the frequencies together and calculate the mean as before.

12. The answer to this question is given in Fig. 16 on page 108 and the accompanying text.

ANSWERS: MEASURES OF VARIATION

CHAPTER VII

1 and 2. See text in Chapter.

3.

	Male	Female
L.Q.	5 years	5 years
Median	28 ,,	26 ,,
U.Q.	56 ,,	54 ,,
Q.D.	25 ,,	24 ,,

The above figures have all been rounded to the nearest year; greater precision is not practicable with this distribution. The calculation of the mean and standard deviation would be difficult by virtue of the large class intervals.

4.

	G.M.Ps.	Dentists
L.Q.	970	1690
Median	1250	2270
U.Q.	1650	3080
Q.D.	340	695

As in Q.3, a degree of approximation is necessary, in this case to the nearest £10. The results indicate that dentists are generally better paid and have a higher salary range.

5. Counties, median = 8·7, Q.D. = 2·9.
   C.Bs., median = 10·1, Q.D. = 1·9.

6. The arithmetic mean = 41 years; standard deviation = 10 years.
   Results to the nearest year.

   Assume the last class has an interval of ten years, but the frequency is so small that any interval within reason would be acceptable.

7.   Males: Mean = 46, S.D. = 11.
                  Females: Mean = 44, S.D. = 12.

Here, too, figures are rounded to the nearest unit. The same point *re* the last class as made in Q.6. applies here.

8. The student will find this question easier to answer if he reverses the tables, i.e., starting with 'under 34' at the top, and accumulating this way.

Consultants: mid 53: median = 46, Q.D. = 6·5
             mid 58: median = 48, Q.D. = 6·1.

The results suggest that on average the 733 additional consultants appointed between 1952 and 1958 have been above the median age at 1952.

9. Arithmetic mean=53 years; S.D.=6 years. Results to nearest whole year.

10. Median=52 years; Q.D.=5 years. Results to nearest whole year.

11. General workers: arithmetic mean=223s., S.D.=39s.
Car men: Arithmetic mean=255s., S.D.=35s.

Results to nearest shilling.

12. Median=18 years; Q.D.=10 years. The method of graphical calculation is shown on page 108.

13. Q.D.=0·65 stone; S.D.=1·85 stone.
The S.D. is preferable since it is based upon all the data and it is especially well suited to this type of symmetrical distribution.

ANSWERS: STATISTICAL AND ARITHMETIC ACCURACY

CHAPTER VIII

1. Calculate the maximum and minimum weight of beef per head for each year. Thus, in 1954 the number of cattle could vary between 215,000 and 224,000, while the quantity of beef could range from 1,217,500 cwts. and 1,218,400 cwts. Dividing the largest quantity of meat by the smallest No. of cattle gives the highest possible figure of beef per head; the minimum figure per head can be derived by using the other two values. The calculations of the maxima and minima show that the range of the annual values per head is never such that one year overlaps with another. Thus the trend is as shown in the rounded figures.

2. The estimate of 31 m.p.g. based on 615 mile run requires petrol consumption of 19·84 gallons, equal to 89·28 litres (1 gall.=4·22 1.) Since the gauge is 10% out, the true consumption lies between 98·21 and 81·35 litres which expressed in gallons is 21·58 and 17·88 g. respectively. The estimated 615 miles converted more accurately at 1·61 kms. per m. gives a true mileage of 611·18. Dividing this mileage by the estimated max. and min. petrol used gives a m.p.g. of 28·32 and 34·18.
If 31 m.p.g. is an exact estimate, then the consumption of 95 litres must be adjusted to between 85·5 and 104·5 representing 18·79 and 22·97 gallons. Given exactly 31 m.p.g. he covers anything between 582 and 712 miles.

3.

	A	B	C	Total
Average wage £	943	881	1,000	948
Error on Wage bill £	±50	±50	±50	±150
Error on numbers employed	± 5	+ 5	± 5	± 15
Error on average wage £	±18	±15	±12	± 15

4.

	Absolute Error	Relative Error %
(i)	± 40	± 10
(ii)	$\pm 1{,}969$	± 21
(iii)	± 40	± 11.4
(iv)	$\pm 3{\cdot}3$	$\pm 22{\cdot}2$

5. Calculate the A.M. and S.D. from the distribution. A.M.$=9{\cdot}09$ and S.D.$= 3{\cdot}75$. The difference in the statistics can only be explained by the rounding to nearest unit of the data in the published distribution, i.e. the original data were given to one decimal place at least, if not two since the answer is given to two decimal places.

6. The neo-natal death rate for Area B is 103% of that of Area A with a possible error of $\pm 19\%$. The total infant death rate for Area B is 107% of that of Area A with a possible error of $\pm 21\%$.

ANSWERS: REGRESSION AND CORRELATION

CHAPTER IX

1. The correlation coefficient is $+0{\cdot}88$ which indicates a close direct relationship. This is supported in the idea that as production increases so the results of that production have to be shifted.

2. The correlation coefficient is $+0{\cdot}05$ which indicates virtually no relationship between the two sets of data. This would support the idea that expenditure on fuel and light represented a basic outlay which would not increase with higher income.

3. The drawing of a scatter diagram is described on page 158. The equations for drawing the regression lines are:
$$X = -0{\cdot}279Y + 49{\cdot}25$$
$$Y = -0{\cdot}337X + 41{\cdot}74$$

4. The correlation coefficient is $+0{\cdot}26$.

5. Regression is described on pages 157-161.

6. The formulae for the two regression lines are:
$$X = 0{\cdot}004Y + 13{\cdot}78$$
$$Y = 134X - 1340$$
$$X \text{ at age of } 40 = \pounds 4020$$

7. The correlation coefficient is $+0{\cdot}76$ which indicates a reasonably close direct relationship between the two sets of data. This supports the idea that the longer the baby is carried in the womb the larger it will grow.

8. The regression equation of weight on gestation time is $Y=0{\cdot}06X-8{\cdot}1$. If the gestation time is 280 days the weight should be $8{\cdot}7$ lb. at birth. A regression line of gestation time on weight could indicate the probable gestation time from a given birth weight (useful in helping to establish paternity).

9. The correlation coefficient is —0·7 which indicates a fairly close inverse relationship. This supports the idea that the infant mortality rate falls as the social level rises.

10. The differences between product moment and rank correlation shown on page 172. The rank correlation coefficient from the data is —0·6.

12. (a) The rank correlation coefficient is +0·65 which indicates a moderate degree of relationship between male and female choice of subjects. On the other hand the correlation is not close enough for the two groups to be considered as one.

    (b) The technique of correlation coefficient by product moment is discussed on pages 162-6.

ANSWERS: INDEX NUMBERS

CHAPTER X

1. See pp. 182 to 185.

2. See pp. 180-183 and Chap. XVI section on Living Standards.

3. Multiply the group indices by their corresponding weights, add the products and divide the total by the sum of the weights. Index for (1) to (4) combined= 141·5; all items index 132·5.

4. See pp. 185-190 stressing the weighting, choice of items and the base year. See Chapter XVI for details of the two indices.

5. An alternative index of quantities is derived by deflating the indices in Col. 1 by reference to the price movements given in Col. 3, i.e. divide latter indices into the former. Year two will be ninety-seven, year three, ninety-four, year six, seventy-seven, etc. The difference between the two indices, i.e. Col. 2 and the alternative arises from the differing composition of the export total for each year as compared with that in year one used for the index in Col. 2.

6. Note the increase in expenditure shown in the first two columns is due in part to price changes and partly to quantity changes. The quantity change is derived by relating the 1954 outlay to the 1960 outlay at 1954 prices, i.e. 14,133/11,950. The price change is the balance, i.e. total change less quantity change or 4,381 less 2,183=2,198 as a ratio of the base year 11,950, or 14,148/11,950.

7. This is similar to the Sauerbeck index illustrated on p.191. Derive the price relatives for each year for the three commodities, weight the individual relatives, sum the products and divide the total by ten, i.e. the sum of the weights.

8. Use the average price per unit values to calculate the price relatives in the two years. The weights to be used with the change in the relatives can be based on the average expenditure data. If the 1956 outlays are used then the resultant index

will be base year 1956 Laspeyre type; alternatively use the 1961 outlays as the weights and obtain a current year weighted index.

9. See p.192 and Chapter on Social Statistics: Living Standards.

10. Derive the logarithms of the prices in each of the two years, multiply them by the weights, sum the products and divide by the sum of the weights. The percentage change is 87·2.

ANSWERS: TIME SERIES

CHAPTER XI

1. Using a two year moving average for each of three series it is found that there is no marked trend, although there does appear to be a two-yearly variation of approximately + or —1·5 cases per 1,000 in England. In Wales the movement is more pronounced, approximately + or — 2·5; in London the trend differs from that in Wales but there is no marked seasonal movement.

2. The method to be used follows that illustrated in Tables 32 and 33. The trend shows a significant upward movement from 38,000 tons to 55,000 over the four year period. If this trend were to be maintained then the four quarter values in 1959 would be 58, 59, 60 and 61. Applying the seasonal variation to these derived values, the expected quarter values would be 86, 58, 36 and 54.

3. Care is required with the calculations. Add the first twelve months, then for the next moving total drop the 116 and add 119, etc. The series is rather short for a twelve month moving average and the variation about the trend is quite small, i.e. between 116 and 119.

4. As in Q.3 calculate the twelve-month average for the trend values which will show a U shaped curve ranging from 140 down to 115 and then up to 150.

5. The chart should resemble that in Fig. 18; use a four quarterly moving average to derive the trend. This displays no marked movement in either direction. The residual fluctuations are also insignificant.

6. The answer to this question is to be found on pp. 199-201.

7. The purpose of trend fitting by moving averages is primarily to remove periodic fluctuations in the series from the trend values. It also requires a fairly marked periodicity, e.g. quarterly, or twelve month, or five yearly movement if the m.a. is to be accurately estimated. Such periodicity is not evident in this series and the method of m.a. is not appropriate. There seems to be a rising trend and initially the best method would be to plot the data and sketch the trend line in. If the trend is linear then it could be defined more exactly by the method of least squares. The main point in the comment to the question is that over this period the infant mortality rate has declined continuously and the fall has been progressively more marked among male infants.

ANSWERS: BASIS OF SAMPLING
CHAPTER XII

1. Half, or 50:50 since it is independent of any previous toss.

2. See pp. 218-222.

3. The s.d. of a normal distribution is the standard error so that the mean of any sample has a 95/100 chance of being within the range of two s.e. about the Mean. The point in the question about four values may be ignored.

4. The s.e. of the proportion of heads in a sample of 200 tosses is equal to 0·035 with a mean proportion (P) of 0·5. Thus we can expect about 95/100 of tosses of 200 coins to yield a result between forty-three and fifty-seven per cent or eighty-six to 114 tosses to fall heads. The probability of at least eighty-five tails (or 42½%) is 95/100 approx. Probability of ninety or fewer heads is given by the probability of forty-five per cent or less of P, i.e. P — 1½ s.e. = seven per cent. If eighty heads or forty per cent, then the observed difference is ten and, since s.e. is 3·5, it is equal to almost three times the s.e. Thus the conclusion is that the coin is not falling randomly but is probably biased, or if 200 coins involved, then the process of tossing is biased.

5. See pp. 217-9.

6. See pp. 223-5.

ANSWERS: SIGNIFICANCE TESTS
CHAPTER XIII

1. See pp. 228-233.

2. Calculate the proportion of cures in each group, i.e. 12/60 and 20/80, then the standard error of the difference between the two proportions, approx. seven per cent. The observed difference between the samples, i.e. five per cent, is smaller than s.e., hence accept null hypothesis.

3. A.M.=112, S.D. 16 and s.e. 7. The first two figures based on assumption that class intervals of first and last classes = 10.

4. Calculate the standard error of the difference between the averages which is about ·07 inches. Since the observed difference (1 inch) is many times larger than the s.e. the difference is statistically significant.

5. Calculate standard error of the difference between proportions of satisfied men and women, i.e. sixty-seven and seventy-seven per cent. Since the observed difference is ten and the s.e. approx. 2·5 the difference between the two sexes in this respect is highly significant.

6. First obtain the number of couples with no children and those with four or more, i.e. eighty and eighty-four. Then calculate the s.e. of the difference between proportions described as 'happily married' in these two samples. The observed difference is thirty and the s.e. approx. 7·4, hence statistically significant. But, note

very small samples of each type and also would need more information on definitions employed.

7. Calculate the s.e. of the percentage given which is equal to 0·96, say one. Hence statement that between ten and fourteen per cent of Fellows in society are between 25-29 is likely to be correct 95/100 times.

8. Calculate the s.e. of the difference between the proportions, i.e. 44·5 and 35·1 which is equal to 3·7. Since the observed difference is 9·4 and almost 2½ times as large as the s.e. we can reject null hypothesis and assume difference between two groups of mothers is statistically significant.

9. Home students: A.M. 29·6 S.D. 16·8
   Lodgings 18·5 14·7

The s.e. of the difference between means is 3·09 but difference of 11·1 is 3·6 times as great so there is a statistically significant difference between the groups.

10. The calculated value of chi-square is 3·9. With one degree of freedom this is significant at the five per cent level.

ANSWERS: VITAL STATISTICS

CHAPTER XVII

1. Need for comparison; explain direct and indirect methods with special reference to S.M.R. and A.C.F.

2. Explain crude birth rate, limitations, etc., then fertility rates including G.R.R. and N.R.R. and finally cohort analysis.

3. As on lines of Q.1.

4. G.R.R.=2·237. N.R.R.=1·348. Mortality of young women was very much higher in 1841.

5. As for Q.2.

6. See appendix on life tables.

7. (a) T_0=4,150. T_{10}=3,405. $\overset{\circ}{e}_{10}$=54·05.
   (b) If this method were applied to data in this life table, we get 815 instead of 745. If we apply formula $\cdot3l_x + \cdot7l_{x+10}$ the result is 741, a close approximation to the figure in the table. The disparity arises from the fact that the assumption on which the first formula is based is not justified for the first year, or first decade of life, i.e. deaths not spread evenly over period.

   (c) crude birth rate $=l_0 \div T_0=100 \div 4150=24\cdot9$ per 1000 general fertility rate $l_0 \div L_{10\text{-}50}$, i.e. number of women in age groups ten to fifty which total 2297. Assuming male and female births equal, then f.r.=200÷2297 or 72 per 1,000. Note 'general' f.r. applies to both sexes as distinct from female f.r.

   (d) Survivorship ratio$=L_{40\text{-}50} \div L_{20\text{-}30}=515 \div 598$ which applied to cohort of 50,000 women gives a total of 43,000 survivors approx.

INDEX

497